SUBMILLIMETRE ASTRONOMY

ASTROPHYSICS AND SPACE SCIENCE LIBRARY

A SERIES OF BOOKS ON THE RECENT DEVELOPMENTS
OF SPACE SCIENCE AND OF GENERAL GEOPHYSICS AND ASTROPHYSICS
PUBLISHED IN CONNECTION WITH THE JOURNAL
SPACE SCIENCE REVIEWS

Editorial Board

R. L. F. BOYD, *University College, London, England*

W. B. BURTON, *Sterrewacht, Leiden, The Netherlands*

C. DE JAGER, *University of Utrecht, The Netherlands*

J. KLECZEK, *Czechoslovak Academy of Sciences, Ondřejov, Czechoslavakia*

Z. KOPAL, *University of Manchester, England*

R. LÜST, *European Space Agency, Paris, France*

L. I. SEDOV, *Academy of Sciences of the U.S.S.R., Moscow, U.S.S.R.*

Z. ŠVESTKA, *Laboratory for Space Research, Utrecht, The Netherlands*

VOLUME 151
PROCEEDINGS

SUBMILLIMETRE ASTRONOMY

PROCEEDINGS OF THE KONA SYMPOSIUM ON
MILLIMETRE AND SUBMILLIMETRE ASTRONOMY,
HELD AT KONA, HAWAII, OCTOBER 3–6, 1988

edited by

GRAEME D. WATT and ADRIAN S. WEBSTER

Joint Astronomy Centre, Hilo, Hawaii, U.S.A.

KLUWER ACADEMIC PUBLISHERS

DORDRECHT / BOSTON / LONDON

Library of Congress Cataloging in Publication Data

```
Kona Symposium on Millimetre and Submillimetre Astronomy (1988 :
  Kailua Kona, Hawaii)
    Submillimetre astronomy : Proceedings of the Kona Symposium on
  Millimetre and Submillimetre Astronomy held at Kona, Hawaii, October
  3-6, 1988.
        p.    cm. -- (Astrophysics and space science library ; v. 158)
    ISBN 0-7923-0614-7
    1. Infrared astronomy--Congresses.  2. Radio astronomy-
  -Congresses.   I. Webster, Adrian S. (Adrian Shaw), 1948-      .
  II. Title.    III. Series.
  QB470.A1K66  1988
  522'.68--dc20                                                89-48930
```

ISBN 0–7923–0614–7

Published by Kluwer Academic Publishers,
P.O. Box 17, 3300 AA Dordrecht, The Netherlands.

Kluwer Academic Publishers incorporates
the publishing programmes of
D. Reidel, Martinus Nijhoff, Dr W. Junk and MTP Press.

Sold and distributed in the U.S.A. and Canada
by Kluwer Academic Publishers,
101 Philip Drive, Norwell, MA 02061, U.S.A.

In all other countries, sold and distributed
by Kluwer Academic Publishers Group,
P.O. Box 322, 3300 AH Dordrecht, The Netherlands.

Printed on acid-free paper

All Rights Reserved
© 1990 by Kluwer Academic Publishers
No part of the material protected by this copyright notice may be reproduced or
utilized in any form or by any means, electronic or mechanical,
including photocopying, recording or by any information storage and
retrieval system, without written permission from the copyright owner.

Printed in the Netherlands

Proceedings of the Kona Symposium on

Millimetre and Submillimetre Astronomy

held at Kona, Hawaii on October 3-6, 1988

We therefore recommend again and again, to the curious investigators of the stars to whom, when our lives are over, these observations are entrusted, that they, mindful of our advice, apply themselves to the undertaking of these observations vigorously. And for them we desire and pray for all good luck, especially that they not be deprived of this coveted spectacle by the unfortunate obscuration of cloudy heavens, and that the immensities of the celestial spheres, compelled to more precise boundaries, may at

ERRATUM

Submillimetre Astronomy

G. D. WATT and A. S. WEBSTER (editors)

Please not the volume number of the Series page (p. ii) should read '158' instead of '151.

TABLE OF CONTENTS

Sponsors and Organisation	xvii
Introduction and Acknowledgements	xix
List of Participants	xxi

Section I: Physics of the Interstellar Medium and Evolved Objects

Structure and Fragmentation of Quiescent Molecular Clouds   3
 Puget J L, Falgarone E

The Self-Similar Structure of Molecular Clouds   9
 Falgarone E, Phillips T G

Extended and Compact CO 7-6 Emission from OMC-1   11
 Schmid-Burgk J

The Size Distribution of Interstellar Dust   13
 Tielens A G G M

High Resolution Observations of Dust Emission   19
 Chini R

The Magnetic Field at the Center of the Galaxy   25
 Hildebrand R H

Hot Molecular Gas in the Protoplanetary Nebula CRL 2688   29
 Smith M G, Geballe T R, Sandell G, Aspin C

CO J=3-2 Observations of IRC+10216   31
 Tauber J A, Kwan J, Goldsmith P F, Snell R L, Erickson N R

CO Observations of Evolved Giant Stars   33
 Knapp G R, Gammie C F, Young K, Phillips T G

Far-Infrared and Submillimeter Analysis of a Dense Core: Temperature and Density Structure of L1489   35
 Ladd E F, Myers P C, Casey S C, Harper D A, Davidson J A, Benson P J, Levreault R M

NGC 2071: A Twin for L1551, and Evidence for Backflow?   37
 Moriarty-Schieven G H, Hughes V A, Snell R L

Section II: Instrumentation and Cosmic Background Experiments

A New Technique for Surface Measurements of Radio Telescopes — 41
Serabyn E, Masson C R, Phillips T G

Gravitational Deflection of the Leighton Telescopes — 43
Woody D

The SMT: A Joint Submillimeter Telescope Project of the Max-Planck-Institut für Radioastronomie and Steward Observatory — 45
Baars J W M, Krügel E, Martin R N

Submillimeter Instrument Development at Steward Observatory — 47
Martin R N, Walker C E, Walker C K, Miller R E

Operation of Bolometric Detectors under Conditions of varying Sky Background — 49
Griffin M J

SCUBA: A Submillimetre Common-User Bolometer Array for the James Clerk Maxwell Telescope — 51
Duncan W D

Submillimeter Cosmology with the Multiband Imaging Photometer for SIRTF — 53
Timbie P T, Bernstein G M, Richards P L, Gautier T N, Rieke G H, Werner M W

Receiver Work at the CFA — 55
Bloemhof E E, Dhawan V

The Coupling of Submillimeter Corner-cube Antennas to Gaussian Beams — 57
Grossman E N

Airborne Heterodyne Receiver Technology for the 100-300 micron Range — 59
Wattenbach R

The Millimeter Multibeam System Project at Onsala Space Observatory — 61
Emrich A

A 60 cm Submillimeter Survey Telescope Project — 63
Hayashi M, Hasegawa T, Sunada K, Kaifu N

Intra-Cavity Pumped FIR Laser System — 65
Chin G, Dave H

Imaging Triple-Fabry-Perot-Spectrometer for Far-Infrared Astronomy — 67
Poglitsch A, Geis N, Haggerty M, Genzel R, Stacey G, Townes C

Small-Area Niobium/Aluminum Oxide/Niobium Junctions for SIS Mixers   69
   Zmuidzinas J, Sharifi F, Van Harlingen D J, Lo K Y

An SIS Receiver for the JCMT   71
   Davies S R, Cunningham C T, Little L T, Matheson D N

SIS Receiver Development at Nobeyama Radio Observatory   73
   Inatani J, Kasuga T, Kawabe R, Tsuboi M, Sakamoto A, Yamamoto M, Watazawa K

Techniques in Small Area SIS NbN Junction Manufacture   75
   Vaneldik J F, Routledge D, Brett M

205 GHz SIS Receiver Development   77
   McGrath W R, Byrom C N, Ellison B N, Frerking M A, Leduc H G, Miller R E, Stern J A

New Measurements of the Spectrum and Anisotropy of the Millimeter Wave Background   79
   Bernstein G M, Fischer M L, Richards P L, Peterson J B, Timusk T

Millimeter and Submillimeter Interferometry   81
   Welch Wm J

The Millimeter Array   87
   Brown R L

The Nobeyama Millimeter Array: New Developments and Recent Observational Results   89
   Ishiguro M, Kawabe R, Morita K -I, Kasuga T, Chikada Y, Inatani J, Kanzawa T, Iwashita H, Handa K, Takahashi T, Kobayashi H, Okumura S K, Murata Y, Ishizuki S

Millimeter VLBI   91
   Wright M C H

The Submillimeter Wave Astronomy Satellite   93
   Melnick G J

Section III: Chemistry of the Interstellar Medium

Spectroscopy of Circumstellar Envelopes with the IRAM 30-m Telescope   97
   Lucas R, Guelin M

Millimeter and Optical Observations of Translucent Molecular Clouds   103
   Van Dishoeck E F, Black J H, Phillips T G

A Submillmeter Line Survey of Sgr B2 — 105
Sutton E C, Jaminet P A, Danchi W C, Masson C R, Blake G A

H_3O^+ Revisited — 107
Wootten A, Boulanger F, Zhou S, Combes F, Encrenaz P, Gerin M, Bogey M

Millimeter and Submillimeter Studies of Interstellar High Temperature Chemistry — 109
Ziurys L M

A Survey of Orion A Emission Lines from 330-360 GHz — 111
Jewell P R, Hollis J M, Lovas F J, Snyder L E

Millimeter-Wave Spectral Line Survey at NRO — 113
Ohishi M

Observations of the CH_2CN $1_{0,1}$-$0_{0,0}$ and $4_{0,4}$-$3_{0,3}$ Transitions — 115
Irvine W M, Madden S C, Ziurys L M, Friberg P, Hjalmarson Å, Matthews H E, Turner B E

Heterodyne Spectroscopy of CII in Molecular Clouds — 117
Betz A L, Boreiko R T, Zmuidzinas J

Ion-Molecule Chemistry of Carbon in Shielded Regions — 123
Langer W D

158 micron [CII] Line Emission from Galaxies — 125
Stacey G J, Genzel R, Lugten J B, Townes C H

Hot Quiescent Gas in Photodissociation Regions: CO and C^+ Observations of NGC 2023 — 127
Jaffe D T, Howe J E, Genzel R, Harris A I, Stutzki J, Stacey G J

DCO^+ in Nearby Dense Cores — 129
Butner H M

Cirrus Cloud Cores: A Deficient Chemistry? — 131
Turner B E, Rickard L J, Lan-Ping X

Section IV: Star Formation

Star Formation in Accretion Disks — 135
Pudritz R E

Molecular Clouds and Bipolar Flows — 141
Sandell G

Submillimeter Emission from Small Dust Grains Orbiting Nearby Stars — 147
Becklin E E, Zuckerman B

Results from 1.4 Millimeter Wavelength Interferometry at the Owens Valley   155
Mundy L G, Sargent A I, Scoville N Z, Padin S, Woody D P

A Survey of Galactic Sources for Submillimeter Emission from High Excitation Molecules   157
Evans N J, Butner H M, Zhou S, Mayer C E, Howe J E, Wall W F, Laycock S C

Mm and Sub-mm Observations of the Protostellar Disks   159
Hayashi S S

Studies of Star Forming Regions in M33   161
Wilson C D

Circumstellar Dense Gas in B5 IRS1   163
Fuller G A

^{12}CO (2-1) Observations of a Complete Sample of Dark Clouds   165
Scott P F, Parker N D, Padman R

Plateau Emission from Orion in the CO J=17-16 Line   167
Boreiko R T, Betz A L, Zmuidzinas J

Structure of Dust Discs Around Young Stars   169
Dent W R F

Probing the Lower Main Sequence with Molecular Clouds   171
Dickman R L, Jarrett T H, Herbst W

1.0-mm Continuum, ^{12}CO, ^{13}CO and C^{18}O Mapping of the NGC 6334 Complex and Comparison with IRAS Observations   173
Gezari D, Blitz L

Observations of HCO$^+$ in B335   175
Hasegawa T I, Rogers C, Hayashi S S

Radio Continuum Activity in Cepheus A   177
Hughes V A, Moriarty-Schieven G H

The Near-Stellar Environment of Cool, Evolved Stars   179
Judge P G, Stencel R E, Linsky J L

Millimeter and Submillimeter Observations of the NGC 7538 Molecular Cloud   181
Kameya O

Far Infrared and Submillimeter Continuum Observations of the Sagittarius B2 Molecular Cloud Core   183
Lis D C, Goldsmith P F, Hills R E, Lasenby J

Synthesis Imaging of the DR21(OH) Protocluster ... 185
Mangum J G, Wootten A, Mundy L G

New High-Velocity Outflows from the Protostar W3 IRS5 ... 187
Mitchell G F, Belcourt K, Maillard J P, Allen M

Solar Mass Clumps in the B5 Core ... 189
Pendleton Y, Davidson J, Casey S, Harper A, Pernic R, Myers P

Infrared and Submillimeter Observations of the Rho Ophiuchi Dark Cloud ... 191
Ward-Thompson D, Robson E I, Walther D, Duncan W D, Gordon M

Outflows from Massive, Pre-Main Sequence Objects: Results from the OVRO Millimeter Wave Array ... 193
Barsony M

Section V: Galaxies

The Evolution of Starburst Galaxies to Active Galactic Nuclei ... 197
Scoville N Z

Submillimetre Observations of Active Galaxies ... 203
Gear W K

Millimeter Observations of Luminous IRAS Galaxies ... 209
Keene J, Carico D P, Neugebauer G, Soifer B T

JCMT Observations of CO (3-2) from M82 ... 211
Lo K Y, Stephens S, Rosenthal E, Eales S, Wynn-Williams C G

CO (2-1) / CO (1-0) Observations of Luminous Infrared Galaxies ... 213
Sanders D B, Sargent A I, Scoville N Z, Phillips T G

Millimetre and Submillimetre Observations of Blazars ... 215
Robson E I

Cold Dust in Galaxies ... 217
Eales S A, Wynn-Williams C G, Duncan W D

CO (3-2) Observations of Spiral Galaxies ... 219
Turner J L, Martin R N, Ho P T P

CO(2-1) Emission from NGC 3256: An Interacting Pair of Galaxies ... 221
Sargent A I, Sanders D B, Phillips T G

CO (2-1) Studies of Centaurus A ... 223
Phillips T G, Sanders D B, Sargent A I

H_2 Emission and CO Absorption towards the Nucleus of Centaurus A: A Circumnuclear Disk?   225
Van Dishoeck E F, Israel F P, Koornneef J, Baas F, Black J H, de Graauw Th

CO 3-2 Observations of NGC 253   227
Bash F N, Davis J H, Jaffe D T, Wall W F, Sutton E C

Dust Emission from Radio-Quiet Quasars   229
Chini R

Photometry of the Hotspots of Cygnus A   231
Eales S A, Duncan W D, Alexander P

Molecular Gas in IC 10   233
Hauschildt M, Fairclough J H, Wright G S, Walker T M

Models of Colliding Galaxies: Kinetic Energy and Density Enhancements   235
Lamb S A, Gerber R A, Miller R H, Smith B F

Millimeter Wave Molecular Line Observations of Galaxies   237
Nakai N

Molecular Spiral Structure in M51   239
Rand R J, Kulkarni S R

CO (3-2) Emission from the Nucleus of NGC 6946   241
Rosenthal E, Eales S, Stephens S, Lo K Y

The Azimuthal Distribution of the ISM in NGC 6946   243
Tacconi L J

Detection of CO J=1-0 Emission from an IRAS Selected Sample of S0 Galaxies   245
Sage L J, Wrobel J M

Radial Distribution of Atomic and Molecular Gas in Disk Galaxies   247
Wang Z, Cowie L L

CO (1-0) in a Newly-Born Elliptical Galaxy: NGC 7252   249
Dupraz C, Casoli F, Combes F, Gerin M

Virgo Galaxies with Extreme CO/HI Ratios   251
Kenney J D P

Section VI: Molecular Studies

Galactic Cloud Spectroscopy   255
White G J

Detection of Far-Infrared ^{13}CO Line Emission Genzel R, Poglitsch A, Stacey G	261
The Rotation Curve of the S106 Molecular Disc Padman R, Richer J	263
A Search for Dense Gas Around Young Stars Masson C R, Keene J B, Mundy L G, Blake G A, Sutton E C, Danchi W, Jaminet P	265
An Unbiased Survey for Dense Cores and Star Formation in the Orion B Molecular Cloud Lada E A	267
Physical Conditions in Molecular Clouds derived from Sub-mm and Far-IR Spectroscopic Observations Stutzki J, Stacey G J, Genzel R, Graf U U, Harris A I, Jaffe D T, Lugten J B, Poglitsch A	269
Vibrationally Excited H_2O in Orion A Petuchowski S J, Bennett C L	275
HCN J=9-8 Emission in the Orion Core and W49 Harris A I, Genzel R, Graf U U, Jaffe D T, Stutzki J	277
Molecular Line Studies of W49A Miyawaki R, Hayashi M, Hasegawa T	279
Atomic and Molecular Outflow in DR21 Russell A P G	281
First Heterodyne Observations at 690 and 800 GHz with the James Clerk Maxwell Telescope Webster A S, Russell A P G, Matthews H E, Watt G D, Hayashi S S, Coulson I M, Genzel R, Harris A I, Stutzki J, Graf U U, Padman R	283
Extensive C^+ 158 micron Line Mapping of W3 and NGC 1977 Howe J E, Geis N, Genzel R, Jaffe D T, Poglitsch A, Stacey G J	285
First SEST Observations of the Large Magellanic Cloud Johansson L E B, Booth R S, Murphy D M, Olberg M	287
Highly Collimated, High Velocity Gas in the Orion B Molecular Outflow Richer J, Hills R E, Padman R, Scott P F, Russell A P G	289

Section VII: Solar and Solar System Studies

One-Millimeter Observations of Asteroids with the James Clerk Maxwell Telescope Redman R O, Feldman P A, Halliday I, Matthews H E	293

Prospects for Submillimeter Observations of the Total Solar Eclipse of 1991   295
  Roellig T L, Lindsey C A

Solar Astronomy on the New Large Submillimeter Facilities on Mauna Kea   297
  Lindsey C A, Roellig T L

Indices:

Author Index   311

Molecule Index   315

Object Index   317

General Index   321

Symposium Sponsors

The International Union of Radio Sciences (URSI)
California Institute of Technology (Caltech)
United Kingdom Science and Engineering Research Council (SERC)
University of Hawaii (UH)
National Research Council of Canada (NRC)

Scientific Organising Committee

D.N.B.Hall, Institute for Astronomy, University of Hawaii, Honolulu, Hawaii, USA
R.E.Hills, Cavendish Laboratory, Cambridge, England, UK
T.G.Phillips, Caltech, Pasadena, California, USA
W.J.Welch, University of California, Berkeley, California, USA

Local Organizing Committee

D.H.Beattie, Joint Astronomy Centre, Hilo, Hawaii, USA
Y.Boyce, Joint Astronomy Centre, Hilo, Hawaii, USA
R.C.Campbell, Joint Astronomy Centre, Hilo, Hawaii, USA
P.Ching, Joint Astronomy Centre, Hilo, Hawaii, USA
M.A.Johnston, Joint Astronomy Centre, Hilo, Hawaii, USA
W.Steiger, Caltech Submillimeter Observatory, Hilo, Hawaii, USA
A.S.Webster, Joint Astronomy Centre, Hilo, Hawaii, USA
C.G.Wynn-Williams, Institute for Astronomy, University of Hawaii, Honolulu, Hawaii, USA

Introduction and Acknowledgements

In recent years the field of millimetre astronomy has blossomed and that of submillimetre astronomy has begun to be explored in earnest. The major topics that have initiated these drives have been the construction of large surface area, high surface accuracy telescopes designed for operation at wavelengths of less than 1mm, the development of extremely sensitive receivers utilizing both heterodyne and bolometric techniques to enable extremely faint signals to be detected to adequate confirmation within in reasonable integration times, and the construction of complex software code running on large mainframes or supercomputers for chemical and physical modelling of individual clouds and entire galaxies. It is a direct result of the developments and progress made in each of these specialized areas that led to the request for this Symposium.

The Symposium titled 'Millimetre and Submillimetre Astronomy' was intended primarily as an introduction for the astronomical community to the two new submillimetre telescopes now in operation on Mauna Kea, namely the James Clerk Maxwell Telescope (JCMT) and the Caltech Submillimeter Observatory (CSO).

The content was not aimed to clash with that of the Summer School held during June 1987 in Stirling, Scotland but the intention was rather that the two meetings should go hand in hand. The Stirling Proceedings are a more technical oriented reference while the current volume is a collection of observations and their reduced and deduced analyses. In order to avoid a little confusion, the Stirling proceedings are called 'Millimetre and Submillimetre Astronomy' while this volume is simply 'Submillimetre Astronomy'.

A great number of people attended the Symposium, probably because of the added attractions of sunshine, swimming, surfing, sunbathing, scuba-diving, etc. Many of them probably went home with aching limbs, over-cooked skin and several cuts and bruises from landing on sharp lava rocks. There is much to be said about holding a meeting in Hawaii even though a major difficulty for the organisers is in persuading the attendees to attend the talk sessions instead of wandering off to the beach or bar.

It is extremely difficult to thank everybody by name who assisted in some way with the smooth running of the events so a communal 'Thank You' is in order if you contributed in any way. More specifically, we must thank the Sponsors for their generous donations towards the costs of the Symposium. We are very grateful to the Management and Staff of the King Kamehameha Hotel in Kailua-Kona for hosting this rather large international gathering. Particular thanks to them for the Luau which served as our Symposium Banquet and was attended by a great many of our participants lured by the delicious food and excellent polynesian dancing displays as well as by the free bar.

The Hawaii Resorts Transportation Company deserve a vote of thanks for their supply of minibuses for the Mauna Kea excursion and for the Volcano Park trip. All but one of their buses made it to Mauna Kea! We also thank the Mauna Kea Support Service Staff for supplying snacks for the visitors at Hale Pohaku waiting while smaller groups were ferried up to the summit for lightning tours around both the CSO and the JCMT.

On a more technical note, thanks are extended to the Scientific Organising Committee for arranging such a varied and comprehensive schedule of talks and posters; to the Local Organising Committee for their enormous effort at dealing with the administration and finances and for the many hours of office duty spent at the Hotel fielding strange questions from attendees about anything from astronomy to volcanoes to pineapples; and to the Local Assistance Team who ran the airport collection vehicles, provided assistance with the Mauna Kea trip, and were generally on hand during the entire time to cope with the plethora of 'odd jobs' that cropped up.

As Editors, we must thank all those who submitted a contribution for these Proceedings and must also apologise to you all for the late publication. It does take an enormous amount of time and effort to edit a volume like this especially when trying to maintain an operational telescope at the same time. A volume with 20 long contributions has got to be simpler to deal with than this one which has 114 short papers.

Most of you managed to get the camera-ready scripts in the correct format which saved us the task of having to retype items. Some of the extra pages were beautifully setup and typed by Donna Delorm (who secretly runs the JCMT single-handed!). Thanks Donna. The headings, indices, etc were set in LaTeX by GDW with necessary instruction and advice from Henry Matthews. We are particularly grateful to Paulette Ching for her dedication towards keeping the whole administrative and secretarial organisation in hand prior to the event.

The Organisers were very pleased with the presentations and seemed to be amazed at how smoothly the whole event ran. We hope that you were suitably impressed too and (should we ever get around to a follow up in a few years!) would be keen to venture off to Hawaii again.

And finally GDW wishes to personally thank ASW for getting him involved with this chaos and for actually getting his own contribution submitted!

Aloha,

Graeme D Watt

Adrian S Webster

September 27, 1989

LIST OF PARTICIPANTS

L. Avery, Herzberg Institute of Astrophysics, Ottawa, Ontario, Canada
J. Baars, Max-Planck-Institut für Radioastronomie, Bonn, West Germany
L. Baath, Onsala Space Observatory, Onsala, Sweden
M. Barsony, Leuschner Observatory, University of California, Berkeley, California, USA
F. Bash, Astronomy Department, University of Texas, Austin, Texas, USA
D.H. Beattie, Joint Astronomy Centre, Hilo, Hawaii, USA
E. Becklin, Institute for Astronomy, University of Hawaii, Honolulu, Hawaii, USA
G. Bellaiche, CNES, Toulouse, France
G. Bernstein, Physics Department, University of California, Berkeley, California, USA
A. Betz, Space Sciences Laboratory, University of California, Berkeley, California, USA
J.H. Black, University of Arizona, Steward Observatory, Tucson, Arizona, USA
G. Blake, Caltech, Pasadena, California, USA
L. Blitz, Astronomy Program, University of Maryland, Maryland, USA
E. Bloemhof, Center for Astrophysics, Cambridge, Massachusetts, USA
A. Boehmar, JRW, California, USA
R. Boreiko, Space Sciences Laboratory, University of California, Berkeley, California, USA
D.R. Brock, Joint Astronomy Centre, Hilo, Hawaii, USA
R.L. Brown, NRAO, Charlottesville, Virginia, USA
J. Burnell, Royal Observatory, Edinburgh, Scotland, UK
H.M. Butner, Department of Astronomy, University of Texas, Austin, Texas, USA
T. Buttgenbach, Caltech, Pasadena, California, USA
J. Carr, Institute for Astronomy, University of Hawaii, Honolulu, Hawaii, USA
M.M. Casali, Joint Astronomy Centre, Hilo, Hawaii, USA
H. Chen, Institute for Astronomy, University of Hawaii, Honolulu, Hawaii, USA
G. Chin, NASA/Goddard Space Flight Center, Greenbelt, Maryland, USA
P. Ching, Joint Astronomy Centre, Hilo, Hawaii, USA
R. Chini, Max-Planck-Institut für Radioastronomie, Bonn, West Germany
C.R. Cordell, Joint Astronomy Centre, Hilo, Hawaii, USA
I.M. Coulson, Joint Astronomy Centre, Hilo, Hawaii, USA
T. Davidge, Canada-France-Hawaii Telescope Corporation, Kamuela, Hawaii, USA
J. Davidson, NASA/Ames Research Center, Moffett Field, California, USA
S. Davies, Electronic Engineering Laboratories, The University, Canterbury, England
J. Davis, Astronomy Department, University of Texas, Austin, Texas, USA
W.R.F. Dent, Royal Observatory, Edinburgh, Scotland, UK

R.L. Dickman, FCRAO, University of Massachusetts, Amherst, Massachusetts, USA
W.D. Duncan, Joint Astronomy Centre, Hilo, Hawaii, USA
A. Emrich, Onsala Space Observatory, Onsala, Sweden
P. Encrenaz, Observatoire de Meudon, Meudon, France
N. Evans, Astronomy Department, University of Texas, Austin, Texas, USA
J.H. Fairclough, Joint Astronomy Centre, Hilo, Hawaii, USA
E. Falgarone, Caltech, Pasadena, California, USA
M. Fich, Department of Physics, University of Waterloo, Waterloo, Ontario, Canada
M.A. Frerking, JPL, Pasadena, California, USA
G. Fuller, Radio Astronomy Laboratory, University of California, Berkeley, California, USA
R.P. Garden, Department of Physics, University of California, California, USA
W.K. Gear, Royal Observatory, Edinburgh, Scotland, UK
R. Genzel, Max-Planck-Institut für Physik und Astrophysik, Garching bei München, West Germany
M.J. Griffin, Queen Mary College, Department of Physics, London, England, UK
T. Groesbeck, Caltech, Pasadena, California, USA
E. Grossman, Department of Astronomy, University of Texas, Austin, Texas, USA
D.N.B. Hall, Institute for Astronomy, University of Hawaii, Honolulu, Hawaii, USA
J.P. Hamaker, Joint Astronomy Centre, Hilo, Hawaii, USA
J. Hamilton, Canada-France-Hawaii Telescope Corporation, Kamuela, Hawaii, USA
A. Harris, Max-Planck-Institut für Physik und Astrophysik, Garching bei München, West Germany
T.I. Hasegawa, Department of Astronomy, Saint Mary's University, Halifax, Nova Scotia, Canada
T. Hasegawa, University of Tokyo, Nobeyama, Minamisaku, Nagano, Japan
M. Hauschildt, Joint Astronomy Centre, Hilo, Hawaii, USA
M. Hayashi, Department of Astronomy, University of Tokyo, Tokyo, Japan
S. Hayashi, Joint Astronomy Centre, Hilo, Hawaii, USA
R. Hayward, Herzberg Institute of Astrophysics, Ottawa, Ontario, Canada
P. Hekman, Joint Astronomy Centre, Hilo, Hawaii, USA
R.H. Hildebrand, The Enrico Fermi Institute, University of Chicago, Chicago, Illinois, USA
R.E. Hills, Cavendish Laboratory, Cambridge, England, UK
J. Howe, Astronomy Department, University of Texas, Austin, Texas, USA
V.A. Hughes, Astronomy Group, Department of Physics, Queen's University, Kingston, Ontario, Canada
J. Inatani, Nobeyama Radio Observatory, Nobeyama, Minamisaku, Nagano, Japan
W.M. Irvine, FCRAO, University of Massachusetts, Amherst, Massachusetts, USA

M. Ishiguro, Nobeyama Radio Observatory, Nobeyama, Minamisaku, Nagano, Japan
D. Jaffe, Astronomy Department, University of Texas, Austin, Texas, USA
P. Jaminet, Space Sciences Laboratory, University of California, Berkeley, California, USA
P.R. Jewell, NRAO, Tucson, Arizona, USA
L.E.B. Johansson, European Southern Observatory, Santiago, Chile
M.A. Johnston, Joint Astronomy Centre, Hilo, Hawaii, USA
P. Judge, Joint Institute for Laboratory Astrophysics, University of Colorado, Boulder, Colorado, USA
N. Kaifu, Nobeyama Radio Observatory, Nobeyama, Minamisaku, Nagano, Japan
O. Kameya, Nobeyama Radio Observatory, Nobeyama, Minamisaku, Nagano, Japan
J. Keene, Caltech, Pasadena, California, USA
J. Kenney, Owens Valley Radio Observatory, California, USA
E.R. Keto, Lawrence Livermore National Lab, University of California, Livermore, California, USA
G.R. Knapp, Department of Astrophysical Science, Princeton University, Princeton, New Jersey, USA
C. Koempe, Herzberg Institute of Astrophysics, Ottawa, Ontario, Canada
K. Krisciunas, Joint Astronomy Centre, Hilo, Hawaii, USA
S. Kulkarni, Caltech, Pasadena, California, USA
E. Lada, Astronomy Department, University of Texas, Austin, Texas, USA
N. Ladd, Harvard-Smithsonian Center for Astrophysics, Cambridge, Massachusetts, USA
S. Lamb, Department of Astronomy, University of Illinois, Urbana, Illinois, USA
W.D. Langer, Plasma Physics Laboratory, Princeton University, Princeton, New Jersey, USA
W. Latter, Steward Observatory, University of Arizona, Tucson, Arizona, USA
T.J. Lee, Royal Observatory, Edinburgh, Scotland, UK
T.H. Legg, Herzberg Institute of Astrophysics, Ottawa, Ontario, Canada
R. Leighton, Caltech, Pasadena, California, USA
C.A. Lindsey, Institute for Astronomy, University of Hawaii, Honolulu, Hawaii, USA
D.C. Lis, FCRAO, University of Massachusetts, Amherst, Massachusetts, USA
K.Y. Lo, Department of Astronomy, University of Illinois, Urbana, Illinois, USA
R. Lucas, Groupe d'Astrophysique, Observatoire de Grenoble, Saint-Martin d'Heres, France
J. Lugten, Institute for Astronomy, University of Hawaii, Honolulu, Hawaii, USA
J. Luthe, Joint Astronomy Centre, Hilo, Hawaii, USA
J.M. MacLeod, Herzberg Institute of Astrophysics, Ottawa, Ontario, Canada
S. Madden, FCRAO, University of Massachusetts, Amherst, Massachusetts, USA

J.G. Mangum, NRAO, Charlottesville, Virginia, USA
R.N. Martin, Steward Observatory, University of Arizona, Tucson, Arizona, USA
C. Masson, Caltech, Pasadena, California, USA
H.E. Matthews, Joint Astronomy Centre, Hilo, Hawaii, USA
J. Maute, Caltech Submillimeter Observatory, Hilo, Hawaii, USA
R. McGrath, JPL, Pasadena, California, USA
G.J. Melnick, Harvard-Smithsonian Center for Astrophysics, Cambridge, Massachusetts, USA
G.F. Mitchell, Department of Astronomy, Saint Mary's University, Halifax, Nova Scotia, Canada
R. Miyawaki, Fukuoka University of Education, Munakata City, Fukuoka, Japan
G. Moriarty-Schieven, Department of Physics, Queen's University, Kingston, Ontario, Canada
L. Mundy, Astronomy Program, University of Maryland, College Park, Maryland, USA
D. Nadeau, Dept. de Physique, Universite de Montreal, Montreal, Quebec, Canada
N. Nakai, Nobeyama Radio Observatory, Nobeyama, Minamisaku, Nagano, Japan
A. Natta, Astronomy Department, University of Texas, Austin, Texas, USA
P. Neill, Caltech, Pasadena, California, USA
D. Neufeld, Department of Astronomy, University of California, Berkeley, California, USA
L. Noreau, Université Laval, Faculté des Sciences et de Génie, Cité universitaire, Québec, Canada
M. Ohishi, Nobeyama Radio Observatory, Nobeyama, Minamisaku, Nagano, Japan
R. Padman, Cavendish Laboratory, Cambridge, England, UK
F. Palla, Observatorio Astrofisica di Arcetri, Firenze, Italy
J. Payne, NRAO, Tucson, Arizona, USA
Y. Pendleton, NASA/Ames Research Center, Moffett Field, California, USA
B. Pernick, Yerkes Observatory, Williams Bay, Wisconsin, USA
S. Petuchowski, Infrared Astrophysics Branch, NASA/Goddard Space Flight Center, Greenbelt, Maryland, USA
T.G. Phillips, Caltech, Pasadena, California, USA
G. Pillbratt, Astrophysics Division, ESTEC, Noordwijk, The Netherlands
K.K. Pisciotta, Joint Astronomy Centre, Hilo, Hawaii, USA
R. Pudritz, Department of Physics, McMaster University, Hamilton, Ontario, Canada
J. Puget, Radioastronomie Laboratoire, L'Ecole Normale Superieure, Paris, France
R. Rand, Caltech, Pasadena, California, USA
R.O. Redman, National Research Council of Canada, Administrative Services and Publications Branch, Ottawa, Ontario, Canada
N. Reid, Caltech, Pasadena, California, USA
S. Remington, Joint Astronomy Centre, Hilo, Hawaii, USA

J. Richer, Cavendish Laboratory, Cambridge, England, UK
E.I. Robson, Lancashire Polytechnic Institute, Preston, England, UK
T. Roellig, NASA/Ames Research Center, Moffett Field, California, USA
C. Rogers, University of Toronto, Toronto, Ontario, Canada
E. Rosenthal, Institute for Astronomy, University of Hawaii, Honolulu, Hawaii, USA
D. Routledge, Electrical Engineering Department, University of Alberta, Edmonton, Alberta, Canada
A. Rudolph, Radio Astronomy Laboratory, University of California, Berkeley, California, USA
A.P.G. Russell, Joint Astronomy Centre, Hilo, Hawaii, USA
L.J. Sage, Department of Physics, New Mexico Tech, Astrophysics Research Center, Socorro, New Mexico, USA
G. Sandell, Joint Astronomy Centre, Hilo, Hawaii, USA
D. Sanders, Caltech, Pasadena, California, USA
A. Sargent, Caltech, Pasadena, California, USA
W.L.W. Sargent, Caltech, Pasadena, California, USA
N.V.G. Sarma, Radiophysics Division, CSIRO, Epping, New South Wales, Australia
D. Sasselov, Department of Astronomy, University of Toronto, Ontario, Canada
A. Schinkel, Caltech Submillimeter Observatory, Hilo, Hawaii, USA
J. Schmid-Burgk, Max-Planck-Institut für Radioastronomie, Bonn, West Germany
P. Scott, Cavendish Laboratory, Cambridge, England, UK
N. Scoville, Caltech, Pasadena, California, USA
E. Serabyn, Caltech, Pasadena, California, USA
P. Shaver, European Southern Observatory, Garching bei München, West Germany
B. Shuter, Department of Physics, University of British Columbia, Vancouver, British Columbia, Canada
M.G. Smith, Joint Astronomy Centre, Hilo, Hawaii, USA
P. Solomon, Astronomy Program, State University of New York, Stony Brook, New York, USA
G. Stacey, Department of Physics, University of California, Berkeley, California, USA
W. Steiger, Caltech Submillimeter Observatory, Hilo, Hawaii, USA
J. Stutzki, Max-Planck-Institut für Extraterrestrische Physik, Garching bei München, West Germany
E. Sutton, Space Sciences Laboratory, University of California, Berkeley,, California, USA
K. Tabor, P.O. Box 2656, Kamuela, Hawaii, USA
L. Tacconi, Radiosterrenwacht Dwingeloo, Dwingeloo, The Netherlands
J. Tarter, NASA/Ames Research Office, SETI Program Office, Moffett Field, California, USA
J.A. Tauber, FCRAO, University of Massachusetts, Amherst, Massachusetts, USA
P. Thaddeus, Harvard-Smithsonian Center for Astrophysics, Cambridge, Massachusetts, USA
C. Thum, Instituto de Radioastronomia Milimetrica, Granada, Spain

A. Tielens, NASA/Ames Research Center, Moffett Field, California, USA
P. Timbie, Department of Physics, University of California, Berkeley, California, USA
B.E. Turner, NRAO, Charlottesville, Virginia, USA
J. Turner, Department of Astronomy, University of California, Los Angeles, California, USA
A. van Ardenne, Radiosterrenwacht Dwingeloo, Dwingeloo, The Netherlands
W. van Citters, Institute for Astronomy, University of Hawaii, Honolulu, Hawaii, USA
E.F. van Dishoeck, Department of Astrophysical Science, Princeton University, Princeton, New Jersey, USA
J.F. Vaneldik, Electrical Engineering Department, University of Alberta, Edmonton, Alberta, Canada
S. Vogel, Department of Physics, Rensselaer Polytechnic Institute, Troy, New York, USA
C. Walker, Steward Observatory, University of Arizona, Tucson, Arizona, USA
T.M. Walker, Joint Astronomy Centre, Hilo, Hawaii, USA
D.M. Walther, Joint Astronomy Centre, Hilo, Hawaii, USA
Z. Wang, Institute for Astronomy, Kula, Hawaii, USA
P.G. Wannier, JPL, Pasadena, California, USA
G.D. Watt, Joint Astronomy Centre, Hilo, Hawaii, USA
A.S. Webster, Joint Astronomy Centre, Hilo, Hawaii, USA
W.J. Welch, Radio Astronomy Lab, University of California, Berkeley, California, USA
G. White, Queen Mary College, London, England, UK
C. Wilson, Caltech, Pasadena, California, USA
G. Winnewisser, I. Physikalisches Institut, Universität zu Köln, Köln, West Germany
D. Woody, Owens Valley Radio Observatory, California, USA
G.S. Wright, Joint Astronomy Centre, Hilo, Hawaii, USA
M. Wright, Radio Astronomy Laboratory, University of California, Berkeley, California, USA
G. Wynn-Williams, Institute for Astronomy, University of Hawaii, Honolulu, Hawaii, USA
K. Young, Caltech, Pasadena, California, USA
L. Ziurys, FCRAO, University of Massachusetts, Amherst, Massachusetts, USA
J. Zmuidzinas, Department of Astronomy, University of Illinois, Urbana, Illinois, USA

Section I

Physics of the Interstellar Medium and Evolved Objects

STRUCTURE AND FRAGMENTATION OF QUIESCENT MOLECULAR CLOUDS

J.L. PUGET[1], E. FALGARONE[1,2]
[1] *Radioastronomie, Ecole Normale Supérieure,*
24 rue Lhomond, 75235 Paris Cedex 05, France
[2] *California Institute of Technology, 320-47, Pasadena, CA91125, USA*

The many physical problems to be studied in the process of star formation can be separated into two categories: firstly, the determination of the small scale structure of molecular clouds which provides the boundary conditions to the collapse process and second, the physics of the collapse itself. This paper concentrates on some of the questions to be solved in the first category.

In the classical picture, when a protostellar cloud of mass comparable to a stellar mass becomes gravitationally unstable, it collapses and the physical questions to be solved are: how does the temperature evolve during the collapse (it depends on the opacity of the cloud), do smaller masses become gravitationally unstable (the Jeans mass decreases as long as the collapse is roughly isothermal), can fragmentation into smaller unstable subunits occur? The answers to these questions are of course critically dependent on the rate at which the angular momentum is taken away from the collapsing gas. This picture of an isolated unstable protostellar cloud has been challenged by Shu, Adams and Lizano (1987) who consider that the collapse starts from the central part of a quasi hydrostatic self-gravitating sphere of gas and propagates outside. In both cases, the initial conditions are given by the structure of the molecular cloud out of which the collapse occurs and must be confronted with data on molecular clouds. The problem might be formulated more precisely in the following way: what is the density structure and velocity field in dense molecular clouds at scales containing a few stellar masses of gas (or less) and furthermore what are the dynamical interactions between the potentially unstable gas and its surroundings. At present, the classical picture of a rather uniform isolated sphere of gas above its Jeans mass collapsing and fragmenting until the fragments become opaque to the radiation carrying away the collapse energy (see for example Silk, 1977) does not fit the observations very well and the initial conditions considered by Shu et al. (1987) need observational backing.

This paper discusses recent observational results on the small scale structure of molecular clouds and the problem of gravitational instability in the presence of supersonic turbulence. We restrict ourselves to clouds which do not form massive stars so that we can hope that the observed structure is not severely affected by the stars already formed. A reasonable criterion for such clouds (referred to as quiescent clouds) is that the mechanical input and luminosity of already formed stars are negligible in comparison to, respectively, the mechanical energy in the form of large scale motions and the total luminosity of the cloud due to heating by the ambient radiation field.

1. The small scale structure of quiescent molecular clouds

The structure of molecular complexes as revealed in the last few years is that of a hierarchy of condensations obeying two scaling laws first noticed by Larson (1981) between their sizes and both their masses and internal velocity dispersions. This has been discussed in many recent reviews (Scalo, 1987; Falgarone and Pérault, 1987). We are concerned here with the physics at the scale of the molecular clouds defined as the entities of a few parsecs in which the average density derived from empirical mass determinations and the actual volume density of hydrogen molecules deduced from the excitation of CO rotational lines are comparable and $\sim 200\, cm^{-3}$. Most of such clouds are roughly in virial equilibrium between their self-gravitating energy and the kinetic energy in their large scale internal motions. Recently, Myers and Goodman (1988a) have compiled data on magnetic field which indicate that its energy density is, when measured, equal within a factor of 2 to the kinetic energy density. An extrapolation of these few measurements leads to the conclusion that clouds are turbulent with a turbulent velocity at scales comparable to the cloud size which are supersonic but just about Alfvenic.

At smaller scales, dense cores of a few solar masses are observed. These cores which follow the mass versus size scaling law ($M \propto R^2$) are thus gravitationally bound entities. An extensive study of such dense cores, selected as maxima of visual extinction in the Taurus complex and detected in the NH_3 line (Myers and Benson, 1983) suggests that they form stars within a time comparable to their free fall time ($\sim 2\, 10^5 yrs$) (Fuller and Myers, 1987). The argument relies upon the frequency of the association of these cores with infrared sources (Beichman et al., 1986). The bias introduced by the selection procedure is not well understood but more importantly low mass dense cores of a few $0.1\, pc$ are now known to be themselves structured down to $\sim 0.02\, pc$ (Guélin and Cernicharo, 1989; Pérault and Falgarone, 1988) and one might ask if the so-called dense cores in the Taurus complex are not concentrations of smaller entities or density peaks out of which only one has collapsed to form a star. In this case the estimate of the cores lifetime by Fuller and Myers would be severely underestimated.

The recent improvement of the angular resolution and sensitivity of millimeter and submillimeter observations has allowed the detection of density peaks at much smaller a contrast above the average density than that of the dense cores and over scales as small as a few $0.01\, pc$. This is clearly a new type of structure. Beside the smallest unresolved structures close to gravitational binding, which are observed within dense cores, either in the $C^{18}O$ line or with high density tracers like HC_3N or HC_5N, condensations at the same scale ($\sim 0.02\, pc$) are found with a local density low enough that they are not gravitationally bound. Instead of masses of a few $0.1 M_\odot$, they have $M < 10^{-3} M_\odot$ and lie well below the mass versus size scaling law. The surprising result is that they roughly follow the other scaling law, between the internal velocity dispersion and the size. It seems to suggest that the velocity dispersion at a given scale is controlled by some turbulent cascade and that the process leading to the bound dense cores is such that the gas loses all memory of what the turbulent velocity field before condensation was.

The detailed properties of the turbulent velocity field at various densities and spatial frequencies within molecular clouds is certainly one of the most crucial observational input to the understanding of star formation. On the one hand, the comparison between the velocity dispersion of structures of similar size but very different densities provides a valuable information on the mechanism which feeds the turbulence and controls the energy transfer between the scales (section 2), on the other hand, the knowledge of the velocity dispersion as a function of scale but at comparable densities is the indispensable tool to discuss the criterion for the onset of gravitational instability (Section 3).

2. The origin of the turbulent velocity field within molecular clouds

Two possible sources have been invoked so far for the turbulent velocity field inside molecular clouds. One is the injection of mechanical energy at small scale by low mass stars through their powerful outflows (Norman and Silk, 1980; Lada, 1988; Myers et al., 1988), the other source is the large scale kinetic energy of the complex, a fraction of which is pumped out of the orbital motions of the clouds into the cloud interiors. These orbital motions can be fed regularly out of the differential galactic rotation (Jog and Ostriker, 1988). They can also be fed by the gravitational potential energy of the slowly contracting complexes (Falgarone and Puget, 1985; Scalo and Pumphrey, 1982), an energy ultimately regenerated once the complexes have formed massive stars which disperse the complexes under the action of HII regions and supernovae.

The critical point here is the mechanism invoked for the transfer of energy and momentum between the scales. Norman and Silk (1980) refer to pure hydrodynamical processes which are likely to be poorly efficient if the density contrasts are large. Hydromagnetic waves avoid this difficulty. It was suggested by Arons and Max (1975). Magnetic interaction between clouds (Clifford and Elmegreen, 1979) is an efficient mechanism to pump kinetic energy from the orbital motions of clouds into kinetic energy of their internal motions which leads naturally to an equipartition between internal kinetic energy and magnetic energy (Falgarone and Puget, 1986).

It should be stressed here that the observable velocity dispersion of hydromagnetic waves in a cloud is not the Alfven velocity v_A. The gas velocity dispersion is a fraction of v_A by a factor dependent on the amplitude of the wave and given by the ratio of the gas displacement to the wavelength. It reaches v_A only in the case of non linear waves for which the perturbation of the magnetic field is comparable to the average field. If we are actually observing hydromagnetic waves with velocity dispersion comparable to v_A as suggested by Myers and Goodman (1988a), this has strong implications. It means that the observed waves have reached the non linear regime and the perturbation of the field is comparable in intensity to the ordered one. Hence, the magnetic field is able to affect the dynamics of the collapse in all directions, parallel to the average field as well as perpendicular to it. The role of an ordered field in increasing the Jeans mass is often discussed and may be of importance (see for example Shu et al., 1987). Nevertheless, the underlying assumption in the determination of the critical magnetic field able to stabilize a cloud is that it has comparable dimensions along and across the field. For such an assumption to be correct, some kind of hydromagnetic turbulence of the type briefly discussed above has to be at work.

We restrict ourselves below to the compressible character of the turbulence to discuss the gravitational stability. The magnetic field is certainly an essential ingredient in the dynamics of protostellar clouds and needs to be included eventually in the discussion which follows.

3. Gravitational instability in presence of supersonic turbulence: the concept of turbulent pressure

Observational data unambiguously show now that the large scale motions and the magnetic field are the two agents which prevent the molecular clouds (which contain hundreds of their Jeans masses) from collapsing in free fall. This absence of collapse at the cloud scale was shown in detailed on one cloud in the Taurus complex by Murphy and Myers (1985). However it is not straightforward to account theoretically for the global stability of clouds in which dense cores and low mass stars form. The gradual loss of magnetic support in the densest and less ionized parts may explain the slow formation of dense cores (Shu et al., 1987; Myers and Goodman, 1988b), but the support of the large scale envelope via hydromagnetic turbulence has to be accounted for simultaneously. We present below an analysis of the possible role of the turbulent pressure. If one compresses a gas of average density ρ enclosed in a box, one can be easily convinced that if this gas is moving with large scale supersonic motions of rms velocity v, these motions create on the walls of the box a force equivalent (on the average) to a pressure larger than the thermal pressure and of the order of ρv^2. Nevertheless the existence of such a turbulent pressure is not so obvious in an open system like the interstellar medium. In order to estimate qualitatively the effect of turbulence on gravitational stability, we consider a situation much simpler than the complex molecular clouds and consider an infinite medium in which an homogeneous supersonic velocity field has been generated. We then create a density perturbation in this medium over a scale $2\pi K^{-1}$ with a given external force. How the gas is going to react to the compression? Is it going to build up a pressure gradient larger than the thermal pressure gradient? Various processes will be at work depending on the scale. The turbulent velocity field in the compressed region dissipates and interacts with neighbouring regions through viscosity and non linear cascade. At scales comparable to $2\pi K^{-1}$, the turbulent velocity field stretches the perturbation. The compressible part of the field randomly increases or decreases the amplitude of the density perturbation. At small scales ($k > K$), the turbulent velocity field (incompressible and compressible) is likely to act as the turbulent field in a closed box and resist the compression (this is what is implicitely assumed when turbulent pressure is mentioned). Each of these processes has a different typical timescale.

An important characteristic of the turbulent pressure is that it is likely to be scale dependent. Any motion at a scale larger than that of the density perturbation considered only carries the perturbation without affecting its stability. Only motions at a scale snmaller or comparable to $2\pi K^{-1}$ are likely to contribute to the turbulent pressure. Bonazzola et al. (1987) have shown in 2D numerical simulations that Jeans unstable scales can be stabilized by supersonic turbulence. An analytical approach to this question is possible using renormalization group techniques (Bonazzola et al., 1989). They write an equation of motion for the velocity field built on Fourier components of spatial frequencies smaller than K. For linear density perturbations, this equation keeps the same form as the Navier-Stokes equation with pressure and viscosity terms renormalized to describe the dynamical effects of the small scales ($k > K$). The form of the turbulent viscosity they find is a rather classical result (see for example Moffat, 1981). The original result of Bonazzola et al. (1989) is that a term proportional to the density gradient appears in the equation of motion (the turbulent pressure gradient) and is given by an integral of the power in turbulent motions over spatial frequencies larger than K. The contribution of the turbulent velocity field to the pressure is given by:

$$V_K^2 = \int_K^{+\infty} dk \left[\frac{2}{3}I^\perp(k) + AI^\|(k)\right]$$

where $0 \leq A \leq \frac{1}{3}$ and $I^\perp(k)$ and $I^\|(k)$ are the power spectra of the correlations of respectively the incompressible and compressible parts of the velocity field. It should be noticed that the incompressible part of the velocity field contributes to the turbulent pressure in the same way as thermal motions of the particles to the kinetic pressure and that the same energy density in the form of a compressible field only would contribute less to the turbulent pressure.

If a linear gravitational stability analysis is carried out on a fluid controlled by this equation of motion, the following dispersion relation is derived:

$$\omega^2 - (c_S^2 + V_K^2)K^2 - 4\pi G\rho_0 = 0$$

(ρ_0 is the average density and c_S the sound velocity; the viscous terms have been omitted for clarity). The stabilisation effect of the turbulence is immediately visible since the pressure gradient term is increased. Furthermore, as shown by Bonazzola et al. (1987), if the turbulent spectrum is steep enough, the usual Jeans criterion might even be reversed, the largest scales being the most stable. The turbulent pressure is only one aspect of the possible effects of supersonic turbulence on gravitational instability. A full analysis must include the generation of non linear density fluctuations by the compressible velocity field and that of their lifetime. This analysis has not been done yet.

Recently, Léorat et al.(1989) have considered the same problem and their 2D numerical simulations show a stabilisation by supersonic turbulence (see also Passot, 1987) but their interpretation of this stabilisation is different from the previous one. They claim that the compressible part of the velocity field is the only one at work in the process. A major improvement in the understanding of the problems mentioned above will occur with 3D simulations including self-gravity over a large range of spatial frequencies.

The elements summarized here indicate a possible explanation of the behaviour of molecular clouds which form dense cores and low mass stars without being themselves globally unstable, which is in contradiction with the classical Jeans criterion. If the turbulent velocity field is fed by a cascade in which the dominant energy source is at large scale, it shows that the dynamical environment of a cloud, over scales much larger than that of the cloud itself, might affect the star formation process inside of it.

References

Arons, J., Max, C.E.: 1975, *Astrophys. J. Letters*, **196**, L77
Beichman, C.A. et al.: 1986, *Astrophys. J.*, **307**, 337
Bonazzola et al.: 1987, *Astron. Astrophys.*, **172**, 293
Bonazzola et al.: 1989, in preparation
Clifford, P., Elmegreen, B.G.: *Monthly Notices Roy. Astron. Soc.*, **202**, 629
Falgarone, E., Puget, J.L.: 1985, *Astron. Astrophys.*, **142**, 157
Falgarone, E., Puget, J.L.: 1986, *Astron. Astrophys.*, **162**, 235
Falgarone, E., Pérault, M.: 1987, *Physical Processes in Interstellar Clouds*, eds. G.E. Morfill and M. Scholer
Fuller, G.A., Myers, P.C.: 1987, *Physical Processes in Interstellar Clouds*, eds. G.E. Morfill and M. Scholer
Guélin, M. and Cernicharo, J.: 1987, *Molecular Clouds in the Milky Way and External Galaxies*, eds. J. Young and R. Snell
Jog, C.J., Ostriker, J.P.: 1988, *Astrophys. J.*, **328**, 404
Lada, C.J.: 1988, *Galactic and Extragalactic Star Formation*, eds. R.E. Pudritz and M. Fich
Larson, R.B.: 1981, *Monthly Notices Roy. Astron. Soc.*, **194**, 809
Léorat, J., Passot, T., Pouquet, A.: 1989, *Astron. Astrophys.*, in press
Moffat, H.K.: 1981, *J. Fluid Mech.*, **106**, 27
Murphy, D.C., Myers, P.C.: 1985, *Astrophys. J.*, **298**, 818
Myers, P.C., Benson, B.J.: 1983, *Astrophys. J.*, **266**, 309
Myers, P.C., Goodman, A.A.: 1988a, *Astrophys. J. Letters*, **326**, L27
Myers, P.C., Goodman, A.A.: 1988b, *Astrophys. J.*, **329**, 392
Myers, P.C. et al.: 1988, *Astrophys. J.*, **324**, 907
Norman, C.A. and Silk, J.: 1980, *Astrophys. J.*, **238**, 158
Passot, T.: 1987, *Thèse d'Etat Université Paris VII*
Pérault, M. and Falgarone, E.: 1987, *Molecular Clouds in the Milky Way and External Galaxies*, eds. J. Young and R. Snell
Scalo, J.M., Pumphrey, W.A.: 1982, *Astrophys. J. Letters*, **258**, L29
Scalo, J.: 1987, *Interstellar Processes*, eds. D.J. Hollenbach and H.A. Thronson
Shu, F., Adams, F.C., Lizano, S.: 1987, *Annual Review of Astron. and Astrophys.*, **25**, 23
Silk, J.: 1977, *Astrophys. J.*, **214**, 152

THE SELF-SIMILAR STRUCTURE OF MOLECULAR CLOUDS

E. FALGARONE[1,2], T.G. PHILLIPS[1]
[1] *California Institute of Technology, 320-47, Pasadena, CA91125, USA*
[2] *Radioastronomie, Ecole Normale Supérieure,
24 rue Lhomond, 75235 Paris Cedex 05, France*

Information on the structure of the interstellar clouds is contained in both spatial maps and spectral profiles of molecular gas emission lines. In this paper, we show that non gaussian aspects of line profiles are found at all scales. This excess wing emission is shown to be generated in dense gas. From a study of the moments of the lines as a function of scale size, we deduce a break-down in the exactness of self-similarity in the hierarchical structure of the interstellar clouds.

We have selected 40 high quality CO spectra from published profiles (Scoville and Young, 1983; Pérault et al., 1985; Blitz and Stark, 1986; Magnani et al., 1988; Falgarone and Pérault, 1988) and unpublished data from the Bordeaux, IRAM-30m and CSO telescopes. Most profiles show a clear departure from a gaussian lineshape (e.g. Fig.1) with strikingly similar slopes of the

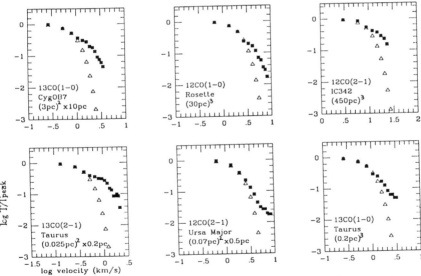

Fig. 1: *Log-log plot of a subset of normalized profiles. Open triangles trace the gaussian of same half power width. The dimensions of the volume sampled by each spectrum are given by its size on the sky i.e. beam size (or size of the area of integration) × estimated length along the line of sight.*

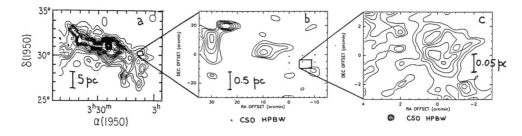

Fig. 2: *(a) Velocity-integrated intensity of ^{12}CO (J=1–0) in the Taurus complex from Ungerechts and Thaddeus (1987). Lowest contour $0.5\,K\,km\,s^{-1}$, step $1.5\,K\,km\,s^{-1}$. (b) T_a^* contours of ^{12}CO (J=2–1) emission. First contour (3σ level) $1.5\,K$, step $0.3\,K$. Dots indicate the sampling. (c) Same as (b) over the small area drawn in (b). First contour (3σ level) $1.K$, step $0.3\,K$.*

wing excesses. The spectra sample regions which are extremely different, not only in their size (the largest are fragments of spiral arms in face-on galaxies, the smallest are unresolved bright spots in nearby clouds) but also in their average volume density.

In order to elucidate the origin of this widespread behavior, we have made high angular resolution CSO observations of the gas responsible for the wing emission in two different galactic complexes (Taurus and CygOB7). We find in both cases that the large scale wing emission results from beam diluted thermalized fragments which have unresolved structure down to the smallest accessible scales ($0.02\,pc$ in Taurus, $0.1\,pc$ in CygOB7) and which have unexpectedly large brightness contrasts above the background emission level. This is illustrated in Fig.2. The $\sim 3\,pc$ fragment (Fig.2a) has an average column density of only $N_{H_2} = 9\,10^{20}\,cm^{-2}$ (and therefore an average density $\sim 100\,cm^{-3}$) but was selected for further CSO mapping on the basis of its $10\,km\,s^{-1}$ shift with respect to the bulk velocity of the neighborhood (Ungerechts and Thaddeus, 1987). The undersampled ^{12}CO (J=2–1) map (Fig.2b) reveals islands ($\sim 0.3\,pc$) of strong emission. Further unresolved emission appears in the small and more fully sampled map (Fig.2c) with brightness contrasts at the $0.04\,pc$ scale, comparable in magnitude to those observed at the ten times larger scale. These maps are roughly consistent with self-similar structure over a range of 100 in scale. However, if we now consider the line profiles as tracers of the statistics of the velocity field, it appears that the self similarity is not exact. This is done by analyzing the variations of the moments of the profiles against the scale of the sampled region. We find a deviation from the early Kolmogorov prediction (1941) which can be interpreted as evidence for intermittency, or unexpectedly large velocity vortices. This will be described in more detail in a future publication.

This work was supported by NSF grant AST83-11849. We thank the crew of the CSO for their unfailing support and R.E. Miller for the SIS junctions used in the receiver.

References

Blitz L., Stark A.A. 1986: *Astrophys. J. Letters* **300**, L89
Falgarone E., Pérault M. 1988: *Astron. Astrophys.* in press
Kolmogorov A.N. 1941: *Dokl. Akad. Nauk.* **26**, 115
Magnani L., Blitz L., Wendel A. 1988: *Astrophys. J. Letters* **331**, L127
Pérault M., Falgarone E., Puget J.L. 1985: *Astron. Astrophys.* **152**, 371
Scoville N., Young J. 1983: *Astrophys. J.* **265**, 148
Ungerechts H., Thaddeus P. 1987: *Astrophys. J. Suppl.* **63**, 645

EXTENDED AND COMPACT CO 7-6 EMISSION FROM OMC-1

J. Schmid-Burgk
Max-Planck-Institut für Radioastronomie
Auf dem Hügel 69
D53 Bonn 1
Germany, F. R.

ABSTRACT. High-resolution spectroscopy of OMC-1 reveals a "spike" source of size 1 pc plus very narrow lines from the photodissociation region, as well as a second hot-outflow source in addition to BN-KL.

We have mapped the central 6' × 8' region of OMC-1 in the 7-6 transition of CO at 806 GHz from aboard the KAO (beam size 98"). A 1.5' × 1.5' cross scan on BN-KL was also made in the same transition from the U of Hawaii 88" telescope (beam 45"). We used the MPIfR submm heterodyne receiver (Roeser et al. 1987) at spectral resolution 0.35 km/s for both experiments.
The main results are as follows (see Schmid-Burgk et al. 1989):
1. A 7-6 "spike" component, of width several km/s, is spread out over the whole area, often ressembling the corresponding CO 2-1 emission in both line shape and brightness temperature. Hence, the gas density must be at least $1...2 \times 10^4 cm^{-3}$ over an area $\approx$ 1 pc in diameter. With standard assumptions, the total dust emission is then estimated to be of order $1 \cdot 10^5 L_\odot$ (if $T_{dust} \approx 50K$), as contrasted to the total of $10 L_\odot$ emitted in CO 7-6.
2. At several of the more central positions, the spike is enhanced by an additional narrower component (Fig. 1) whose width, of order 2 km/s, appears similar to low-J transitions of rare CO isotopes (^{13}CO 2-1) rather than of ^{12}CO. This component is shown to originate from the thin, hot photodissociating layer that separates the Trapezium HII region from the ambient molecular cloud. Its intensities indicate gas temperatures of several hundred K for this layer, which requires an efficient mechanism for CO self-shielding to be operative against UV destruction.

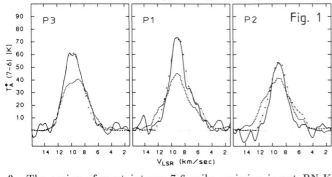

Fig.1. Bright narrow profiles from several positions on the UV dissociation sheet. Heavy lines: CO 7-6; step functions: CO 2-1; dots: ^{13}CO 2-1. Distance to the Trapezium is about 0.5 pc for all three positions.

3. The region of most intense 7-6 spike emission is not BN-KL but an area $\approx$ 2' South of it. At that position, high-velocity wings over a 30 km/s velocity interval are seen as well (Fig. 2); these must stem from $\approx 1 M_\odot$ of hot outflow gas at temperatures 500...1000 K. The wings are not visible in low-J CO lines but have been detected in SiO 2-1 (Ziurys and Friberg 1987) where they are very

similar in shape to CO 7-6. Thus besides BN-KL, a second hot-outflow source is clearly present in OMC-1. Because its dynamical properties differ strongly from BN-KL, comparative chemical studies are warranted. The upper limit to the SiO:CO abundance ratio is $4 \cdot 10^{-4}$ in the Southern source. It remains to be understood why, of the very few hot-outflow sources known for the Galaxy, two are found so close together in a fraction of space where the more common cold outflows are not seen at all.

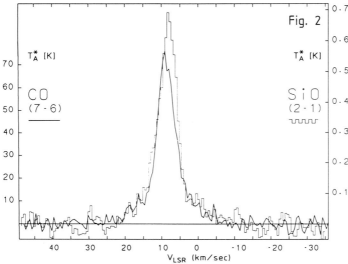

Fig. 2. The new source of hot outflow, about 1.5' to the South of BN-KL. Note the similarity of wing profiles for CO and SiO inspite of an optical depth ratio around 150. The decline of intensities with v_{LSR} is nearly three times as steep as in the BN-KL plateau.

4. 7-6 emission from BN-KL is dominated, even in our 98" beam, by the high-velocity outflows of the plateau. Outside the narrow v_{LSR} regime of the spike, the profile cannot in practice be distinguished from the published CO profiles of lower J transitions. In contrast, 2-1 profiles of the rare isotopes (both ^{13}CO and $C^{18}O$) are seen to differ completely from this standard main-isotope profile, presumably because their major contributions stem from the hot core (whose ^{12}CO lines are optically very thick) rather than from the plateau.

The dependence of plateau intensity on J strongly suggests the outflow to be optically thin for all values of v_{LSR} between ≈ 0 and -30 km/s, with the sole exception of CO 1-0. Thus our 7-6 observations indicate plateau gas temperatures of order (at least) 250...300 K. Such values, albeit higher than what recent CO 17-16 measurements (Boreiko et al. 1989) would suggest - provided these can be interpreted in terms of LTE at all - find support from our CO 12-11 and 14-13 observations of September 1988 which give brightness temperatures close to those of 7-6. Here again, profile differences to other transitions are at most marginal. Hence, main-isotope CO plateau lines with J_{up} from 2 up to 17 should all emanate from the same phase of the outflow gas. Our 14-13 high-resolution spectroscopy shows the plateau source to vary in geometrical properties with the value of v_{LSR}, in support of previous 7-6 results.

References

Boreiko, R.T., Betz, A.L. and Zmuidzinas, J. (1989), Astrophys. J. in press

Roeser,H.P., Schaefer,F., Schmid-Burgk,J., Schultz,G.V., van der Wal,P. and Wattenbach,R. (1987), J. IR mm Waves, vol. 8, no. 12

Schmid-Burgk,J., Densing,R., Krügel,E., Nett,H., Roeser,H.P., Schaefer,F., Schwaab,G., van der Wal,P. and Wattenbach,R. (1989), Astron. Astrophys. in press

Ziurys, L.M. and Friberg, P. (1987), Astrophys. J. 314, L49

The Size Distribution of Interstellar Dust

A.G.G.M. TIELENS
Space Sciences Division, NASA Ames Research Center
and Space Sciences Laboratory, UC Berkeley.

1. Introduction

The interstellar extinction curve is a particularly useful tool for probing the grain size distribution (Greenberg 1978). Typically, the extinction at a wavelength, λ, is dominated by grains with sizes, $a \approx \lambda/2\pi$ and extinction cross sections, $C_{ext} \approx \pi a^2$. Thus, the NIR-visible extinction is dominated by grains with a size of about 3000Å, while the FUV extinction results from about 100Å grains. The steep FUV rise of the extinction curve implies that small grains dominate the total geometric cross section of the dust, a major feature of all dust models (Greenberg and Hong 1974; Mathis et al. 1977). Conversely, since extinction below 1000Å has not been measured and since IR extinction is small and difficult to measure, the size distribution outside of these limits is not constrained very well. Here the small size limit of the interstellar grain size distribution will be reviewed.

2 The Molecular Domain

Based on an analysis of IR emission observations rather than FUV extinction, it is now generally accepted that the grain size distribution extends well into the molecular domain (ie., $a \approx 5$Å; cf., Allamandola et al. 1987). Many objects show strong infrared emission features at 3.3, 6.2, 7.7 and 11.3 μm, which are very characteristic for carbonaceous materials having a planar aromatic carbon structure with H atoms bonded to its edges. Figure 1 illustrates some relevant structures. Polycyclic Aromatic Hydrocarbon molecules (ie., PAHs) consist of C atoms arranged in a planar, honeycomb structure. The coordination of the edge C atoms is completed by peripheral H atoms. Larger PAHs can be made by adding successively more fused benzene rings, ultimately leading to a single graphite sheet. Such PAHs can be stacked together in the form of a cluster and finally a soot particle. PAHs in a cluster are often linked by tetrahedral C chains either interstitial or at the peripheries.

There is presently little direct spectroscopic evidence favoring free-flying PAHs over clustered PAHs (ie., amorphous carbon grains) in the ISM and probably both types are present. However, relatively intense emission in the 3.3 μm feature has been observed far from the illuminating star in reflection nebulae, where equilibrium temperatures of classical ($a>100$Å) dust grains are far too low for them to be responsible. The 3μm feature, and by inference the others as well, is therefore attributed to emission by small

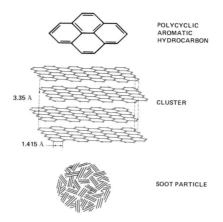

Fig. 1: Structural relationship between different carbonaceous materials (see text).

"particles" transiently heated through the absorption of a single UV photon to temperatures above 500 K (Sellgren 1984). Since the total excitation energy available is limited to less than 13.6 eV, an upper size limit on the emitter can be placed of about 5-10Å. This size range falls actually in the molecular domain and the excitation and emission process should be properly described as UV pumped, IR fluorescence of PAH molecules (cf., Allamandola et al. 1987). Nevertheless, the overall conclusion remains the same: the emission carrier contains 20 to 50 C atoms.

Since ≈5% of the total IR flux is emitted in the IR emission features, the far UV opacity of the carriers is ≈5% relative to that of the dust. Using a typical UV absorption cross section (≈10^{-17} cm^2/C-atom) and a standard dust absorption cross section per H atom (cf., Savage and Mathis 1979), this requires that the emission carrier contains about 1% of the elemental carbon. With a size of 20 C atoms per molecule this translates to an abundance of PAHs per H atom of 2x10^{-7}.

It has been suggested that vibrational excitation energy could be highly localized in an individual aromatic moiety, which is loosely bound to an amorphous carbon grain (Duley and Williams 1988). In this picture, such a PAH island stuck in or on a carbon grain would be thermally isolated and thus emit as a "free flying" molecule. Low-temperature studies of vibrationally excited molecules isolated in inert gas matrices have shown that, under certain conditions, IR vibrational decay is favored over multi-phonon energy transfer to the thermal bath (cf., Legay 1977). However, energy transfer through dipole-dipole interaction to vibrational modes close in energy of neighboring molecules is very rapid (≈10^{-11}s; Legay 1977). Thus, two PAH islands separated by 100Å will still share their vibrational excitation (ie., relax non radiatively) before IR emission takes place (≈0.1s). Moreover, PAH islands are bonded to a grain by carbon chains, whose fundamental vibrations are very close in energy, facilitating rapid vibrational energy transfer even between distant islands. Note that the experimental evidence cited to support localized vibrational excitation (Duley and Williams 1988) actually refers to local storage of electronic excitation energy. Indeed, the thermal diffusion timescale for localized electrons is $\approx 10^{-3}$s at 300K in a semi-conductor with a bandgap of 1eV (Kastner 1985). However, the high energy vibrations excited by decay of this electronic excitation thermalize with the phonon bath on a timescale of 10^{-10} s (Malinovsky 1987), much faster than the IR emission timescale. Thus, since vibrational energy transfer is rapid in a solid, the 3µm feature has to result from a small carrier.

3 Carbon Clusters

The observed IR spectra show evidence for a broad plateau underlying the 6.2 and 7.7µm features which varies independently (Cohen et al. 1988). This plateau has been attributed to CC stretching modes in PAH clusters. Emission around these wavelengths requires temperatures of about 400K. Assuming that its specific heat is similar to that of graphite, the temperature of a cluster containing N_c C atoms is given by $T \approx 2 \times 10^3 \, (h\nu(eV)/N_c)^{0.4}$ K. Thus these observations imply a size of ≈ 500 C atoms ($a \approx 10$Å). Typically, $\approx 5\%$ of the energy is in this plateau emission. Assuming the same UV absorption cross section as for PAHs, this translates into a cluster abundance of 10^{-8} per H atom ($\approx 1\%$ of C abundance). Probably, the galactic 12µm cirrus observed by IRAS is dominated by emission from PAHs and these clusters. Indeed, the 12 µm IRAS flux of reflection nebulae, the high flux equivalent of cirrus, is dominated by the IR emission features (Puget 1987).

4 Small Carbon Grains

Most of the 60 µm cirrus emission observed from IRAS is also due transiently heated grains, as evidenced by the dependence of the 60-100µm color temperature on the intensity of the incident UV field, G_o (fig. 2; Puget 1987; Ryter et al. 1987). For a λ^{-1} dust emissivity law, the equilibrium dust temperature, T_d, is proportional to $G_o^{0.2}$. A semi-empirical fit of the emissivity to reflection nebulae observations predicts a dust temperature in the average interstellar radiation field of about 10K (cf., fig. 2). In contrast, the dust temperature of the cirrus, corrected for a λ^{-1} dust emissivity law, is typically ≈ 25 K. Using the MRN model results in slightly higher temperatures (≈ 20K), due to the lower IR emissivity of graphite and silicates (fig 2; Draine and Anderson 1985), but the dependence on G_o is steeper than observed and the reflection nebulae data is not fitted very well. In contrast, models incorporating transiently heated grains seem to agree with all observations (cf., Fig 2; Draine and Anderson; Ryter et al. 1987). Typically, 60µm emission requires a temperature of ≈ 50K ($T \approx 0.3/\lambda$) and this translates into a 45Å grain size ($N_c \approx 4 \times 10^4$) for a 10eV photon. About 15% of the total IR emission is emitted in the 60µm band, thus implying 2.5% of all the C in 40Å grains ($\approx 3 \times 10^{-10}$ /H atom). Finally, note that these color temperatures are not directly related to physical grain temperatures.

5 The Grain Size Distribution

Thus, the grain size distribution extends, probably continuously, into the molecular domain (fig 3). Equilibrium emission by classical dust grains is

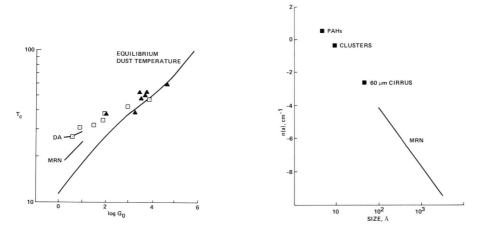

Fig. 2 (left): The 60/100μm color temperature as a function of the incident UV field (in units of the average interstellar radiation field). Typical G_0's for cirrus, reflection nebulae and HII regions are ≈ 1, 10^3, 10^5, respectively. IRAS (squares) and KAO (triangles) observations have been taken from the literature. Equilibrium dust temperatures have been calculated assuming a λ^{-1} emissivity law and a simple radiative transfer code. MRN and DA refer to calculations based on the MRN dust model without and with the effects of transient heating included (Draine and Anderson 1985). Figure adapted from Ryter et al. (1987).

Fig. 3 (right): The derived interstellar carbon grain size distribution.

expected to dominate only for $\lambda > 300\mu m$ ($T_d \approx 10K$) and submm observations might probe the grain size distribution out to about 250Å. Previous studies of the IR cirrus have suggested that the grain size distribution steepens considerably for a<30Å (Draine and Anderson 1985), reflecting the observed higher 12μm (T≈300K; a≈10Å) than 25μm (T≈120K; a≈25Å) flux. This difference with our analysis results from the incorrect use of bulk graphite optical properties for small grains. In fact, for small carbon grains the IR emissivity is dominated by the C-C vibrations rather than electronic intraband transitions (cf., Tielens and Allamandola 1987). These vibrations are intrinsically much stronger in the 5-15μm than in the 20-30μm region (Allamandola et al. 1987), thus, explaining the increase in flux to shorter wavelength.

The size distribution of interstellar grains probably reflects their evolutionary interrelationship. Processes that affect the grain size distribution are shattering and coagulation upon grain-grain collisions. Since only a relative grain velocity of 1 km/s is required for the former, besides SN shocks, it can also be of importance in molecular clouds (Tielens 1988). Sticking of grains will only occur for very low grain velocities, v<3 m/s for 1000Å carbon grains (Chokshi et al. 1988). Extreme high densities (>10^9 cm^{-3}) are then required before coagulation becomes important in molecular clouds. Coagulation can proceed at much lower densities if grains are ice-coated. But unless these icy grain mantles are rapidly processed into a more refractory grain mantle, such conglomerates will not be able to withstand the rigors of the diffuse ISM and coagulation will have no lasting effect on the grain size distribution (Tielens 1988).

Finally, it should be emphasized that the abundance of extremely large grains (a>>3000Å) is not well known. A simple extrapolation of the MRN distribution towards larger grain sizes would provide some difficulty with the observed NIR extinction curve in the diffuse interstellar medium. Moreover, extrapolation beyond about 1 μm is limited by cosmic abundance constraints. Interestingly, while the small grain population can be studied at long wavelengths, large grains can be best probed at short wavelengths, ie., through the diffraction pattern of X-ray point sources (cf., Catura 1983). Note that the size distribution of interstellar SiC grains isolated from meteorites extends to micron-size scales (Anders et al. 1988). Also, although generally considered unlikely, it has sometimes been argued that cm-sized Ca,Al rich inclusions in chronditic meteorites have an interstellar origin, (cf., Wood 1985). In any case, meteorites may have preserved a record of the size distribution of large grains, albeit altered by protoplanetary processes. Possibly, the best source of information on interstellar grains in general may reside in comets and a comet sample-return mission may revolutionize this research field.

References

Allamandola, L.J., Tielens, A.G.G.M., and Barker, J.R., 1987 in *Physical Processes in Interstellar Clouds*, eds. G.E. Morfill and M. Scholer, (Reidel, Dordrecht), p.305.
Anders, E., Lewis, R.S., Tang, M., and Zinner, E., 1988, in *Interstellar Dust*, ed. L.J. Allamandola and A.G.G.M. Tielens, (Kluwer, Dordrecht), in press.
Catura, R.C., 1983, *Ap.J.*, **275**, 645.
Chokshi, A., Tielens, A.G.G.M., and D. Hollenbach, 1988, in preparation.
Cohen, M., et al., 1988, *Ap. J.*, in press.
Draine, B.T. and Anderson, N., 1985, *Ap. J.*, **292**, 494.
Duley, W.W., and Williams, D.A., 1988, *M.N.R.A.S.*, **231**, 969.
Greenberg, J.M., 1978, in *Cosmic Dust*, ed. J. McDonnell, (Wiley,New York), p.187.
Greenberg, J.M., and Hong, S.S., 1974 in *Galactic Radio Astronomy*, ed. F.J. Kerr and S.C. Swans (Reidel, Dordrecht), p.155.
Kastner, M.A., 1985, in *Physical Properties of Amorphous Materials*, ed. D. Adler, B.B. Schwartz and M.C. Steele, (Plenum Press, New York), p.381.
Legay, F., 1977, in *Chemical and Biochemical Applications of Lasers*, ed. C.B. Moore, (Academic Press, New York), p.43.
Malinovsky, V.K., 1987, *J. Non-Cryst. Sol.*, **90**, 37.
Mathis, J.S., Rumpl, W., and Nordsieck, K.H., 1977, *Ap. J.*, **217**, 425.
Puget, J.L., 1987, in PAHs and Astrophysics, eds. A. Leger, L. d'Hendecourt, and N. Bocarra, (Reidel, Dordrecht), p.303.
Ryter, C., Puget, J.L., and Perault, M., 1987, *Astr. Ap.*,**186**, 312.
Savage, B.D. and Mathis, J.S., 1979, *Ann. Rev. Astr. Ap.*, **17**, 73.
Selgren, K.S., 1984, *Ap. J.*, **277**, 623.
Tielens, A.G.G.M., and Allamandola, L.J., 1987, in *Physical Processes in Interstellar Clouds*, eds. G.E. Morfill and M. Scholer, (Reidel, Dordrecht), p.333.
Tielens, A.G.G.M., 1988, in *Interstellar Dust*, ed. L.J. Allamandola and A.G.G.M. Tielens, (Kluwer, Dordrecht), in press.
Wood, J.A., 1985 in *Protostars and Planets II*, eds., D. Black and M. Mathews, (Univ. Arizona Press, Tucson), p.687.

HIGH RESOLUTION OBSERVATIONS OF DUST EMISSION

Rolf Chini
Max-Planck-Institut für Radioastronomie, Auf dem Hügel 69,
5300 Bonn 1, F.R.G.

ABSTRACT. This review primarily presents high resolution continuum observations at 1300μm as obtained with the MPIfR bolometer attached to the IRAM 30m MRT. Some outstanding measurements will be shown in the following, more to demonstrate the possibilities of this system rather than to discuss an individual object in detail. In particular the capability of mapping extended galactic and extragalactic sources at a resolution of 11" is shown with examples.

1. Observational details

The MPIfR bolometers (Kreysa, 1985) operate with ^{3}He-cooled Ge-crystals at a temperature of 0.27K. Each cryostat is optimized for one of the atmospheric windows at 350, 730, 870, and 1300μm; the individual bandpasses are defined by interference mesh filters; their transmission at shorter wavelengths was determined to be below 10^{-6}. During the observations, atmospheric transmission is frequently checked by skydips. The calibration is done via the planets; from the repeatability of observations of flat spectrum radio sources an absolute accuracy of order 10-20% is expected for results with signal to noise $\geq 10\sigma$ decreasing to factors 2 or 3 for 3-4σ detections.

Mapping at MRT and SEST is done continuously at a speed of 8"-15"/s and with scan direction parallel to chopping direction; the scan separation is 1/3 HPBW. This procedure yields dual beam maps in Nasmyth- or Az, El- coordinates which then are converted and restored by the algorithm of Emerson, Klein and Haslam (1979). Flux density measurements obtained from such maps are of the same accuracy as quoted above. So far most prominent galactic star forming regions and strong galaxies have been mapped on the northern (MRT) as well as on the southern (SEST) hemisphere at our most efficient wavelength of 1300μm. A few examples are given below.

2. High resolution maps of galactic regions

2.1 NGC2024

Six so far unknown condensations of cold (≤ 16K) dust and gas have been detected towards the dark lane across NGC2024 (Fig.1). From their physical properties (n_{H_2} ~10^8-10^9cm^{-3}, m ~$10-60 M_\odot$, linear sizes $10^{16}-10^{17}$cm) we suggest that they are isothermal protostars at the end of their contraction phase. A comparison with low resolution (90") observations show that the condensations in Fig.1 contain only 23% of the total mass of the cloud which amounts to ~$800 M_\odot$ (for details see Mezger et al., 1988).

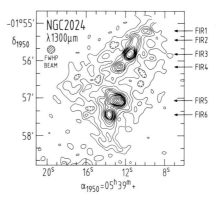

2.2 M17

The molecular cloud SW of M17 has been mapped with resolution of 90" (IRTF) and 11" (MRT) at 1300μm and was partly observed at 350μm with a 30" beam (IRTF). Fig.2 shows the high resolution map at 1300μm (0,0:$18^h17^m28^s$, $-16°14'$) which still includes free-free emission from the underlying HII region. Like in many other regions we find increasingly clumpy structure when going to higher resolutions.

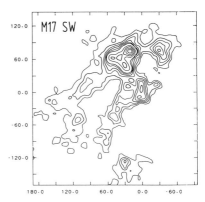

2.3 The galactic center

Fig.3 shows the galactic center at 1300μm as seen with the MRT (Mezger et al. 1989). A ring of dust with a total gas mass of ~$6 \ 10^4 M_\odot$ and a highly clumpy structure surrounds the synchrotron shell source Sgr A East (0,0:$17^h42^m29^s$, $-28°59'19"$).

3. Continuum emission from external galaxies

The 1300μm emission from galaxies is generally rather weak and only a few, outstanding objects can be mapped at that wavelength. Fig.4 shows as examples the starburst galaxies M82 and NGC253, as well as the prominent radio galaxies Virgo A and Cygnus A (Salter et al. 1989). Generally, however, our knowledge of the 1300μm extent of galaxies will be restricted to cross scans or multi-aperture observations. It has been shown that the emission from active Mkn-galaxies is below 40mJy at 1300μm in an 11" beam (Krügel et al. 1988 a,b). Recent re-observations of this Mkn-sample

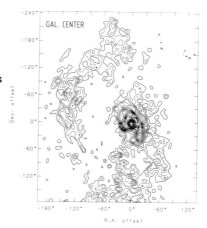

at SEST with a 30" beam indicate that the major fraction of gas is concentrated within the nucleus of these galaxies, without any evidence for an extended component.

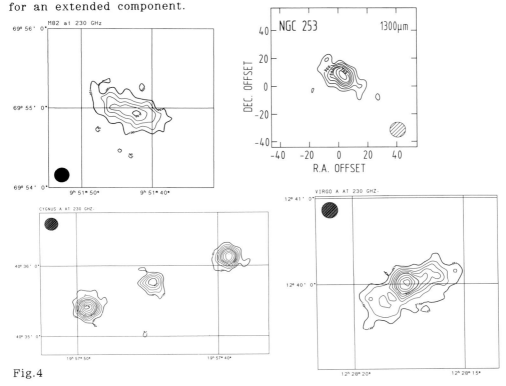

Fig.4

The spatial extent of mm/submm emission from Arp 220 is heavily debated among continuum observers as well as among CO-line people. Agreement could be achieved that this galaxy contains a central source of 130mJy (5") (Scoville), 290mJy (11") (Chini), 280mJy (16") (Robson), 240mJy (Keene) and 350mJy (30") (Chini); these numbers were quoted by the corresponding authors at this conference and indicate little dependence on the angular size of the observed region. Whether there is an additional extended component has to be investigated by low resolution observations.

Fig.5 shows a map of Arp 220 as obtained at the MRT; here only small, probably unresolved feature is to be seen. The CO-data presented for Arp 220 during this conference indicate that besides a strong nuclear component (Scoville) also extended emission seems to be present (Solomon). The distribution of gas and dust within galaxies will remain a major challenge for submm-observations during the next years.

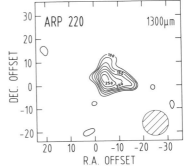

4. Mass determinations

It is well known that the optically thin thermal emission from heated dust basically allows to estimate the amount of radiating grains. Using the canonical conversion factor of 100 between gas and dust one can even derive H_2 masses for various objects. Like other methods, gas masses derived from dust emission have their uncertainties; the latter mainly arise from our poor knowledge of dust properties and chemical abundances. The other possibility for mass estimates comes from CO observations. Recently Chini et al. (1989a) have started a systematic comparison between the two methods by observing all Mkn galaxies for which dust masses are available (Krügel et al. 1988a,b).

Fig.6 shows a preliminary result of this study. The surprisingly good correlation emphasizes that there are no systematic errors with either approach.

Closely related to the above problem is the conversion of optical depth at 1300μm into a visual extinction. Although there may be problems in the individual galactic regions (arising from abnormal extinction at optical wavelengths), reasonable agreement can be found for external galaxies:

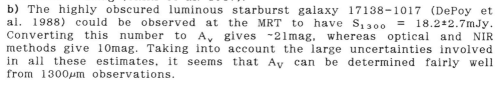

a) The IRAS quasar 13349+2438 (Beichman et al. 1987) suffers from 2.5mag A_V. Converting the observed 1300μm optical depth into a visual extinction one finds 5.2mag (Chini et al. 1987).

b) The highly obscured luminous starburst galaxy 17138-1017 (DePoy et al. 1988) could be observed at the MRT to have S_{1300} = 18.2±2.7mJy. Converting this number to A_V gives ~21mag, whereas optical and NIR methods give 10mag. Taking into account the large uncertainties involved in all these estimates, it seems that A_V can be determined fairly well from 1300μm observations.

5. The wavelength dependence of dust opacity

The MRN dust model (Mathis et al. 1977) and the dust properties given by Draine and Lee (1984) have been very successful in reproducing the interstellar extinction curve from UV to submm and in describing most observations in this wavelength range. These models suggest a square law dependence of dust opacity on frequency.

There is also overwhelming observational evidence for ß=2:
a) From a sample of ~90 compact HII regions we derived ß≥1.7 combining IRAS and 1300μm data (Chini et al. 1986a,b).
b) A subset of this sample was observed at 350, 800, 1100 and 3300μm (Gear et al., 1988) corroborating our result and quoting ß=1.8±0.2.
c) Spectra of 23 Mkn galaxies can be fitted by a single dust temperature from 60 to 1300μm, indicating that ß=2 (Krügel et al. 1988a,b).
d) Spectra of 25 radio-quiet quasars may also be fitted by a single blackbody curve from 60 to 1300μm and ß=2. The steepness of the

spectra (mostly upper limits at 1300μm) clearly rules out a linear wavelength dependence of dust opacity (Chini et al. 1989b and contribution in these proceedings).
e) Energy distributions of a number of individual galaxies also suggest ß~2. As an example Fig.7 shows the spectrum of NGC4945 as given by Brock et al. (1988). These authors propose a fit with ß=1 (dashed line) and therefore predict a 1300μm flux density of 3160mJy. We have observed an unresolved source of 750mJy (dot in Fig.7) towards the center of that galaxy - again favouring a square law dependence of dust opacity.

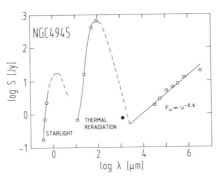

References

Beichmann, C.A., Soifer, B.T., Helou, G., Chester, T.J., Neugebauer, G., Gillett, F.C., Low, F.J.: 1987, Ap.J. **308**, L1
Brock, D., Marshall, J., Lester, D.F., Harvey, P.M., Ellis, H.B.: 1988, Ap.J. **329**, 208
DePoy, D.L., Wynn-Williams, C.G., Hill, G.J., Becklin, E.E.: 1988, A.J. **95**, 398
Chini, R., Kreysa, E., Mezger, P.G., Gemünd, H.P.: 1986a, Astron.Astrophys.**157**, L1
Chini, R., Kreysa, E., Krügel, E., Mezger, P.G.: 1986b, Astron.Astrophys. **166**, L8
Chini, R., Kreysa, E., Salter, C.J.: 1987, Astron.Astrophys.**182**, L63
Chini, R., Krügel, E., Steppe, H.: 1989a, in prep.
Chini, R., Kreysa, E., Biermann, P.: 1989b, Astron.Astrophys., in press
Draine, B., Lee, H.: 1984, Ap.J. **285**, 89
Emerson, D.T., Klein, U., Haslam, C.G.T.: 1979, Astron.Astrophys. **76**, 92
Gear, W.K., Robson, E.I., Griffin, M.J.: 1988, MNRAS **231**, 55p
Kreysa, E.: 1985, Proc. URSI Int.Symp."mm- and submm-wave Astronomy", Granada, Spain, p.153
Krügel, E., Chini, R., Kreysa, E., Sherwood, : 1988a, Astron.Astrophys. **190**, 47
Krügel, E., Chini, R., Kreysa, E., Sherwood, : 1988b, Astron.Astrophys. **193**, L16
Mathis, J., Rumpl, W., Nordsiek, K.: 1977, Ap.J. **217**, 425
Mezger, P.G., Chini, R., Kreysa, E., Wink, J.E., Salter, C.J.: 1988, Astron.Astrophys. **191**, 44
Mezger, P.G., Zylka, R., Salter, C.J., Wink, J.E., Chini, R., Kreysa, E., Tuffs, R.: 1989, Astron.Astrophys. in press
Salter. C.J., Chini, R., Haslam, C.G.T., Junior, W., Kreysa, E., Mezger, P.G., Spencer, R.E., Wink, J.E., Zylka, R.: 1989, Astron.Astrophys., in press

THE MAGNETIC FIELD AT THE CENTER OF THE GALAXY

ROGER H. HILDEBRAND
The University of Chicago
5640 South Ellis Avenue
Chicago, IL 60637 U.S.A.

ABSTRACT. The polarization of the 100 μm emission from the 3 pc dust ring at the center of the Galaxy has been measured at five points. The magnetic field inferred from the measurements lies in the plane of the ring but does not appear to be toroidal. We estimate a field $B \approx 10^{-2}$ gauss.

Yusef-Zadeh and Morris (1987) have found long non-thermal arcs and filaments in their 6 cm continuum observations of the central 50 pc of the Galaxy. These features run along magnetic field lines perpendicular to the plane of the Galaxy. Within the region of the 6 cm observations is a molecular cloud with a ridge of submillimeter thermal emission extending ~30 pc parallel to the Galactic Plane (Hildebrand et al. 1978 and Mezger et al. 1988). The central portion of the cloud radiates strongly at 100 μm (Fig. 1a). That portion, the "3 pc dust ring," is heated from within by a luminous source ($>10^7$ L$_\odot$; Becklin, Gatley, and Werner 1987).

Werner et al. (1988) and Hildebrand et al. (1988) have shown that the 100 μm emission from the dust ring is polarized with the E-vector normal to the ridge of emission (Fig. 1a). The inferred direction of the magnetic field is in the plane of the ring: approximately normal to the field in the 50 pc arcs and filaments. Contrary to what one would expect for a toroidal field in a toroidal dust ring seen edge-on, the polarization seen at the flux density peak at the northern edge of the ring is greater than that seen at the middle of the ring.

We estimate the magnitude of the field in the dust ring by methods introduced by Chandrasekhar and Fermi (1953). For the first method we assume that the angular dispersion, $\alpha \approx 0.16$ rad, of the polarization vectors is due to a hydromagnetic wave. (Angles and errors are tabulated in Hildebrand 1988). The field is given by $B = (4\pi\rho/3)^{1/2} v/\alpha$ where $\rho = 7 \times 10^3$/cm^3 is the beam averaged hydrogen volume density (Genzel et al. 1985) and $v = 50$ km/s is the r.m.s. velocity dispersion (Güsten et al. 1987). The result is $B \approx 7 \times 10^{-3}$ gauss. This result is of course based on fewer points (5) than would be desirable.

The second method is to use the relationship (magnetic pressure) + (kinetic pressure) = (gravitational pressure). Using a central mass M = 5×10^6 M$_\odot$, a (clump) density $n = 10^5$/cm^3, and a rotational velocity $v = 110$ km/s (at $R \approx 1.5$ pc) we have $B = 18 \times 10^{-3}$ gauss (at equilibrium). The values chosen for the parameters entering into these estimates are debatable, but we shall assume on the basis of the estimates that the field is of order 10^{-2} gauss.

The relatively tenuous region within 1 pc of the center (within our central beam) contains features known as the "northern arm" and the "bar" which appear in the near-IR continuum (Fig. 1b). Smith, Roche, and Aitken (1988) have measured the 12.5 μm polarization from near-IR sources in these features. Their results, corrected for selective absorption in foreground clouds, indicate a magnetic field along the ridge of emission in the northern arm (*i.e.* approximately parallel to the field in the ring as shown in Fig. 1b).

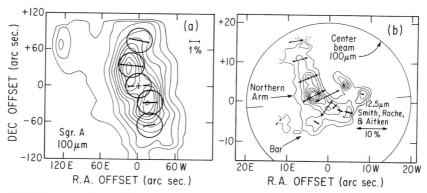

FIGURE 1. (a) Polarization of the 100 μm emission from the dust ring surrounding Sgr A (Hildebrand *et al.* 1988). The map of flux densities is by Davidson *et al.* (1988). The cirlces show the positions of the beams used in the polarimetry. The enclosed bars show the degree and direction of the polarization (E-vector) as measured with the University of Chicago far-infrared polarimeter on the Kuiper Airborne Observatory. The inferred field is perpendicular to the polarization vectors. Offsets are measured from $17^h 4^m 29.4^s$, $-28°59'19''$ (1950). (b) Comparison of 12.5 μm and 100 μm polarimetry. The contours show the flux density at 8-13 μm and the arrows the polarization at 12.5 μm (Smith, Roche, and Aitken 1988). The large circle and radial tick marks show the central beam and position angle for the 100 μm observations (Fig. 1a). Again the inferred field is perpendicular to the polarization vectors.

These observations raise the following questions: What sustains the non-thermal emission from the arcs? (See discussion by Yusef-Zadeh and Morris 1987.) Is the magnetic field confined within the arcs? If so, why don't they expand? If not, why do we see large-scale motion perpendicular to these features? Is the field in the dust ring toroidal? If so, why is the polarization greater at the northern flux density peak than in the center? If not, if the large-scale field normal to the plane is not wound up by the rotating ring, then why is the field in the plane of the ring? Why is the polarization in the northern arm relatively strong and uniform compared to the field in the bar?

J.A. Davidson, M. Dragovan, D. Gonatas, M. Morris, G. Novak, S.R. Platt, M. Werner, and X. Wu participated in the far-infrared measurements presented here. I wish to thank D. Aitken and R. Rosner for many discussions. I am grateful to C.H. Smith, P.F. Roche, and D. Aitken, and to J.A. Davidson, P. Harvey, D. Lester, M. Morris, and M. Werner for permission to show their data. This work was supported by NASA Grant No. NSG 2057 and NSF Grant No. AST 8513974.

References

Becklin, E.E., Gatley, Ian, and Werner, M.W. (1982) 'Far-infrared observations of Sagittarius A: the luminosity and dust density in the central parsec of the Galaxy,' Astrophys. J. **258**, 135-142.

Chandrasekhar, S. and Fermi, E. (1953) 'Magnetic fields in the spiral arms,' Astrophys. J. **118**, 113-11 .

Davidson, J.A., Harvey, P.M., Lester, D.F., Morris, M., and Werner, M.W. (1988) (in preparation).

Genzel, R., Watson, Dan M., Crawford, M.K., and Townes, C.H. (1985) 'The neutral-gas disk around the Galactic Center,' Astrophys. J. **297**, 766-786.

Güsten, R., Genzel, R., Wright, M.C.H., Jaffe, D.T., Stutzki, J., and Harris, A.I. (1987) 'Aperture synthesis observations of the circumnuclear ring in the Galactic Center,' Astrophys. J. **318**, 124-138.

Hildebrand, Roger H. (1988) 'Polarized infrared emission from dust,' IAU Symposium 135 on Interstellar Dust, July 1988 (EFI preprint 88-57).

Hildebrand, R.H., Whitcomb, S.E., Winston, R., Stiening, R.F., Harper, D.A., and Moseley, S.H. (1978) 'Submillimeter observations of the Galactic Center,' Astrophys. J. **219**, L101-L104.

Hildebrand, R.H., Davidson, J.A., Gonatas, D., Morris, M., Novak, G., Platt, S.R., Werner, M.W., and Wu, X. (1988) (The 100 μm results presented here will be incorporated in a latter publication).

Mezger, P.G., Zylka, R., Salter, C.J., Wink, J.E., Chini, R., and Kresa, E. (1988) 'Continuum observations of Sgr A at mm/submm wavelengths,' (preprint, submitted to Astronomy and Astrophysics).

Smith, C.H., Roche, P.F., and Aitken, D.K. (1988) (in preparation).

Werner, M.W., Davidson, J.A., Morris, M., Novak, G., Platt, S.R., and Hildebrand, R.H. (1988) 'The polarization of the far-infrared radiation from the Galactic Center,' Astrophys. J. (in press) .

Yusef-Zadeh, Farhad, and Morris, Mark (1987) 'The linear filaments of the radio arc near the Galactic Center,' Astrophys. J. **322**, 721-728.

HOT MOLECULAR GAS IN THE PROTOPLANETARY NEBULA CRL 2688

M. G. Smith, T. R. Geballe, G. Sandell, and C. Aspin
U.K./Canada/Netherlands Joint Astronomy Centre.

A protoplanetary nebula is the result of the mass loss that occurs as a red giant star evolves into the core of a planetary nebula. The mass loss produces a cool, massive, and slowly expanding envelope around the dying star. Fast and hot winds, emanating from the central star as it approaches the white dwarf stage, overtake and interact with the more slowly-moving envelope (e.g. Kwok 1982, Ap.J., 258, 280). CRL 2688 is a particularly interesting case of this phenomenon, which we have been studying at the JCMT and the UKIRT.

The profile of CO J=3-2 emission from the central 13" (0.06pc) of CRL 2688 is shown in Fig. 1a. The profile has 3 noteworthy features: (a) a roughly parabolic central core, the signature of the red giant wind expanding at about 20 km/s; (b) wide blue wings extending from -55 km/s to about -90 km/s (LSR), identifiable with the fast wind; (c) a self-absorption feature near -54 km/s. The high-speed wind extends over a region about 20" across, implying a dynamical time scale for it of ~1000 yr (assuming a distance of 1 kpc and an inclination of 30 deg).

An absorption feature is also seen in the CO J=1-0 line profile (Kawabe et al. 1987, Ap.J., 314, 322) and in the CO J=2-1 profile shown by Knapp at this conference. Similar features are seen in low J CO transitions in a number of other objects, as illustrated here by Lucas. The absorption most likely arises in the cold, outer envelope as it expands away from the central, hot core. In CRL 2688 the absorption dips reported for the 1-0 and 2-1 lines are narrower than that of the 3-2 line (FWHM ~ 8 km/s). Given the temperature gradient in the material surrounding the central core, it is understandable that the line width increases at higher J levels. The value of T_A^* for the J=3-2 profile observed with the 15-m JCMT is everywhere about twice that for the J=1-0 profile observed by Kawabe et al. with the 45-m Nobeyama telescope (i.e. with similar angular resolution). We estimate that T(ex) is about 80K from models calculated by Stutzki, in good agreement with independent estimates by Kawabe et al. CO J=1-0 and possibly even J=2-1 are optically thin at this temperature, so the larger width of the J=3-2 line may be due to a first glimpse of self-absorption by the fast wind or by material directly affected by it.

We have brought higher spatial resolution to bear on the core of CRL 2688 by using the imaging camera (IRCAM) with 0.6" pixels, along with Fabry-Perot and grating spectrometers, on UKIRT. The images in the

H$_2$ S(1) line and in the near and mid-infrared continuum have allowed us to build up a composite picture of the region (Fig. 1b). The two clumps of excited H$_2$ along the lines of sight to the optical (i.e., N and S) scattering lobes had already been mapped by Beckwith et al. (1984, Ap. J., 280, 648); the clumps to the E and W (Fig. 1c) are a new discovery. Our spectra covering the 2-micron region show that the detected H$_2$ is predominantly shock-excited in all lobes.

Profiles of the H$_2$ S(1) line, obtained at 12 km/s resolution, reveal a surprising situation (Fig. 1d): the N and E lobes are both approaching at similar velocity, whereas the peaks of the S and W lobes are more redshifted. The axis of the bipolarity is therefore less clearly the line joining the lobes of scattered light. The obvious disk-like morphology revealed in other nebulae such as M 2-9 by Aspin et al. (1988, Astron. Ap., 196, 227) seems at first sight to occur (faintly) in the K image of CRL 2688; however, our more detailed study has left the situation less clear. Perhaps a large cone-opening angle will still provide consistency with a simple bipolar outflow.

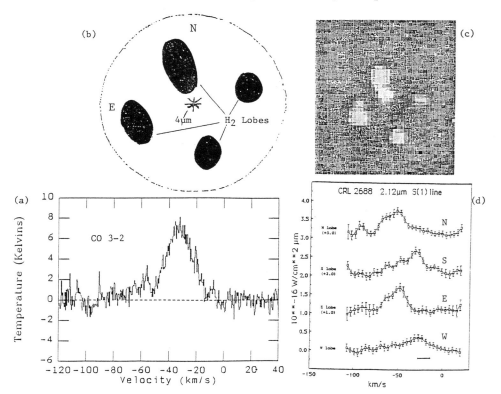

Figure 1 (a) CO(3-2) line profile of CRL 2688.
(b) Schematic composite of the observations.
(c) IRCAM+FP (FWHM = 100 km/s) image of H$_2$ S(1) emission.
(d) S(1) line profiles of different lobes at 12 km/s res.

CO J=3→2 Observations of IRC+10216

J.A. TAUBER, J. KWAN, P.F. GOLDSMITH, R.L. SNELL,
and N.R. ERICKSON
Five College Radio Astronomy Observatory
University of Massachusetts
Amherst, MA 01002

The circumstellar envelope of the old star IRC+10216 has become an important testing ground of theoretical models which seek to determine the physical properties of stellar winds. It is in particular important to evaluate accurately the mass loss rate from this class of stars, as they may play an important role in the replenishment and enrichment of the interstellar medium. The mass loss rate is one of the parameters that goes into radiative transfer models which calculate the thermal structure of the gas together with molecular line excitation and emission self-consistently. The prototype of these models is that of Kwan and Linke (1982) (KL), which has been applied to the case of CO emission in its lower two rotational transitions, J=1→0 and J=2→1. This model can explain adequately the main features of the observed lines; however, there are some small discrepancies in the variation of the line intensities with impact parameter. The resulting best fit parameters are a compromise solution, which somewhat overestimates the intensities at the center of the source, and underestimates them in the outer envelope. KL point out that one of the possible reasons for the discrepancies is that the beam patterns of the telescopes used may have contained close-in sidelobes that picked up power from the central region of the source, even when the telescope was pointed quite far away from the central position, and thus lead to an enhanced antenna temperature when the outer regions of the source are observed. In the work of KL, the nature of these sidelobes was sufficiently uncertain that the corrections deduced were very uncertain. There is therefore an incentive to observe this source using antennas with very accurately known beam patterns; optical telescopes fulfill this requirement, since their surfaces are virtually perfect at millimeter wavelengths. Unfortunately the beam sizes obtainable at millimeter wavelengths with optical telescopes are usually too large to constitute useful probes; one therefore has to observe at the highest possible frequency to offset this problem.

In this work, we have observed the CO J=3→2 transition at 345 GHz, using the 2.3 m dish of the Wyoming Infrared Observatory. The receiver used was built at the Five College Radio Astronomy Observatory and has been described elsewhere (Tauber 1989). The beam shape was determined quite accurately by scanning the limb of the Moon, and can be described by two gaussians, the first with HPBW of 87 arcseconds containing 60% of the total power, and the second with HPBW 7.5 arcminutes containing 25% of the power. The source was strip mapped in the N-S and E-W directions at 0.5 arcminute spacing. In Figure 1 we present the observed spectra, after azimuthal averaging of the offset positions. The falloff in intensity is quite rapid, and at 1.5 arcminutes from the center the observed line is already below the confusion limit.

We have modeled the source as in KL, using the original code and the same input parameters. The results that we obtain are in good agreement with our observations (see Figure 2), as well as with previous single point observations of this source (Phillips et al. 1982; Jewell, Snyder and Schenewerk 1988). The discrepancies noted in the lower two transitions are not present in our strip map. However, the model also shows that the CO J=3→2 emitting region is very small, so that our large beamsize barely resolves it. Higher spatial resolution observations will be needed to determine conclusively whether beam pattern effects can be the cause of the differences between model predictions and observations of the J=1→0 and J=2→1 transitions.

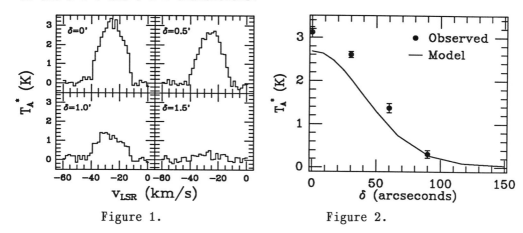

Figure 1.   Figure 2.

References

Jewell, P.R., Snyder, L., and Schenewerk, M. 1988, in preparation.
Kwan, J., and Linke, R.A. 1982, *Ap.J.*, **254**, 587.
Phillips, J.P., White, G.J., Ade, P.A.R., Cunningham, C.T.,
    Richardson,K.J., Robson, E.I., and Watt, G.D. 1982,
    Astron.J., **116**, 130.
Tauber, J.A., Kwan, J, Goldsmith, P.F., Snell, R.L., and Erickson, N.R.
    1989, *Astron.J.*, in press.

CO OBSERVATIONS OF EVOLVED GIANT STARS

G.R. Knapp and C.F. Gammie
Princeton University Observatory

K. Young and T.G. Phillips
California Institute of Technology

The copious mass loss which takes place at late stages of stellar evolution, while a star is on the AGB, causes the formation of a cool extended circumstellar envelope expanding at speeds like 15 km s^{-1}. These envelopes are observed in many molecular lines, which give the stellar velocity, the wind velocity and an estimate of the mass loss rate.

A program has recently begun using the Caltech Submillimeter Observatory (CSO) to study the thermal emission from the rotational lines of CO in these envelopes. At first, we are concentrating on surveying evolved stars with a wide range of properties, to establish mass loss parameters at different evolutionary stages. To date, 47 new detections have been made, increasing the number of detected evolved stars by 25%.

The typical emission line from a circumstellar envelope is somewhere between parabolic and flat-topped in shape (depending on optical depth and resolution effects) and the emission descends sharply to zero intensity at the positive and negative values of the outflow velocity. A particularly interesting recent finding is the presence of a second, faster wind in objects which are evolved past the AGB stage, and are becoming, or have become, planetary nebulae. Such winds have previously been seen in OH231.8+4.2, V Hya, NGC7027 and CRL2688, and the CSO observations have turned up several more. The terminal outflow velocities of both the fast and slow winds are listed in Table 1; all data except those for OH231.8+4.2 and V Hya are from CSO CO(2-1) observations.

Star	V_o(slow)	V_o(fast)
V Hya	15	29
OH231.8+4.2	23	70
CRL618	21	76
CRL2688	20	37
NGC7027	18	23
M2-56	19	63
M1-92	23	53
CPD $-$ 56°8032	23	74

CO(2-1) line profiles for CRL618, CPD $-$ 56°8032 and M2-56 are shown in

Figure 1. The fast winds are found to be much more prominent in the CO(2-1) line than in the (1-0) line, and their profiles drop smoothly into the noise (unlike the line profiles for slow winds), making it difficult to assign a maximum velocity. The objects in Table 1 show signatures of post AGB evolution, such as hot central stars, optical emission lines, radio continuum emission etc. Several bright AGB giants have been searched for such wings, so far without success.

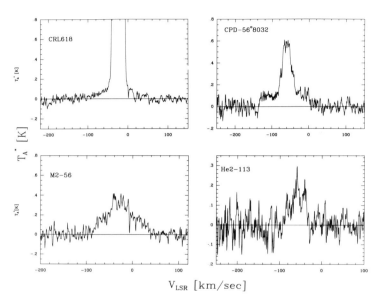

Figure 1. CO(2-1) line profiles for CRL618, CPD − 56°8032, M2-56, and He2-113

The cool molecular winds from AGB stars are likely to be driven by radiation pressure on dust - for a sample of about 60 such stars, we find $\dot{M}V_0 \leq L_*/c$, with more than half of the stars having the wind and luminosity momenta approximately equal. This is not the case for the fast winds, however, for which $\dot{M}V_0$ is found to be tens of times higher than L_*/c, and indeed the energy flux in the winds is found to be several percent of the stellar luminosity. The properties of these fast molecular winds are very similar to those seen around young stars embedded in molecular clouds - there is even some evidence that these winds are bipolar.

A final point of interest is that CPD − 56°8032 is a Wolf-Rayet (WC) star, i.e. a very hot star showing emission lines from several ionization stages of carbon. It is one of four stars considered to be an extension of the WR sequence to cooler, later types (the WC11 stars). We have detected weak CO emission from a second of these, He2-113 (also shown in Figure 1), and the IRAS spectra of both show emission from the "PAH" features. Several other cool WR stars were observed, but all were close to the galactic plane and confused by strong emission from galactic CO.

Research at the CSO is supported by the N.S.F. under grant AST83-11849. The N.S.F. also supported this work via grant AST87-62945 to Princeton.

FAR-INFRARED AND SUBMILLIMETER ANALYSIS OF A DENSE CORE: TEMPERATURE AND DENSITY STRUCTURE OF L1489

E.F. Ladd, P.C. Myers
Harvard-Smithsonian Center for Astrophysics, Cambridge, MA
S.C. Casey, D.A. Harper
Yerkes Observatory, University of Chicago, Williams Bay, WI
J.A. Davidson
NASA/Ames Research Center, Moffett Field, CA
P.J. Benson
Wellesley College, Wellesley, MA
R.M. Levreault
Wesleyan University, Middletown, CT

ABSTRACT. We present far infrared and submillimeter wavelength images of the dark cloud core Lynds 1489, taken with the Kuiper Airborne Observatory (KAO) and the Infrared Telescope Facility (IRTF). Using these images we have constructed maps of the temperature and dust density structure in the vicinity of the forming protostar. In brief, the observed structure is consistent with that of a hot (3000 - 6000 K) point source located on the edge of a cold (10 - 15 K) dense (10^5 cm^{-3}) core. Our results indicate that the protostar, in concert with the interstellar radiation field, plays a large role in radiatively heating its surroundings including the associated dense core.

1. Introduction

Lynds 1489 is a relatively isolated clump of gas and dust on the western edge of the Taurus molecular cloud complex (Lynds 1962). It is a small cloud, covering about 0.5 square parsecs in extent, with a maximum visual extinction (measured from star counts) of 3.8 magnitudes and an inferred mass of 92 M$_\odot$ (Cernicharo, Bachiller, and Duvert 1985). The dense core contained within this cloud is considerably smaller than the cloud itself. Measured by NH$_3$ emission from the (J,K) = (1,1) inversion transition, the core extends over 0.05 x 0.08 pc with a maximum density of n(H$_2$) = 6 x 10^4 cm^{-3} (Gaida, Ungerechts, and Winnewisser 1984). The thermal and nonthermal linewidths of these observations are exceedingly small, indicating that the dense core is very cold and quiescent. Gas kinetic temperatures derived fom these data are of order 10 K (Benson 1983).

2. Observations

The IRAS source associated with this core (04016+2610) is located some 0.1 pc to the west of the ammonia peak. It is the prime radiator in both of our KAO broadband observations. The 100μm and 160μm maps both peak at the IRAS position, but are slightly resolved. We interpret this spatial broadening to be due to warmed dust in the immediate vicinity of the protostar.

The IRTF observations show much more structure. While one peak coincides with the IRAS position, there are two other emission peaks in the 360μm map. The more extended peak to the east is located within the half power contour of the ammonia emission and is interpreted to be emission from the cold, high density dust within the core. The second peak, approximately 1 arcminute north of the IRAS source position is a surprise. Ammonia has been detected at this position, though not at the strength measured in the eastern core.

3. Analysis

We have calculated temperatures and densities as function of map position by fitting a blackbody to any two of the three wavelengths measured. Dust column densities were then calculated by determining the optical depth given the derived temperature at each map position. The temperature map derived from comparison of the 100μm and 360μm observations is shown at left below and the 100μm opacity map is at right. The protostar appears to sit in a region of relatively flat temperature structure, though there is a real gradient from southwest to northeast. Toward the ammonia emission in the north and east, the temperature drops significantly. The opacity map also clearly mimics the observed ammonia structure, with very high column densities to the north and east. From this analysis it appears that the northern peak 360μm is both colder and denser than the eastern core.

To the southwest, however, the density appears to be very low, and the temperature reaches a maximum. This low temperature, high density region is roughly coincident with the blue lobe of L1489's ^{12}CO (2 $\rightarrow$ 1) outflow measured with the Five College Radio Astronomy Observatory 14m telescope. The high temperature, low density region detected with the far-infrared/submillimeter analysis may be a conduit through which the bulk outflow from the star passes. Contour intervals on the maps are 5 K in temperature and 0.0026 in 100μm opacity.

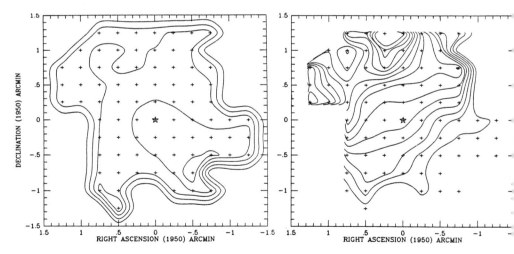

4. References

Benson, P.J. 1983. Ph.D. thesis, Massachusetts Institute of Technology.
Cernicharo, J., Bachiller, R., and Duvert, G. 1985. *Astron. Astrophys.*, **149**:273.
Gaida, M., Ungerechts, H., and Winnewisser, G. 1984. *Astron. Astrophys.*, **137**:17.
Lynds, B.T. 1962. *Ap. J. Suppl.*, **7**:1.

NGC 2071: A TWIN FOR L1551, and EVIDENCE FOR BACKFLOW?

G. H. Moriarty-Schieven, V. A. Hughes
Dept. of Physics, Queen's University
Kingston, Ontario K7L 3N6 Canada

R. L. Snell
FCRAO, University of Massachusetts
Amherst, MA 01003

We present ^{12}CO J=2-1 maps of the high velocity emission of the NGC 2071 bipolar molecular outflow (Figure 1) obtained at FCRAO. Despite enormous differences in the luminosities of the sources driving the L1551 and NGC 2071 outflows (39 $L_\odot$ and 10^3 $L_\odot$ respectively), the two outflows themselves are remarkably similar. Both exhibit a similar expanding shell morphology and have a similar amount of mass involved in the outflow. NGC 2071 is much more highly inclined out of the plane of the sky than L1551, and if this greater inclination is taken into account, then the momentum, energy and physical sizes of the two outflows are virtually identical (Moriarty-Schieven and Snell, 1988).

Another feature in common with L1551 is the presence of red-shifted emission on the limb of the blue lobe, and blue-shifted emission on the opposite limb of the red lobe. This was suggested by Uchida et al. (1987) in L1551 to be evidence for corotation of the outflow with a magnetized rotating disk around IRS-5. We can compare the angular momentum of such a disk (assuming the disk reported by Kaifu et al. (1984) which has an angular momentum of $\sim 3.5 \times 10^{-2}$ $M_\odot$ km/s pc) with the angular momentum of the outflow. The latter is estimated from only the red-shifted emission on the limb of the blue lobe, and vice versa. For NGC 2071, this value is 2×10^{-3} $M_\odot$ km/s pc and for L1551 is 3.2×10^{-2} $M_\odot$ km/s pc, a value which is nearly identical to that of the "disk" of Kaifu et al. This disk has been called into question, however (Batrla and Menten, 1985; Moriarty-Schieven et al., 1987), and so the disk angular momentum is an upper limit. The "outflow" angular momentum, on the other hand, is an extreme lower limit since it was calculated using only a small fraction of the emission. Thus the feature in question cannot be due to rotation.

An alternative explanation for the peculiar emission is that it is caused by a jet emanating from the central source and which is not aligned with the outflow axis. The emission arises where the jet is blasting into the cavity wall, causing extreme turbulence. This possibility explains the antisymmetry of the emission, but does not account for the extension of the emission outside of the outflow lobe. It is possible that this emission is the signature of backflow of molecular material or some form of large-scale circulation within the molecular cloud. Detailed modelling is required.

References

Batrla, W., and Menten, K.M. 1985, Ap.J.(Letters), **298**, L19.
Kaifu, N., et al. 1984, Astr.Ap., **134**, 7.
Moriarty-Schieven, G.H., et al. 1987, Ap.J.(Letters), **317**, L95.
Moriarty-Schieven, G.H., and Snell, R.L. 1988, Ap.J., **332**, 364.
Uchida, Y., et al. 1987, in M. Peimbert and J. Jugku (eds)., IAU Symposium 115, Star Forming Regions, Dordrecht: Reidel, p. 297.

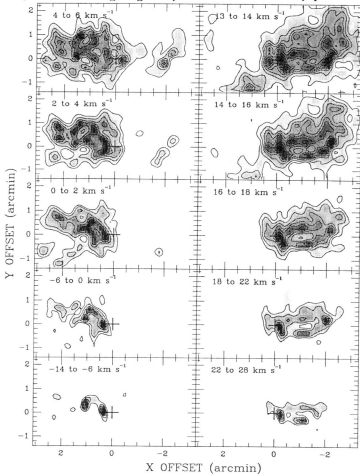

Figure 1: The average intensity within each of ten velocity intervals in the NGC 2071 outflow was found, and the maps were reconstructed using Maximum Entropy (Moriarty-Schieven and Snell, 1988).

Section II

Instrumentation and Cosmic Background Experiments

A NEW TECHNIQUE FOR SURFACE MEASUREMENTS OF RADIO TELESCOPES

E. Serabyn, C. R. Masson, and T. G. Phillips
Division of Physics, Mathematics and Astronomy
California Institute of Technology
Downs Laboratory of Physics, 320-47
Pasadena, CA 91125

In order to optimize the surface figure of the parabolic primary mirror of the Caltech Submillimeter Observatory, holographic measurements of the dish surface are being carried out. A new holographic technique is currently under developement, which has a number of advantages over existing techniques. This measurement technique is based on the use of a shearing Fourier Transform Spectrometer (Malacara 1978; Steel 1983) to examine the shape of the wavefronts from a point-like astronomical source.

The new measurement technique is similar to that used at sites with more that one radio telescope available. In this case, one telescope (the reference telescope) is always pointed directly at an astronomical source, while a second telescope (the one being measured) is pointed at off-center positions, in order to determine the far-field response pattern of the telescope (Scott and Ryle 1977; Hills, this volume). These far-field measurements are then spatially Fourier transformed to yield the electric field distribution in the aperture plane of the telescope, from which the surface deviations are derived.

The technique of "shearing holography" differs from this scenario primarily in that two images of a single telescope are used, instead of two individual telescopes. The way this is done is shown in Fig. 1. After the telescope focuses the collimated millimeter and submillimeter wave radiation from a distant point source onto the Cassegrain focal surface, the radiation reexpands to a tertiary off-axis paraboloid (P1). This mirror recollimates the radiation, and forms an image of the primary mirror about one focal length beyond the tertiary. Insertion of a mylar beamsplitter (B) between the tertiary and the image of the primary then splits the radiation to form two images of the primary. Flat mirrors M1 and M2, positioned at these image surfaces, reflect the radiation back to the beamsplitter, where the two halves recombine before being focused onto a detector by a second off-axis parabola (P2).

The optical arrangement described constitutes a Fourier Transform spectrometer (FTS), and so can be used for astronomical and atmospheric spectroscopy, if the end mirror M1 is translated normal to its plane. However, making use of the fact that the end mirrors of the FTS contain images of the primary, one can see that tipping mirror M2 out of its plane is equivalent to tipping the telescope. Thus, by putting M2 onto a gimbal mount, M2 can be used to steer its beam on the sky off-axis. The two image mirrors are then analogous to the two telescopes in the interferometric technique of Scott and Ryle (1977), with mirror M1 always pointing a beam at the center of the source, while M2 directs the second beam off-axis, to map a grid of far-field positions.

At the beamsplitter, the radiation from the off-axis and on-axis directions are combined, resulting in a superposition of the two electric fields on the detector. The resultant intensity contains a cross-term which is proportional to one quadrature component of the desired off-axis electric field. Thus a power

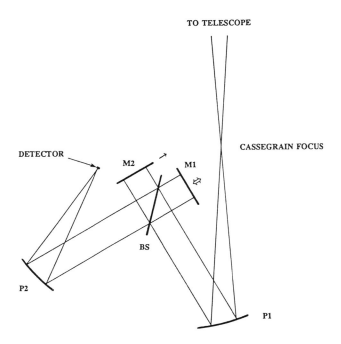

Fig. 1 Optical arrangement of the shearing Fourier Transform Spectrometer. Mirror M1 translates as in a conventional FTS. Mirror M2, mounted on a gimbal mount, is free to rotate out of the plane normal to M1.

detector, such as a bolometer, can be used to measure the field. Moving the translatable mirror by an eighth of a wavelength would yield the other quadrature component, if the radiation (or detector response) were monochromatic. However, with broadband radiation, the far-field pattern, which is a function of wavelength, can still be determined, if mirror M1 is scanned to produce a full interferogram. This interferogram can then be inverted to yield the far-field electric field as a function of frequency. Thus, the far-field, and so also the aperture plane field, can be determined at a number of frequencies simultaneously.

The primary advantages of this new technique of "shearing holography" can be summarized as follows: First, the broadband nature of the technique allows holography to be done at a number of frequencies simultaneously. This allows not only a check of the procedure, but also a measurement of the efficiency of the telescope as a function of frequency. Second, holography can be performed at the high frequencies at which the telescope will be used. Since the surface phase errors are a larger fraction of a wavelength at high frequency, the requirements on signal-to-noise ratio are less severe. Third, by using planets as sources, the deformation of the dish with changing elevation angle can be determined. Finally, the instrument can also be used as a normal Fourier Transform Spectrometer to study atmospheric and astronomical spectra.

References
Malacara, D. 1978, *Optical Shop Testing*, Wiley.
Steel, W. H. 1983, *Interferometry*, Cambridge Univ. Press.
Scott, P. F., and Ryle, M. 1977, *MNRAS*, **178**, 539.

Gravitational Deflection of the Leighton Telescopes

David Woody
Owens Valley Radio Observatory
California Institute of Technology
Big Pine, CA 93513

Abstract The gravitational deflection of the 10.4 meter telescopes used for the Owens Valley Radio Observatory millimeter-interferometer array have been measured using radio holographic techniques. The measurements show that the telescopes are very stiff with less than 20 microns RMS deviation from homology over the zenith angle range from 15 to 70 degrees.

Introduction

The performance of a large radio telescope is ultimately limited by the zenith angle dependent gravitational deflections. Even for telescopes with active control of the panel positions, the accuracy of the gravitational deflection model will determine the surface errors. We have measured the surface errors of the 10.4 meter telescopes built by Bob Leighton as a function of zenith angle.

Three of the Leighton telescopes are at the Owens Valley Radio Observatory (OVRO) where they are used in the millimeter-wave interferometer and a fourth telescope is used by the Caltech Submillimeter Observatory. The telescope surface maps used for this paper were made with the interferometer using standard holographic techniques [1] and using the quasar 3C84 as a point source.

Observations and Model

Telescope maps were made covering the zenith angle range from 4 to 70 degrees. Figure 1 is a example of one of the maps. The resolution is approximately 1 meter. The gravitationally induced distortions between 15 and 65 degrees were only 27 microns RMS.

The maps were used to construct a model of the deflections of the telescopes based upon the theory of the elastic distortions of a structure under the influence of gravity [2]. The deflections at any angle are simply of the form A*SIN(ZA)+B*COS(ZA). The focus and coma aberrations agreed with Leighton's original calculations to within the measurement uncertainty. The astigmatism is several times larger than the predicted value. Higher order perturbations were handled by fitting the displacement of each grid point

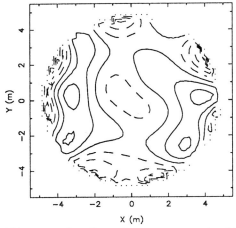
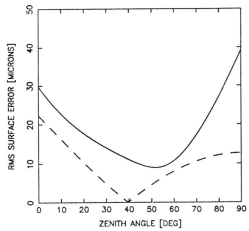

Figure 1. Aperture map at 15 degrees with focus and coma aberrations removed. Contour spacing is 20 microns.

Figure 2. Calculated surface errors. The astigmatism contribution is shown as a dashed line.

to an equation of the form given above. The RMS deviation of the data from this model was only 6 microns.

This model of the telescope was then used to calculate the expected RMS surface errors as a function of zenith angle for different panel settings. The results for a strategy which minimized the average RMS error over the full zenith angle range is shown in figure 2. The focus and coma aberrations are assumed to be removed by positioning the secondary at the optimum position for each zenith angle.

Conclusions

Our study shows that a simple static tuning of the surface of the 10.4 meter diameter Leighton telescopes will yield better than 20 microns RMS surface error over the zenith angle range from 15 to 70 degrees. This also means that only a relatively small range of active panel tuning would be required to make these telescopes useful into the far-infrared wavelength region.

References

[1] P. F. Scott and M. Ryle, "A rapid method for measuring the figure of a radio telescope reflector," Mon. Not. R. astr. Soc., vol. 178, pp. 539-545, 1977.

[2] S. von Hoerner and W. Wong, "Gravitational deformation and astigmatism of tiltable radio telescopes," IEEE Trans. Antennas and Prop., pp.689-695, Sept. 1975.

THE SMT: A JOINT SUBMILLIMETER TELESCOPE PROJECT OF THE MAX-PLANCK-INSTITUT FÜR RADIOASTRONOMIE AND STEWARD OBSERVATORY

J. W. M. Baars, E. Krügel
Max-Planck-Institut für Radioastronomie, 5300 Bonn, W. Germany

R. N. Martin
Steward Observatory, University of Arizona, Tucson AZ 85721, USA

ABSTRACT. We present a summary of the main features of the SMT project with emphasis on the site and the organisational aspects. Technical features of the telescope are presented in the proceedings of the Zermatt Symposium (Baars and Martin, 1989).

1. Introduction

During the construction of the 30 m Millimeter Radio Telescope, planning began at the MPIfR on a smaller, high precision telescope which would be dedicated to ground based observations at the shortest accessible wavelengths (300-1000 μm). To fully exploit the submillimeter windows of radio astronomy, we needed: 1) A high dry site, since less than 50% transmission is expected at 350 μm wavelength under even the best conditions. 2) An extremely accurate telescope which has full efficiency at the shortest wavelength (i.e. 15 μm rms surface under operational conditions). 3) Improved detector technology.
Existing collaboration between the MPIfR and Steward Observatory (SO) was formalized in 1982. These two institutes are jointly designing, building, equipping and operating the 10 m Submillimeter Telescope (SMT) facility.

2. Project Organisation

The project was set up in a way to use the existing expertise of both institutes in an optimal way. The design and construction of the telescope is carried out under the responsibility of the MPIfR by the firms Krupp and M.A.N, which also built the 30 m telescope. The hard- and software of the control system is developed at the MPIfR.
    Design and construction of a protective enclosure is in the hands of SO. It is a novel, simple and cheap design. When closed it looks like a barn; it is depicted open in Figure 1. The development of the site is also the responsibility of Steward Observatory.
    The accurate reflector panels are replicated from molds, made of pyrex glass and ground to 3 μm accuracy at the Optical Sciences Center of

the University of Arizona. SO (under contract to MPIfR) designed and constructed a focusing and chopping system for the secondary reflector. The measurement of the reflector surface is the responsibility of SO.

A first set of facility receivers has been built at the MPIfR. Both ^{3}He-cooled bolometers and cooled Schottky mixer receivers will be available. Further receiver development, e.g. SIS mixers, is being carried out at both partner institutes.

The capital investment of the project amounts to about $ 6M and is being shared by both institutes. The operation will also be a joint venture. Observing time will be divided equally between both partners.

3. The Site

The site chosen for the SMT is Mount Graham at an altitude of 3200 m and is 220 km SE of Tucson. This new site was proposed by the University of Arizona in 1984 for several future large telescopes in the submillimeter, optical and infrared regime. More than four years were used to resolve environmental aspects. Finally an area of 60 hectares have been set aside for an observatory on Mt. Graham by an act of Congress in October 1988.

For 30 months we have measured the precipitable water vapor with a 22 GHz radiometer. A comparison with radio-sonde data for Tucson, which have been gleaned over a period of 22 years (Fig.2), shows excellent agreement. For eight months of the year the 350 and 450 µm windows will be "open" for 10-40% of the time. Since we see no indication of a diurnal effect and as the telescope operates in daytime with unimpaired parameters, we plan to observe 24 hours per day during suitable weather.

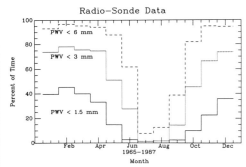

Fig.1 The SMT in the open "dome".   *Fig.2 Distribution of water vapor.*

We plan to prepare the site and erect the enclosure during 1989. The telescope will be installed and commissioned in 1990. We hope to begin full astronomical operation in 1991.

Reference: Baars, J.W.M., Martin, R.N. (1989) 'The 10 m Submillimeter Telescope (SMT)', in G. Winnewisser and T. Armstrong (eds.), Physics and Chemistry of Interstellar Clouds, Springer Verlag, Berlin.

SUBMILLIMETER INSTRUMENT DEVELOPMENT AT STEWARD OBSERVATORY

R. N. MARTIN, C. E. WALKER, C. K. WALKER
Steward Observatory,University of Arizona, Tucson, AZ 85721

R. E. MILLER
AT & T - Bell Labs, Murray Hill, NJ 07974

ABSTRACT. Receiver instrumentation for the UA/MPIfR Submillimeter Telescope (SMT) is being assured through three different mechanisms. (1) There will be standard facility instruments. Initially, these include 460 GHz and 345 GHz Schottky waveguide frontends, a bolometer, and two AOS backends (supplied by the MPIfR). (2) The MPIfR and UA are both, independently, working on supplementary instrumentation. These instruments are more experimental in nature. And (3) arrangements have been made (and encouraged) with other institutions to bring unique instruments. In this paper, we will describe submillimeter instrumentation developments underway at Steward Observatory.

1. SIS Heterodyne Receiver

For low noise temperature and broad frequency coverage, we have begun development of an open structure SIS heterodyne receiver system at Steward Observatory. We are currently using a bowtie as the broadband antenna element, similar to that used by Wengler *et al.* (1985) except that we have scaled down the size of the bowtie to provide optimum performance at 345 GHz. We are also investigating the broadband characteristics of other types of antenna structures which may provide better performance than the bowtie. The SIS junctions and the associated antenna structure are fabricated by R. E. Miller at Bell Laboratories.

We have tested our receiver both in the lab and on the NRAO 12 meter telescope. At 345 GHz the receiver has a double sideband noise temperature of about 350 K. The cryostat is a hybrid dewar using a CTI-22 cold head and a 4 ℓ liquid He reservoir, with a 10–12 day hold time.

2. Fourier Transform Spectrometer (FTS)/Bolometer System

The FTS/bolometer system is designed for broadband spectroscopic, continuum and polarization measurements at submillimeter wavelengths. The FTS is of the Martin-Puplett type with freestanding wire grids as beam splitters. It is usable over the wavelength range of 2 to 0.3 mm with a resolution of 250 MHz (unapodized). Four ^{3}He cooled (T = 0.3K) bolometers are being used as detectors, two optimized for 450 μm wavelength and the other two optimized for 0.8 and 1.2 mm wavelength. It is useful for spectroscopic studies of atmospheric transmission and calibration, absolute source calibration, and searches for strong broad lines. With the FTS set at the central fringe, the system can be used as a broadband continuum detector and polarimeter.

3. Broad Bandwidth Analog Delay Autocorrelation Spectrometer

We have begun the development of a 2 GHz wide autocorrelation spectrometer which uses analog delay lines. Conventional autocorrelation spectrometers sample the incoming signal at the Nyquist rate and do the delay and multiplication in digital hardware. However, the speed of digital sampling and logic limits this technique to spectrometers of about 100 MHz total bandwidth. To reach broad bandwidths, we have gone to analog delays with lengths of cable and circuit board stripline. The signals are digitized after the summation in analog circuitry.

4. SMT Secondary Mirror System

The chopping secondary system has been designed and constructed at UA under contract by the MPIfR. The system has both a high performance chopper and a focus-translation stage. The chopping system nutates the subreflector in one dimension. Design goals are 4' beam throw at 10 Hz and 25" beam throw at 30 Hz with an 80% duty cycle. Initial tests of the system indicate that these goals will be exceeded.

5. SMT Surface Measuring Instrumentation

The position of each SMT panel will be adjusted from the rear by means of differential adjusting screws. The panels will initially be set mechanically utilizing fiducial marks replicated onto the panels. Subsequently, we will measure the errors and refine the setting using the method developed by Ulich, et al. (1986). The method we have developed utilizes phase retrieval by amplitude interferometry.

The modulator box, transmitter, and most of the receiver system have been constructed. Tests with a prototype version were successfully completed in 1984 using the Texas 5 m telescope. Tests with the current version are scheduled for November 1988 at the Texas 5 m telescope.

REFERENCES

Ulich, B. L., Walker, C. K., Philips-Walker, C. E., Davison, W. B., Davis, J. H., and Mayer, C. E. 1986, *SPIE Opt. Align.*, **608**, 55.

Wengler, M. J., Woody, D. P., Miller, R. E., and Phillips, T. G., 1985, *Intl. J. IR MM Waves*, **6**, 8.

OPERATION OF BOLOMETRIC DETECTORS UNDER CONDITIONS OF VARYING SKY BACKGROUND

M. J. GRIFFIN
Physics Dept.
Queen Mary College
Mile End Rd.
London E1 4NS
England

1. Introduction:

The next generation of ground-based submillimetre photometers will use very sensitive bolometric detectors, whose performance will be strongly influenced by varying thermal background radiation from the sky. In this paper, we consider the effects of this background on the detector responsivity, taking the SCUBA bolometer array for the JCMT as an example. A method is proposed to monitor and calibrate out the changes in detector responsivity caused by changes in the sky background

2. The effect of background loading on the detector responsivity:

The total thermal background loading, Q, on the SCUBA detectors will be dominated by the sky contribution, and, in the 855 µm band, will vary by a factor of about 3 - from around 32 pW under good conditions to around 92 pW under bad conditions. This represents a new regime of operation for bolometric detectors. In previous instruments, the background power in any one waveband has been essentially constant regardless of the sky emissivity, being determined largely by the local contribution. The variable background means that the bolometer is now a significantly non-linear device, as the responsivity and NEP both depend on the background. And, since the sky emissivity can vary considerably from night to night, from one time of night to another, from zenith to horizon and from one azimuth to another, the responsivity changes will pose some calibration problems.

Using a theory of the bolometer as an ideal thermal device, it is possible to model the influence of the background radiation (Griffin and Holland, *Int. J. Infrared and Millimeter Waves*, 9, 861, 1988). Figure 1 shows the variation of the normalised responsivity of a model SCUBA detector with γ, a parameter representing the amount of background power ($\gamma = Q/G_0 T_0$, where T_0 is the bath temperature and G_0 is the thermal conductance at T_0). The range of γ for the SCUBA 855 µm channel is indicated. Responsivity curves are plotted for three values of ϕ_a, the normalised operating temperature at $Q=0$. ϕ_a is effectively set by the bias current: $\phi_a = 1.1$ is close to the theoretical optimum,

1.3 is moderately overbiased, and 1.5 is heavily overbiased.
Note the large change in responsivity over the anticipated range of γ - a factor of 2 for optimum biasing. By over-biasing, which does not seriously degrade the NEP, we can reduce this to around 25%. In operation, therefore, it will be necessary to calibrate out responsivity variations at the level of 25% or more.

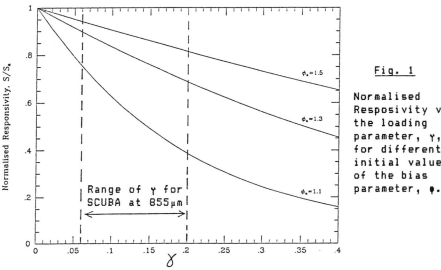

Fig. 1

Normalised Resposivity vs the loading parameter, γ, for different initial values of the bias parameter, ϕ.

3. Correcting for the responsivity variations:

For each of the SCUBA detectors, a correction will have to be applied to the data to remove the dependence of responsivity on the exact sky background level at the time of the observation. There are two basic ways of calculating the correction factor. The first is to measure the responsivity as a function of the DC operating voltage for each detector, and to monitor the voltage during receiver operation.

A second approach, which is more flexible and attractive, is to monitor the responsivity directly by means of an internal calibration source in the receiver. This feature will be incorporated in SCUBA. The internal calibrator (a small electrically heated thermal radiation source) can illuminate the detector array with a power level which is constant for any given detector (but need not be the same for every detector, so that simple optics may be used). The power level must be small compared with the total background. For SCUBA, a value of around 10^{-14} W will be sufficient to give high signal-to-noise. If the source is modulated at a frequency different to the sky-chopping frequency, then it can even be operated while the array is taking data.

In addition, the internal calibrator will serve as a useful diagnostic feature for performing calibration and sensitivity checks on the detectors during testing of the instrument before installation at the telescope and during daytime work on the reciever.

SCUBA: A Submillimetre Common-User Bolometer Array for the James Clerk Maxwell Telescope

W. D. Duncan
Joint Astronomy Centre
665 Komohana Street
Hilo, Hawaii 96720
U.S.A.

1. INTRODUCTION

The large diameter (15m) and high surface accuracy (~38microns) of the James Clerk Maxwell Telescope (JCMT) have allowed submillimetre measurements of unprecedented sensitivity and spatial resolution[1] to be made with the JCMT's continuum photometer UKT14[2]. In order to exploit the new sensitivity limits and high spatial resolution made possible by the JCMT and Mauna Kea site a new submillimetre continuum array instrument has been designed.

2. OVERALL SYSTEM DESIGN

Design Parameter and Explanation

a) SINGLE MODED (DIFFRACTION LIMITED) OPTICS.
   Maximise coupling to Airy pattern and minimise background power.
b) NARROW BAND FILTERS (Q~10).
   Minimise background from wings of telluric lines and the variations of effective wavelength with observing conditions.
c) BOLOMETERS COOLED TO .1K.
   To achieve photon noise limit set by the sky background determined by the single moded optics/filters.
d) MAIN WAVELENGTHS 855 AND 438 MICRONS.
   Gain of antenna peaks at ~450 microns for thermal sources. The 800 micron window good poor weather channel. The nearly two to one ratio of wavelength allows pixel centres of one array to overlap the other in simultaneous observing using dichroic.
e) 855 MICRON ARRAY USED @ 750 AND 600 MICRONS. 438 MICRON ARRAY USED AT 350 MICRONS. ALSO 1.3 & 1.1 mm CHANNELS.
   Flexible wavelength choice for scientific reasons (better estimates of temp., density & emissivity).
f) INTERNAL CALIBRATOR
   .1K bolometers are non-linear with changes in sky background. Responsivity variations removed using temperature modulated internal source.

g) FOCAL PLANE CALIBRATOR
   Allows the local sky transmission to be measured. Important because of the inconstancy of the submillimeter sky.
h) FIELD OF VIEW ~ 2 ARC MINS. (91 pixels in 438 micron array and 37 at 855 microns).
   Field of view set by the dynamics of the chopping secondary and by cost considerations.
i) OPTICAL DESIGN BASED ON BACK TO BACK GAUSSIAN BEAM TELESCOPES[3].
   Gaussian Beam telescopes provide frequency independent beam waist locations and alternate images of sky and telescope aperture fields (ideal for filter and dichroic locations).
j) ALL REFLECTING DESIGN.
   Minimum losses. Design by a combination of geometrical ray tracing and wave optics coupling calculations.

3. SYSTEM SENSITIVITY

The zenith photon noise limited equivalent flux densities are estimated to be 20 mJy at 855 microns and 200 mJy at 438 microns for 1mm of precipitable water vapour. The system is designed to achieve these sensitivities in the absence of '1/f' fluctuations from the sky[4]. The amount of the low frequency fluctuations (sky noise) present in the data will depend on the observing conditions and the effectiveness of any suppression techniques employed. Single element systems employ 3 position chopping[4] and an array can make use of the high degree of spatial coherence of sky noise across the telescope focal plane and employ cross-correlation techniques. Thus, the system may approach the quoted PNEFD's under some observing conditions. This is a great advantage in speed per pixel over previous systems and gives factors of 10^5 or more improvement in mapping speed. This will mean a new era in submillimetre continuum science as the detection and mapping of objects previously considered too faint becomes possible in 1991/2. The system is being constructed, in collaboration with Queen Mary College, by a group at the Royal Observatory, Edinburgh led by Dr. W. K. P. Gear. The optical calculations are due to Anthony Murphy (Maynorth College, Ireland).

REFERENCES

1. Weintraub, D. A, S., Sandell, G., Duncan, W. D. Submitted to Ap.J. Letters.
2. Duncan, W. D., Robson, E. I., Ade, P. A. R., Griffin, M. J., and Sandell, G., "Millimetre/Submillimetre Common User Photometer for the James Clerk Maxwell Telescope", in preparation.
3. Goldsmith, P. F., "Quasi-Optical Techniques at Millimeter & Submillimeter Wavelengths", in "Infrared and Millimeter Waves", Vol. 6, pp. 277 - 343, Academic Press, 1982.
4. Papoular, R., Astron. Astrophys. 117, 46 1983.

SUBMILLIMETER COSMOLOGY WITH THE MULTIBAND IMAGING PHOTOMETER FOR SIRTF

P.T. Timbie, G.M. Bernstein, and P.L. Richards
Department of Physics, University of California
Berkeley, California 94720

T.N. Gautier
Infrared Processing and Analysis Center
California Institute of Technology, Pasadena, California 91109

G.H. Rieke
Steward Observatory
University of Arizona, Tucson, Arizona 85721

M.W. Werner
NASA/Ames Research Center, Moffett Field, California 94035

ABSTRACT. The Multiband Imaging Photometer for SIRTF (MIPS) will provide background-limited sensitivity from 3μm to 700μm. The submillimeter channels will reach $NEP \lesssim 5 \times 10^{-17} W/\sqrt{Hz}$ using bolometers cooled to 100mK by an adiabatic demagnetization refrigerator. These bands will have an angular resolution of ~2' and will achieve a sensitivity of 4-9mJy. These limits are 5 × better than for existing telescopes observing point sources and ~100 × better for observing diffuse sources.

One of the three instruments on the Space Infrared Telescope Facility (SIRTF) is the Multiband Imaging Photometer (MIPS), (Rieke et al. 1986). Here we will focus on two cosmological observations planned for the 200μm-700μm portion of the MIPS.

Submm Background Radiation: Radiation from sources at high redshifts (protogalaxies, massive stars, quasars, etc.) may be absorbed by intervening dust and reradiated in the submillimeter (Bond, Carr, and Hogan 1986). The recently discovered submillimeter excess (see figure) (Hayakawa et al. 1987), here fit to a dust model, may in fact be this predicted background. Bond et al. predict that the submm radiation should be anisotropic on the scale of a few arcminutes. This anisotropy can be measured by SIRTF as is shown in Fig. 1.

Sunyaev-Zel'dovich Effect: Photons in the CBR are expected to undergo inverse Compton scattering when passing through ionized gas in clusters of galaxies. The SIRTF submm channels should detect an increase in flux from the cluster relative to the background. For example, looking at the

Coma cluster, SIRTF's 700μm channel would see an enhancement of ~0.5% of the microwave flux at that wavelength. This effect should also be detectable by SIRTF as is shown in Fig. 1.

This work was supported by NASA.

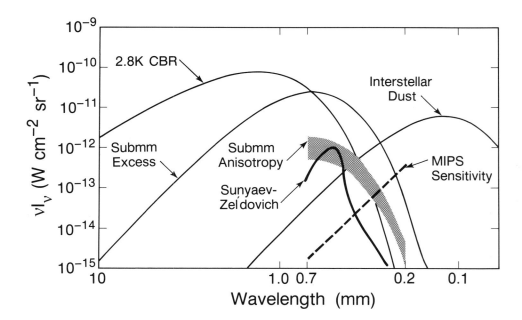

Figure 1. Expected signal levels in the MIPS bolometer channels. Limiting flux is for a 500 second integration and includes detector and photon noise and confusion from galaxies and IR cirrus.

REFERENCES

Bond, J.R., Carr, B.J., and Hogan, C.J. (1986) 'Spectrum and Anisotropy of the Cosmic Infrared Background', Ap.J. 306, 428-450.

Hayakawa, S., Matsumoto, T., Matsuo, H., Murakami, H., Sato, S., Lange, A.E., Richards, P.L. (1987) 'Cosmological Implication of a New Measurement of the Submillimeter Background', Publ. Astron. Soc. Japan 39, 941-948.

Rieke, G.H., Werner, M.W., Thompson, R.I., Becklin, E.E., Hoffman, W.F., Houck, J.R., Low, F.J., Stein, W.A., and Witteborn, F.C. (1986) 'Infrared Astronomy after IRAS', Science 231, 807-813.

RECEIVER WORK AT THE CFA

E. E. BLOEMHOF and V. DHAWAN
Harvard-Smithsonian Center for Astrophysics
60 Garden Street
Cambridge, MA USA 02138

ABSTRACT. We describe initial plans for the construction of submillimeter heterodyne receivers at the Harvard-Smithsonian Center for Astrophysics.

1. Introduction

An interferometric array of submillimeter telescopes has been proposed for construction by the Smithsonian Astrophysical Observatory of the Harvard-Smithsonian Center for Astrophysics. Initial plans (Moran *et al.* 1984) envisioned 6 dishes each of perhaps 6 meters diameter, with surfaces usable to 1000 GHz, and with baseline coverage in the range of a few hundred meters; final specification of these parameters will be made in a formal design study that has been funded to start in late 1988. An interferometer would offer unparalleled spatial resolution at submillimeter wavelengths, permitting the study of a variety of exciting scientific problems. As part of the initial effort on the submillimeter array, a laboratory has been established at the Center for Astrophysics for the construction of heterodyne submillimeter receivers.

2. Requirements for Interferometry

The need for simultaneous operation of an array of front-ends places a premium on ruggedness and simplicity of receiver systems, and this will influence the choices between rival technologies in a number of areas. Solid-state local oscillators, consisting of Gunn diodes followed by one or more diode multipliers, are reliable and compact, although somewhat limited in power and tunability. The use of Gunn-oscillator LO systems would make SIS junctions, with their low power requirement, an attractive mixer option. SIS junctions lend themselves to monolithic focal-plane arrays of junctions coupled to quasi-optical antennas, allowing full exploitation of valuable sky signal and permitting a solution, in principle, to the field-of-view problem facing a high-frequency interferometer. The alternative to the quasi-optical approach is the traditional waveguide mixer block, which presents extreme fabrication challenges at frequencies above a few hundred GHz.

Quasi-optical SIS receivers have demonstrated impressive broad-band response (Büttgenbach *et al.* 1988). The possibility of spanning an octave or more of the submillimeter spectrum with a single mixer structure in a single cryostat is extremely attractive when numerous telescopes must be outfitted with parallel receiver systems.

Practical receiver systems to be used on an array should have closed-cycle cryogenic cooling to minimize operational maintenance, and efficient diplexers to preserve sky and LO signal, both of which become precious at very high frequencies.

3. Early Directions

Our first efforts have been directed towards studying SIS mixer designs based on niobium junctions with non-native oxides. Although most SIS junctions used for astronomy have been fabricated from soft metal such as lead, the greater robustness and higher T_c of niobium make it an attractive electrode material.

Use of a non-native oxide, such as that of aluminum, can partially overcome the otherwise inconveniently high dielectric constant limiting high-frequency operation. The sputter-deposition technique used with niobium and other refractory materials complicates some potentially useful schemes for small-area junction manufacture, such as the hanging photoresist bridge technique exploited by Dolan (1977) working with lead. Nonetheless, there is some hope that this technique can be made to work if proper care is taken to cool the substrate. Edge junctions and variants are another attractive class of small-area junctions; we are currently fabricating SAIL structures (Koch 1987), which have been used for SQUIDs and may be adaptable for mixing applications.

Small-linewidth chrome-on-glass photolithographic masks, with resolution ~ 0.3 μm, are being made at the MIT Microsystems Research Center. Larger-scale masks for quasi-optical antennas, with features no smaller than 10 μm, are made for us by a commercial firm from CAD drawings conveniently transmitted in machine-readable form on floppy disk. Integration into a monolithic quasi-optical mixer/antenna structure is straightforward with a sub-micron mask aligner. For operation at a few hundred GHz, junction areas can be made small enough to achieve $\omega R_n C$ products of order unity, yielding broad-band operation.

Mask designs have been drawn up for more complicated schemes in which monolithic tuning structures rather than extremely small junction areas permit operation to high frequencies. Such structures may consist of short lengths of stripline presenting an appropriate inductance to tune out the junction capacitance over a relatively narrow design bandwidth.

4. References

Büttgenbach, T. H., Miller, R. E., Wengler, M. J., Watson, D. M., and Phillips, T. G. (1988) 'A Broadband Low Noise SIS Receiver for Submillimeter Astronomy', CSO preprint.

Dolan, G. J. (1977) 'Offset Masks for Lift-Off Processing', Appl. Phys. Lett., **31**, 337.

Koch, H. (1987) 'Self-Aligned In-Line Junction–Fabrication and Application in DC-SQUIDS', preprint, ISEC '87, Tokyo.

Moran, J. M., Elvis, M. S., Fazio, G. G., Ho, P. T. P., Myers, P. C., Reid, M. J., and Willner, S. P. (1984) 'A Submillimeter-Wavelength Telescope Array: Scientific, Technical, and Strategic Issues', Report of the Submillimeter Telescope Committee of the Harvard-Smithsonian Center for Astrophysics.

THE COUPLING OF SUBMILLIMETER CORNER-CUBE ANTENNAS TO GAUSSIAN BEAMS

E.N. GROSSMAN
Dept. of Astronomy, RLM 15-308
University of Texas at Austin
Austin, TX 78712

ABSTRACT. We have calculated a number of experimentally useful properties of 90° corner-cube antennas. These are the orientation and waistsize of the best-matched Gaussian beam, the power coupling efficiency to this beam, and the radiation resistance. The coupling calculation incorporates all known effects, including main beam efficiency, polarization mismatch, aberrations. It is found that the conventional prescription for antenna length, L = 4λ, and spacing from the dihedral, d = 1.2λ, does not yield optimal coupling. Exploring a grid in the L-d plane, we find optimal coupling to be obtained for L = 1.35λ, d = 0.9λ. An improvement of 1.5 db in coupling efficiency, from 55% to 78% is thereby achieved. This modification increases the width of the best-matched Gaussian beam from $w_0/\lambda = 1.82$ (FWHM = 12°) to $w_0/\lambda = 1.24$ (FWHM = 17°), and reduces the radiation impedance from 130Ω to 65Ω.

1. Introduction and Assumptions

Despite their ubiquitous use in radioastronomical receivers and other applications, the total coupling efficiency of a corner-cube antenna to a specific incident beam has never been calculated or measured. Therefore, the optimum design values for the whisker length, L, and spacing from the dihedral, d, have never been established. The values L = 4λ, d = 1.2λ became deeply entrenched on the basis of early measurements[1] of one-dimensional cuts through the beam's principal planes. However, the main sidelobes of a corner-cube antenna do not lie in the principal planes[2], so the early measurements are of limited relevance. In the present work, we fill this gap by calculating from first principles, as a function of L and d, the power coupling efficiency of a corner-cube antenna to the best-matched fundamental (m = 0) Gaussian-Hermite radiation mode. The Gaussian beam formalism is by far the most widely used framework for submillimeter optics design. In passing, we have also calculated the radiation resistance of the corner-cube antenna. This is equal to the real part of the antenna driving-point impedance if Ohmic losses may be neglected. Our assumptions are :

1a *Current distribution* : We assume the RF current is a pure traveling wave along the whisker plus three image currents due to the reflector. Current along any horizontal section of wire, and the effects of imperfect termination at the whisker bend are ignored. (See reference 3.)
1b *"Ground plane"* : We ignore the effects of any ground plane at z=0. This is based on the small size of the rearward radiation lobes of a traveling wave current distribution.
1c *Incident Gaussian beam* : We assume the incident Gaussian beamwaist coincides with the dihedral angle at the level of the whisker tip. We assume the polarization of the incident beam coincides with that of the corner-cube in the corner-cube's E-plane.

2. Results

The electric field radiated by the corner-cube, is calculated analytically, preserving the complex phase. The expression's squared absolute magnitude is numerically integrated over solid angle to yield the radiation resistance. For a given Gaussian waistsize and angle of incidence, the power coupling efficiency is calculated from the vector overlap integral of the two fields, and for each value of L/λ and d/λ, the waistsize and angle of incidence of the Gaussian beam are then varied so as to maximize the coupling efficiency. The highest coupling efficiency, 78%, is obtained for L=1.35λ, d=0.9λ. It consitutes a 1.5 db improvement over the conventional corner cube's efficiency, 55%.

Table 1 – Comparison of Optimal and Conventional Designs

$(L/\lambda, d/\lambda)$		(4.0,1.2)	(1.35,0.9)
Best-matched Gaussian beam			
	inclination θ	25°	35°
	waistsize w_0/λ	1.82	1.24
	corresponding to FWHM	11.8°	17.3°
	f-number to -35 db	f/1.37	f/.93
Power coupling to above Gaussian			
	η	55%	78%
Radiation resistance			
	R_{rad}	130Ω	65Ω

We gratefully acknowledge useful conversations with M.J. Wengler, A.I. Harris, and D.T. Jaffe. This work was partially supported by the National Science Foundation under grant no. AST86-11784 to the Electrical Engineering Research Laboratory of the University of Texas at Austin.

4. References

1. H. Krautle, E. Sauter, and G.V. Schultz, *Infrared Phys.* **17**, 477 (1977)
   H. Krautle, E. Sauter, and G.V. Schultz, *Infrared Phys.* **18**, 705 (1978)
2. P. Goldsmith, "Quasi-Optical Techniques at Millimeter and Submillimeter Wavelengths" in *Infrared and Millimeter Waves*, Vol. 6, p. 277 Academic Press, New York (1982)
3. B. Vowinkel, *Intl. J. of Infrared and Millimeter Waves* **5**, 451 (1984)

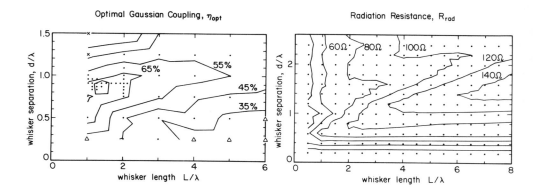

AIRBORNE HETERODYNE RECEIVER TECHNOLOGY FOR THE 100-300 µm RANGE

R. Wattenbach
Max-Planck-Institut für Radioastronomie
Auf dem Huegel 69
5300 Bonn 1, FRG

ABSTRACT. A submillimeter heterodyne receiver has been developed for airborne observations of molecular transitions in the 100-300 µm range. The system and the application at the KAO-telescope are described.

1. Airborne heterodyne receiver

The complete heterodyne receiver has been used several times aboard the KAO and on ground based telescopes (Röser et al., 1987). An extensive study of the receiver technology in the 100-300 µm range was done resulting in improvements of the system performance.
The diplexer is a Martin-Puplett system with freestanding wire grids as beam splitters. Losses in the range 370 µm - 180 µm were measured to be lower than 5 %. A baseline wobbler is used to avoid standing waves between mixer and secondary. Adjustable phase shift and wobbling time reduces baseline ripples more than 12 dB at 214 µm. The detector consisting of a GaAs Schottky barrier diode mixer mounted in a new designed open structure system formed by a whisker long wire antenna and a corner cube was tested down to 118 µm with a maximum i.f. of over 15 GHz. No significant changes in conversion losses were measured at these high intermediate frequencies.
The frequency coverage of a laser as local oscillator (LO) in the 800 GHz to 2200 GHz range allowing an i.f. of ± 20 GHz is shown in Figure 1. All laser lines in Table 1 have already been used as LO for CO detections or will be used in the near future.
Hybrid metal meshes with cross-shaped apertures were used as output couplers for our optically pumped FIR laser. In contrast to conventional hybrid meshs, these new couplers show a uniform reflectivity over a wide spectral range (Densing (1988)) allowing single mode operation and sufficient output power for use as a local oscillator in Schottky mixers (≳ 1 mW for laser line [arrows] in Figure 1).
The acousto-optical spectrometer has a bandwidth of 1 GHz (≈ 200 km/sec at 200 µm) and 1024 channels of resolution. A compact and rigid design allows airborne application. Long term stability was tested by Allan variance and allows chopping cycles of about 30 sec.

2. Airborne observations

The heterodyne receiver was used to detect CO (J=12-11) and CO (J= 14-13) lines in Orion and W3 in September 1988. A map of CO (J=12-11) in OMC-1 showed intensive lines of $T_A^* \gtrsim 150$ K in BN/KL and extended emission of very narrow CO lines over several arcminutes. These results are similar in CO (14-13) at 184 μm. All spectra will be published soon.

Table 1. Laser lines as local oscillators for observation of CO-transitions

$^{12}C^{16}O$	FREQUENCY (GHz)	WAVELENGTH (μm)	LO MOLECULE	FREQUENCY (GHz)	IF (GHz)
6 - 5	691.5	433.6	HCOOH	692.9	1.4
7 - 6	806.7	371.7	$^{15}NH_3$	803.0	3.7
9 - 8	1,036.9	289.1	CH_2F_2	1,042.2	5.3
12 - 11	1,382.0	216.9	CH_2F_2	1,397.1	15.1
14 - 13	1,611.8	186.0	CH_2F_2	1,626.6	14.7
16 - 15	1,841.3	162.8	CH_3OH	1,838.8	2.5
17 - 16	1,956.0	153.3	$^{15}NH_3$	1,962.8	6.8

Table 2. Receiver sensitivity and bandwidth

800 GHZ	6000-8000	K (SSB), IF: 4 GHz, BW: 1-1.5 GHz
1400 GHz	25.000	K (SSB), IF: 15 GHz, BW: 1.5 GHz
1630 GHz	40.000	K (SSB), IF: 15 GHz, BW: 1.5 GHZ
1960 GHz	21.000	K (SSB), IF: 6 GHz [cooled mixer, Betz et al.]

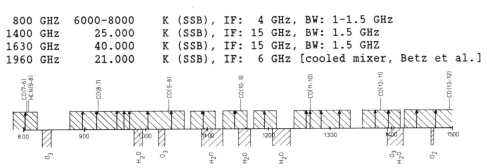

Figure 1: Frequency coverage of submm laser lines (800-2200 GHz)

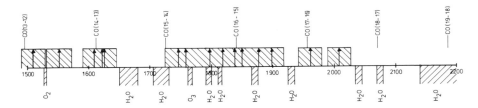

3. References

Densing R. (1988), in Proc. of the 4th Intern. Conf. on IR Physics, pp 313
Zumidzinas, J., Betz, A.L., Boreiko, R.T., to be published in Infr. Physics
Roeser, H.P., Schäfer, F., Schmid-Burgk, J., Schultz, G.V., van der Wal, P., Wattenbach R., (1987), Int. J. of IR and MM Waves 8, 1541

THE MILLIMETER MULTIBEAM SYSTEM PROJECT AT ONSALA SPACE OBSERVATORY

Anders Emrich
Onsala Space Observatory
S-43900 Onsala, Sweden

ABSTRACT. A focal plane array receiver system is now under development at Onsala Space Observatory. The aim is for a 4x4 array of endfire receptors with integrated SIS mixers for the band 80-115 GHz. Each receiver will be equipped with a flexible hybrid spectrometer covering 400 MHz bandwidth in its widest setting and with a resolution of 8 kHz in the highest resolution mode (with 20 MHz bandwidth). Coupling of the antenna to the array is handled by a quasioptical system which includes a beamrotator, beamswitch, local oscillator injector and single sideband filter.

1. Introduction

The sensitivity of state of the art millimeter receivers are mainly limited by the atmospheres contribution to the system temperature and to increase the throughput of radio telescopes, multibeam receiver projects has been started at several places. The multiple samples can be taken in two planes, in the aperture plane or in the focal plane (or anywhere in between) and after some investigation, the focal plane sampling was chosen for the Onsala project.

We are also aiming at arrays for higher frequencies with more elements and have therefore chosen a compact tapered slot line antenna design with integrated SIS-junctions. Specifications:

Frequency coverage	90-115 GHz (usable down to 80 GHz)
System temperature	100 K SSB (at least in the 110-115 GHz band)
Feed	16 beam, polarization interleaved, undersampling<2
Optics	beam derotation, beamswitch, pathlength variator, sideband filter
Instantaneous bandwidth	1 GHz
Spectrometer bandwidth	400 MHz-16 beams, 800 MHz-8 beams, (Δb=2 MHz)
Spectrometer resolution	8.125 kHz, (B=25 MHz)
Data storage	>2 maps of 256x256 pixels, 1024 channels
Unattended operation	>12 hours

2. System design

Quasioptics. The optics will be described starting from the subreflector going to the feed elements. The Onsala Cassegrain antenna has a focal number of 4.5 which is convenient for quasioptical design. The first waist is situated in a beamswitch, probably using a rotating polaristion grid instead of a chopping wheel. After the switch, the beam package will be refocused and inserted in a K-mirror image derotator that compensates for the sky´s apparent rotation in an alt-az mounted antenna also including a path-length variator. Selection of receiver system will then be made by means of a rotating mirror (22 GHz and 43 GHz VLBI receivers will also be present), and the beam package is then inserted into the receiver.

Local oscillator system. The oscillator system will be based on a Gunn diod, phaselocked in an ordinary way. The basic difference is that care is taken so that the local oscillator can be completely computer controlled and that some kind of beam shaper is used so that the receiver elements get the same amount of LO-power. Insertion is made through a 20 dB coupler just outside the dewar.

Frontend receiver design. A cooled single sideband filter terminates the unwanted sideband into a 4°K load. The beam package is then refocused onto the feed array with a focal number of ≈ 1.5. Tapered slot antennas with the configuration 4x4 is used, either situated in one polarization plane or interleaved in two polarizations. An aperture efficiency of at least 0.8 times that shown by a single corrugated horn is expected for an undersampling factor of 2.5 in one dimension. The SIS-junctions will be implemented in an edge type design with the layers: Nb/AlOx/PbBi. HEMT amplifiers centred around 4 GHz will be used in the first stage after the mixers.

Backend receiver design. The hybrid spectrometer backend is based on the Bos chip in the autocorrelator part, preceded by a few channel filterbank. The backend can easily be expanded and reconfigured to fit other requirements but the basic design specifications is 400 MHz bandwidth for extra galactic work and a resolution of 10 kHz for dark clouds, all 16 receivers working at the same time. This has been implemented in two 19" triple eurocard height boxes and a Macintosh II™ is used for control, readout, processing, display and storage. Worth mentioning is perhaps the filterbank design based on a modified type of quadrature mixers, using standard Mini-Circuits filters and no phaseshifters, and the software planned for reducing the channel to channel coupling. Both these details simplifies the spectrometer and reduce the cost considerably.

Control system and data processing. Rather then trying to modify the old control system it was decided to design a new one. The system is based on a structured design in three layers with as much distributed processing as possible. The software for processing will include a new map based data format, optimised for fast storage and access, and all operations on the data file is done through library functions to hide the real implementation. The processing package will use a Macintosh interface and routines for both spectral and maplevel commands will be available, even routines for MEM type beam deconvolution.

3. References

Johansson, Joakim F, (1988) 'Tapered slot antennas and focal plene imiging' Technical report No. 184, Departement of Radio and Space Science, Chalmers University of Technology.
Bos, Albert, (1985-88), Internal technical reports at NFRA, The Netherlands.
Goldsmith, Paul, (1988), private communication.
Weinreb, Sander, (1985), 'Analog-filter Digital-Correlator Hybrid Spectrometer' IEEE Transaction on instrumantion and measurment, IM 34, NO 4, December 1985.

(The project is a joint effort between the Departement of Radio and Space Science and Onsala Space Obsrvatory and is supported by Knut & Allice Wallenberg foundation.)

A 60 CM SUBMILLIMETER SURVEY TELESCOPE PROJECT

M. HAYASHI, T. HASEGAWA, and K. SUNADA
University of Tokyo
and
N. KAIFU
Nobeyama Radio Observatory, National Astronomical Observatory

ABSTRACT. We are now making a 60 cm submillimeter survey telescope at Nobeyama, Japan. The primary purpose of this telescope is to complete a Galactic plane survey in the CO (J=2-1) emission. Elementary heating processes of molecular gas will be investigated by comparing our data with the Columbia CO (J=1-0) survey.

I. SCIENTIFIC GOALS

The main scientific goal of this project is to make a Galactic plane survey in the ^{12}CO (J=2-1) transtition, determining the temperature distribution of molecular gas in the Galaxy by comparing our ^{12}CO (J=2-1) data with the ^{12}CO (J=1-0) Columbia survey data. Large-scale temperature variation of molecular gas is related to the cosmic ray density, mean ultraviolet interstellar radiation, Galactic shocks, and so on.... Investigation of the temperature variation on a global scale in the Galaxy is therefore important to understand these elementary physical processes for the heating of molecular gas. The ^{12}CO (J=1-0) Columbia survey provides a nice view of the distribution of molecular clouds in the Galaxy, but the variation of the molecular gas temperature along and across the Galactic plane will be better understood through the comparison of the J=2-1 and J=1-0 data, because the complicated temperature structure often seen in molecular clouds makes it difficult to infer the typical temperature inside these clouds only from the one lowest transition of carbon monoxide. The large scale variation of the molecular gas temperature in the Galaxy may be closely related to the variation of the cosmic ray density which has been estimated from γ-ray observations.

The 60 cm submillimeter survey telescope will then be used for the ^{13}CO (J=2-1), ^{12}CO (J=3-2) observations. The high surface accuracy noted below will enable us to observe lines with much higher frequencies like CI ($^3P_1 - {}^3P_0$), as long as the telescope is set at a proper site for submillimeter observations.

II. TELESCOPE

The main reflector is an offset paraboloid in order to achieve high telescope efficiencies. Its diameter is 60 cm yielding a similar angular resolution to the 1.2 m Columbia telescope, which is now operated at Center for Astrophysics, so that the direct comparison between the J=2-1 and J=1-0 data will be available. The surface accuracy of the main reflector will be less than 30 μm (rms), giving high efficiencies at 230 GHz. Parameters of the 60 cm submillimeter survey telescope are listed in Table 1.

Figure 1 shows one of the drawings of the 60 cm submillimeter survey telescope. It has a Cassegrain-Coudé optics, which makes the moving portion light ($\sim$ 100 kg). This enables us quick operation of the telescope, so that the time loss for position switchings can be minimized even if off positions are far away. Receivers, local oscillators, and IF systems are set at a fixed small cabin fully air-conditioned located just below the main reflector, which keeps the output signal stable and makes the access to the receiving system easy. We will use a Schottky mixer receiver whose noise temperature is 800 K (SSB) at the initial stage of the survey, but will install an SIS receiver in a few years.

The telescope will be set at Nobeyama 1,300 m in elevation. Cold ($\sim -10\,^\circ C$) and dry climate during the winter season will provide sufficient conditions ($\tau_{\text{zenith}} < 0.3$) for observing ^{12}CO ($J=2-1$) for more than 3 months in a year. The initial survey of the first quadrant of the Galaxy will be finished in 2-3 years. The telescope has a sliding antenna cover to protect the main and sub reflectors from snow falls and winds at Nobeyama.

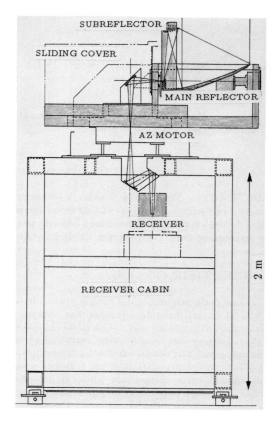

Figure 1. A drawing of the 60 cm submillimeter survey telescope.

Table 1
Parameters of the 60 cm Submillimeter Survey Telescope

OPTICS	OFFSET CASSEGRAIN COUDÉ
DIAMETER (main reflector)	60 cm
SURFACE ACCURACY	$\leq 30\,\mu m$ (rms)
BEAM SIZE (HPBW)	8' at 230 GHz
TRACKING ACCURACY	within $\pm$ 30"

Intra-Cavity Pumped FIR Laser System

Gordon Chin
Code 693
NASA/Goddard Space Flight Center
Greenbelt, MD 20771

Hemant Dave
Applied Research Corporation
Landover, Md 20785

ABSTRACT. We have constructed a new type of FIR laser system for submillimeter heterodyne receivers and we are in the process of evaluating its performance. This new FIR LO laser system makes use of the high circulating power of a CO_2 cavity as the optical pump for the FIR laser. In this manner we hope to achieve several technical improvements over the conventional use of extra-cavity pumping. Improvements can be made such as a reduction in size of the LO system, ability to pump more lines because of the high available pump power, and the possibility of achieving some tunability of the FIR laser with high pump power density. First results have been obtained for the 118 μm CH_3OH FIR transition.

1. **Introduction.** Commercial and scientific far infrared (FIR) lasers use a separate DC high voltage gas discharge CO_2 laser system to provide the necessary 10 μm pump power to invert the FIR gas medium. The CO_2 laser power is extracted by an output coupler which transmits a fraction of the circulating power which resides within the CO_2 laser cavity. In order to provide sufficient power to pump the FIR gas, about 50 W of CO_2 power is needed. This type of CO_2 laser is usually about 2 meters in length. The CO_2 beam is then directed and focused into a small aperture which forms an entrance optic of a second laser cavity. The total FIR system is really two laser systems operating in tandem, a CO_2 laser and a FIR laser. Commercial FIR laser systems are about 2 meters in length, while all astronomical groups use FIR laser cavities which are 1 meter in length. A review of existing FIR heterodyne receiver groups and the state-of-the art of FIR lasers used for astronomical work can be found in Chin 1987, *International Journal of Infrared and Millimeter Waves*, vol. 8 No. 10., p. 1219.

2. **Intra-Cavity Pumping.** The intra-cavity pumped FIR laser system employs the circulating power within a CO_2 laser cavity as the pump for the FIR laser system. The components of the system include a CO_2 gain cell (i.e. high voltage DC discharge cell or alternatively an RF excited waveguide laser gain path), high voltage power supply for the CO_2 gain cell, a FIR vacuum laser cavity consisting of a Pyrex glass tube, input and output optics, a curved grating for CO_2 line selection, a grating mount which allows zero-th order beam monitor, a voice-coil translation for CO_2 mode control, a micrometer translation stage for FIR laser mode control, Invar rods for thermal stability bounded by aluminum machined endplates as the optical bench. The important aspect of this configuration is that the CO_2 laser does not have an output coupler. Instead, the FIR laser is inside of the CO_2 laser cavity (thus intra-cavity pumped) and can therefore make use of the total CO_2 circulating power. The laser configuration can be seen schematically in Figure 1. In Figure 1 elements (1) and (3) form the optical cavity for the CO_2 pump while elements (2) and (3) form the optical resonator for the FIR laser. The CO_2 lasing pump transition and power are monitored via the zero-th order beam from the grating. Other important innovations include the use of a curved grating (1) and focusing optics (2,3) to form stable Fabry-Perot laser resonators for both the CO_2 and FIR laser beams. In addition, consideration was also made to optimize the volume overlap and

collinearity of the FIR and CO_2 beams. With our system, alignment of the CO_2 laser guarantees alignment of the FIR pump with the optical axis of the FIR resonator. This is unlike other (extra-cavity pumped) FIR lasers which require separate alignment processes for injecting the CO_2 beam into the FIR cavity, and for making it collinear with the FIR resonator. Alignment of the FIR resonator is then achieved by adjusting a single element (2).

In our working system we use a high voltage DC gas discharge cell as the gain medium for the CO_2 laser. However, RF excited waveguide lasers are more efficient and more compact in size. The same principle employed in our working system can be used to form a compact, integrated CO_2 and FIR waveguide intra-cavity system. In addition, pumping lasers featuring other lasants (e.g. isotopic species of CO_2, CO, N_2O, etc.) could be substituted. In our system, active length control of the CO_2 resonator is achieved with element (3), and for the FIR resonator with elements (2) and (3). Element (3) affects primarily the 10 μm cavity due to the small amount of its motion. In an alternative embodiment, active length control for the 10 μm cavity can be achieved by mounting element (1) to the voice coil and making element (3) passive, thereby decoupling the pump and FIR cavity controls completely. Other schemes for coupling out the FIR power are also possible. In our working system element (3) is a hybrid quartz or silicon mirror with a gold annulus to reflect the FIR and 10 μm, and with an overcoating of dielectric film to effect high 10 μm reflection. The annulus hole allows FIR transmission. Element (3) may be replaced with all reflecting optic to minimize FIR cavity losses then output of the FIR laser may be achieved by coupling the power through element (2) and monitoring the power reflected off the input ZnSe window through a quartz or mylar vacuum window.

3. Summary. We have constructed a new type of FIR laser system as the prototype for submillimeter LO systems. Performance tests for this new laser are now being carried out to assess improvements in pump efficiency and the optimum parameters for both the CO_2 and FIR laser cavities. First results of intra-cavity lasing has been achieved at 118 μm with CH_3OH.

We acknowledge support from NASA's OAST RTOP 584-01-21.

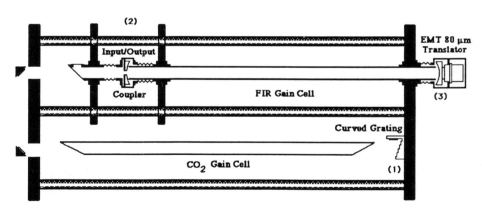

Figure 1. Schematic Configuration for the Intra-Cavity Pumped FIR Laser System

IMAGING TRIPLE-FABRY-PEROT-SPECTROMETER FOR FAR-INFRARED ASTRONOMY

A. Poglitsch, N. Geis, M. Haggerty, and R. Genzel
Max-Planck-Institut für Physik und Astrophysik, D8046 Garching

G. Stacey and C. Townes
University of California, Berkeley CA 94720

ABSTRACT. The development of an imaging Fabry-Perot spectrometer for the wavelength range 40-200μm employing a 5x5 detector array is reported.

For far-infrared/sub-millimeter astronomy, present spectrometers are based on one or a few (coherent or incoherent) detector elements, thus having restricted mapping capabilities. We are now building and testing an imaging spectrophotometer for the wavelength range 40-200μm for operation on NASA's Kuiper Airborne Observatory where observation efficiency is particularly important.

In our experiment we will implement a 5x5 array of Ge:Be, Ge:Sb, and Ge:Ga photoconductive detectors, both stressed and unstressed, mounted on integrating cavities to increase the quantum efficiency of the weakly absorbing detectors. Light cones will provide area-filling light collection. A three-element linear detector array (without cones) has already shown excellent performance. A five element linear array (with cones) is now under test; five of these modules will form the 5x5 array (fig. 1). The detectors are read out by transimpedance amplifiers with cryogenic FET input stages; the signals are processed by a 25 channel digital lock-in amplifier, providing identical response for all channels.

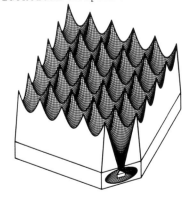

Fig. 1
5x5 detector array with light cones. Cut away shows a detector mounted in its integrating cavity.

These detectors are background-noise limited; therefore, all bandwidth-limiting elements are kept either at LN_2 or at LHe temperature. We expect an NEP of $\approx 10^{-15}$ W/$\sqrt{Hz}$ on the telescope. The spectral resolution will be $\approx 10^5$ in the center and $\approx 10^4$ on peripheral pixels. This resolution is obtained by three Fabry-Perot-interferometers (FPI) in series, set to different orders. Metal-mesh plates provide both a high finesse (≈ 50) and a relatively broad bandwidth. The high-order FPI at LN_2 temperature is designed for both scanning and frequency-switching modes. The dewar unit which contains all elements described above is shown in figure 2. The optics is based on a double-Czerny-Turner configuration to minimize image distortions. By rotating carrousels (10) & (11) to move a chosen pair of mirrors into place, the observer can switch between two f-converters during observation; overall, f-converters corresponding to 10, 20, or 40 arcsec/pixel are available. A combination of reststrahlen filters in the filter wheel (5) and the scatter filter (6) are used as low-pass filters.

An additional calibration unit provides both wavelength calibration (with a gas absorption cell) for the FPI's and intensity calibration (flat-fielding) for the detector array. It also contains a K-mirror assembly to compensate image-rotation during observation and visible light optics for focal-plane tracking. The instrument shall become observational in early 1989.

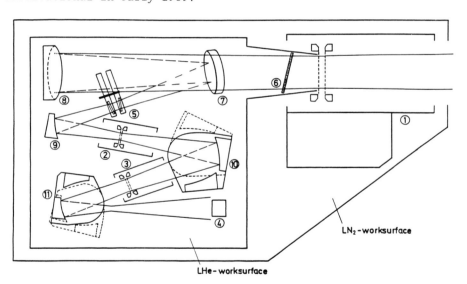

Fig. 2. Schematic diagram of cryogenic optics.
    (1) High order scanning FPI   (6) Scatter filter
    (2) Medium order tunable FPI (7) Plane mirror
    (3) Low order tunable FPI   (8),(9),(10) Off-axis parabolic mirrors
    (4) Detector array
    (5) Reststrahlen filter wheel (11) Off-axis elliptical mirrors

SMALL-AREA NIOBIUM/ALUMINUM OXIDE/NIOBIUM JUNCTIONS FOR SIS MIXERS

J. Zmuidzinas[1], F. Sharifi[2], D.J. Van Harlingen[2], and K.Y. Lo[1]
Astronomy Department[1] and Materials Research Laboratory[2]
University of Illinois, Urbana IL 61801

1. Introduction

Submillimeter-wave SIS receivers require junctions with small RC products. The RC product can only be reduced by decreasing the thickness of the insulating barrier; however, this is generally accompanied by an increased subgap leakage current which adversely affects mixer performance. Thus, it is advantageous to choose a junction fabrication technology which produces inherently high quality barriers and low subgap leakage currents. Junctions fabricated from Nb/Al-Oxide/Nb trilayers (Gurvitch et al. 1983; Lichtenberger et al. 1988) are particularly well known in this regard, and have shown good performance when used in mm-wave mixers (Inatani et al. 1987; Pan et al. 1987). However, junction areas must be reduced to ≤ 1 μm^2 in order to meet impedance-matching requirements at submm wavelengths. Here we report Nb/Al-Oxide/Nb junctions which are among the smallest fabricated to date (see also Imamura and Hasuo 1988).

2. Fabrication

Our fabrication process is quite similar to that described by Yuda et al. (1987); the reader is referred to that paper for more detail. Briefly, a 10x10 μm^2 Nb/Al-Oxide/Nb sandwich is deposited by DC magnetron sputtering through a three-layer photoresist lift-off stencil (Dunkleberger 1978). The junction area is masked with photoresist and defined by reactive ion etching (RIE) of the top Nb layer in a CF_4 atmosphere. The junction area is then isolated by self-aligned evaporation and liftoff of an SiO insulating film. This procedure is actually performed twice in our process. First, a ≈ 1 μm wide photoresist <u>line</u> is patterned on top of the 10x10 μm^2 sandwich, such that the RIE and SiO deposition steps define a 1x10 μm^2 junction. Next, another 1 μm photoresist line is patterned in the orthogonal direction. Repeating the RIE and SiO deposition steps produces a ≈ 1 μm^2 junction. Defining the junction area in two steps eases the lithography considerably. In the final step, the top Nb layer of the junction is cleaned by light Ar ion milling, and a Nb connection layer is then deposited and patterned using liftoff.

3. Results

The I-V curve of a 0.7x0.7 μm^2 junction is shown in Fig. 1. The Josephson current is suppressed by an applied magnetic field. Table 1 lists various parameters obtained from I-V measurements of several samples at 4.2 K. The parameters listed include the ratio of sub-gap to normal state resistances R_{sg}/R_N (at 2 mV) and the roll-off frequency $(2\pi R_N C)^{-1}$. A capacitance of 45 fF/μm^2 is assumed

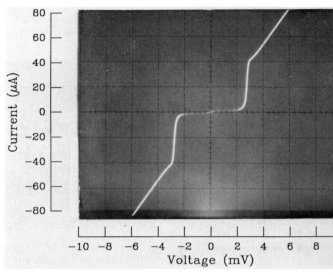

Fig. 1. I-V curve of a 0.5 μm^2 junction.

(Lichtenberger et al. 1988). The decrease in R_{sg}/R_N with decreasing RC product is clearly seen. "J_c" refers to the critical current density expected from the Ambegaokar-Baratoff relationship: $J_c A = I_c \approx \pi V_g/4R_N$. The <u>measured</u> critical currents are smaller than these values typically by a factor of 3, for reasons not yet clear. However, this suppression of the critical current should not hurt mixer performance.

TABLE 1. SIS junction parameters at 4.2 K.

Area (μm^2)	R_N (Ω)	R_{sg}/R_N	$(2\pi R_N C)^{-1}$ (GHz)	"J_c" (A cm^{-2})
1.	15.	6.6	236.	1.7×10^4
1.	24.	11.	147.	1.1×10^4
0.5	77.	15.	90.	$6. \times 10^3$

References

Dunkleberger, L.N. 1978, J. Vac. Sci. Technol., <u>15</u>, 88.
Gurvitch, M., M.A. Washington, and H.A. Huggins 1983, Appl. Phys. Lett., <u>42</u>, 472.
Imamura, T. and S. Hasuo 1988, J. Appl. Phys., <u>64</u>, 1586.
Inatani, J., T. Kasuga, A. Sakamoto, H. Iwashita, and S. Kodaira 1987, IEEE Trans. Magn., <u>MAG-23</u>, 1263.
Lichtenberger, A.W., C.P. McClay, R.J. Mattauch, M.J. Feldman, S.-K. Pan, and A.R. Kerr 1988, IEEE Trans. Magn., submitted.
Pan, S.-K., A.R. Kerr, J.W. Lamb, and M.J. Feldman 1987, National Radio Astronomy Observatory, Electronics Division Internal Report No. 268.
Yuda, M., K. Kuroda, and J. Nakano 1987, Jpn. J. Appl. Phys., <u>26</u>, L166.

AN SIS RECEIVER FOR THE JCMT

S.R. DAVIES [1], C.T. CUNNINGHAM [2], L.T. LITTLE [1], D.N. MATHESON [2]
[1] *Electronic Engineering Laboratories, University of Kent, Canterbury CT2 7NT*
[2] *Rutherford Appleton Laboratory, Chilton, Didcot, Oxon OX11 0QX*

ABSTRACT. A 220–280 GHz prototype SIS receiver system is currently being constructed, which will be tested on the JCMT in January 1989. The prototype receiver will act as a test-bed for developing a dual-channel common-user SIS receiver for the JCMT, tunable from 280–360 GHz, in a collaborative project between UKC, RAL and the Herzberg Institute of Astrophysics, Ottawa.

1. Introduction

In a collaborative project between UKC and RAL, a 230 GHz laboratory receiver has been built using a lead alloy SIS junction as the mixing element, mounted in 4:1 reduced-height waveguide, terminated at one end with a circular cross-section, adjustable backshort [1]. Best results indicate a double-sideband receiver noise temperature of T_R (DSB) ≈ 120 K, at 230 GHz. A preliminary measurement using the same mixer block indicated a receiver noise temperature of T_R (DSB) ≈ 350 K at 345 GHz. Results at both frequencies are far from optimised, and it is expected that improved noise performance will be achieved.

A prototype 220–280 GHz SIS telescope receiver system, based upon the laboratory receiver, is currently being constructed (Fig. 1), and is briefly described here.

2. Receiver Configuration

The Pb-alloy SIS junctions are manufactured with areas of $\leq 1.0\,\mu m^2$. The junctions are formed integrally with a stripline RF choke structure on each electrode, and are fabricated on z-cut crystalline quartz of thickness 0.1 mm. The quartz substrate is machine diced into individual junction chips of size ~ 4 mm × 0.3 mm. A suitable junction is mounted in a rectangular channel milled across the reduced-height waveguide of the electroformed copper mixer block. The 4:1 reduced-height waveguide (of cross-section 0.97 mm × 0.12 mm) is terminated in a half-height waveguide section containing an adjustable, quarter-wave choked, non-contacting backshort. Radiation is coupled into the mixer block using a corrugated feedhorn and a circular to reduced-height rectangular waveguide transition.

An IF impedance transformer, fabricated in microstrip, is situated between the junction and the subsequent 50 Ω components, increasing the IF load impedance presented to the junction from 50 Ω to 200 Ω. The IF signal is passed through a cooled isolator to a cooled 1.5 GHz GaAs FET amplifier which provides 30 dB of gain and has an equivalent noise temperature of T_{IF} ≈ 12 K. These components are mounted on the cold plate of a liquid helium cryostat. The local oscillator signal is provided by frequency tripling the output of a Gunn oscillator. The L.O. signal is coupled to the mixer by a taut, thin piece of mylar film, which acts as a ~ 1% dielectric beam-splitter. The 1.5 GHz I.F. signal from the FET amplifier is filtered, further amplified and passed to a square-law detector. These latter components are all at room temperature.

3. 3-Port Mixer Models

Apart from the physical parameters of the SIS junction, the three features which are to some extent in the control of the designer are : (a) the backshort, (b) the R.F. filter structure, and (c) the I.F. transformer. We have analysed the importance of these parameters [2], developing an equivalent circuit for the waveguide mixer mount, allowing for loss in the backshort, and using the Tucker quantum theory of mixing [3], which permits gain calculations for SIS mixers. Typical plots of gain as a function of the series reactance, X_0 (due to the mounting structure), the backshort VSWR, junction capacitance and I.F. load impedance are shown here. All impedances are normalized relative to the waveguide characteristic impedance Z_0. For the 4:1 reduced-height waveguide described here, $Z_0 \approx 120\ \Omega$. Following scaled model measurements of the embedding impedance, suitable choke structures are now being fabricated with the junctions to provide a series reactance of zero ohms.

Initial measurements in the laboratory with the prototype receiver indicate a receiver noise temperature of T_R (DSB) ≈ 300 K at 245 GHz. The noise performance of the receiver appears to be fairly constant between 230–270 GHz. Further optimisation of the noise performance is being carried out.

References

1. Davies, S.R., Little, L.T., and Cunningham, C.T. : Electronics Letters **23**, 946 (1987).
2. Davies, S.R., and Little, L.T. : (in preparation).
3. Tucker, J.R., and Feldman, M.J. : Rev. Mod. Phys. **57**, 1055 (1985).

Fig. 1

220–280 GHz Prototype Receiver for the JCMT.

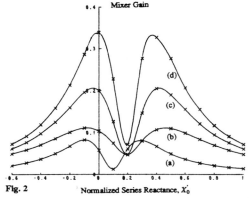

Fig. 2 Normalized Series Reactance, X_0'

SIS mixer gain plotted as a function of series embedding reactance, X_0.
All impedances are normalized relative to the waveguide impedance Z_0.
Normalized junction normal-state resistance is 0.5.
Curves (a), (c) and (d) are for a backshort VSWR of 30.
Curve (b) is for a backshort VSWR of 10.
Curve (a) is for a junction with an area of 1 μm².
Curves (b) - (d) are for a junction with an area of 0.5 μm².
Curves (a) - (c) are for normalized I.F. load impedance = 0.4.
Curve (d) is for normalized I.F. load impedance = 2.0.

SIS RECEIVER DEVELOPMENT AT NOBEYAMA RADIO OBSERVATORY

J.INATANI, T.KASUGA, R.KAWABE, M.TSUBOI, A.SAKAMOTO
Nobeyama Radio Observatory
Nobeyama, Minamisaku, Nagano 384-13, Japan

M.YAMAMOTO, K.WATAZAWA
Sumitomo Heavy Industries, Ltd.
Tanashi, Tokyo 188, Japan

ABSTRACT   We have constructed a very reliable SIS receiver on the base of the Nb/AlOx/Nb junction and the 4-K closed-cycle helium refrigerator. Six SIS receivers are now successfully used in the 45-m telescope and in the Nobeyama Millimeter Array(NMA) for 35-50 GHz and 85-115 GHz.

1. SIS JUNCTION

We have developed a small SIS junction on the base of the Nb/AlOx/Nb technology. This is fabricated by means of a continuous sputter deposition of Nb and Al, and a subsequent reactive ion etching(1). A mixer unit is composed of four SIS junctions connected in series and a millimeter wave choke filter. Four junction device is effective to increase the mixer saturation level, which is about 3000 K for the white noise RF input. Junction size is 2.5 x 2.5 μm^2 for a 40 GHz mixer and 1.8 x 1.8 μm^2 for a 100 GHz mixer.

The Nb/AlOx/Nb junction has a sharp dc I-V curve as shown in Fig.1, which is for a 40 GHz mixer(2). Uniformity among the four junctions in a mixer is quite well. One advantage of this junction is that such a sharp I-V curve is obtained at 4.2 K. Actually it is not necessary to cool the junction to 3 K or less for a mixer application. Furthermore, this junction is quite stable for heat cycling and for a long storage at room temperature.

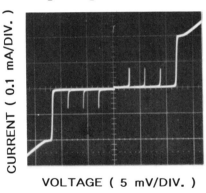
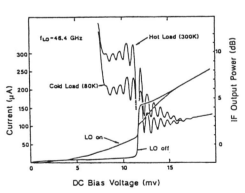

(LEFT) Fig.1 I-V curve of four Nb/AlOx/Nb junctions in series.
(RIGHT) Fig.2 Response of 40 GHz SIS receiver.

2. SIS RECEIVERS

We have constructed several SIS receivers using such a Nb/AlOx/Nb junction. Fig.2 shows a typical response of our 40 GHz SIS receiver. The junction is installed in a waveguide mixer mount which has two mechanical tuners(3). One is a conventional backshort and the other is a waveguide stub tuner. These two realize a wide tuning range(35-50 GHz and 85-115 GHz) of our mixers. We have adopted an extremely low-height waveguide(1/8-reduced one for 40 GHz and 1/7-reduced for 100 GHz), which has been effective to optimize the source impedance of the mixer.

Our 40 GHz SIS receiver is shown in Fig.3. This has been successfully used in the 45-m telescope since 1987. Fig.4 shows our dual channel SIS receiver for 40 GHz and 100 GHz, which has been used at five antennas in the Nobeyama Millimeter Array. All these receivers are cooled by 4-K closed-cycle helium refrigerators. The receiver noise temperatures(DSB) are shown in Fig.5.

(LEFT) Fig.3 SIS receiver for 35-50 GHz used in the 45-m telescope.
(RIGHT) Fig.4 SIS receiver for 40-50 GHz /85-115 GHz used in the NMA.

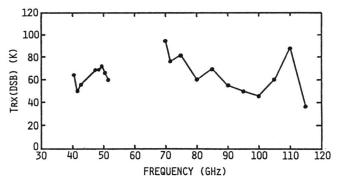

Fig.5 The noise temperatures(DSB) of SIS receivers used at Nobeyama Radio Observatory.

REFERENCES

(1) J.Inatani,T.Kasuga,A.Sakamoto,H.Iwashita and S.Kodaira: IEEE Trans., MAG-23, 1037(1987)
(2) M.Tsuboi,J.Inatani,T.Kasuga,R.Kawabe,A.Sakamoto,H.Iwashita and K.Miyazawa: Int.J.Infrared Millimeter Waves, 8, 1(1987)
(3) M.Tsuboi,J.Inatani,T.Kasuga and S.Kodaira: ISEC'87, 293(1987)

TECHNIQUES IN SMALL AREA SIS NbN JUNCTION MANUFACTURE

J.F. VANELDIK, D. ROUTLEDGE, M. BRETT
University of Alberta
Edmonton, Alberta
Canada, T6G 2G7

ABSTRACT. This paper discusses progress made in the development of techniques for manufacture of submicron area superconductor/insulator/superconductor junctions executed in niobium nitride. The junctions are meant for use as mixers in integrated planar multi-beam array receivers in the 100 to 800 GHz frequency range. Sputtered thin film NbN/MgO/NbN metallized sandwiches are plasma etched into mesa junctions. Standard 1.5 μm UV lithography is used to deposit the smallest possible junctions. Selective over-etching of the upper NbN layer, with MgO acting as the etch-stop, further reduces the area from that produced by the lithography. The smallest junctions produced so far have areas of 3 μm^2, but areas as small as 0.2 μm^2 appear possible.

1. INTRODUCTION

Millimetre and sub-millimetre telescope observing speed may be greatly increased by use of focal plane antenna feed arrays. But, efficient image plane sampling will require densely packed array elements. Each beam-forming array element may, itself, be an array. Potentially, the high packing density requirements are most easily met by implementing the feeds as planar arrays of end-fire stripline horns.
   High critical temperature superconductor/insulator/superconductor (SIS) junctions in small junction area configurations are well suited to design of very low noise and very broad-band RF mixers for receivers in the 100 to 800 GHz frequency range. If properly arranged physically, such junctions may be directly integrated with the stripline horns and associated circuitry to form completely integrated multi-beam focal plane receiver front ends.
   Investigations were undertaken to develop techniques for producing the arrays of very small horns, junctions and circuitry required.

2. TECHNIQUES

NbN with its high (16K) critical temperature and its outstanding chemical and mechanical stability was chosen as the superconductor for this work. This material is useful to frequencies of at least 1.5 THz.

Because of its low losses and low dielectric constant, MgO is the initial insulator selected. The crystal lattice mismatch between MgO and NbN is about 4%. High insulator film quality is, therefore, possible. MgO is also an excellent etch-stop in the SF_6 plasma etching of NbN.

Of the metals investigated as contact layers, aluminium was found to have the best adhesion with NbN and makes a good ohmic contact. Aluminium may even be ultrasonically bonded directly to NbN. Common aluminium etchants do not attack NbN, making selective etching possible.

To make a large number of junctions (and circuitry), a composite film of Al, NbN, MgO, NbN and Cr, in that order, is first sputtered onto a substrate in a multi-target sputtering chamber without ever breaking the vacuum. The primary pattern is laid down on the film using photoresist and UV exposure. A secondary pattern is then chemically etched into the chromium (NbN not affected) and the resist removed. The secondary chromium patterning is essential because resist is damaged by plasma etching. All primary patterning is done with chromium-on-quartz contact masks exposed in a commercial mask aligner.

The upper NbN layer is now plasma etched in a pure SF_6 atmosphere. Slight over-etching is purposely employed here. After acid etching of the exposed MgO (NbN unaffected), the lower NbN layer is plasma etched, automatically stopping at the Al contact layer. The second plasma etch purposely over-etches the upper NbN layer, thereby reducing the effective mesa junction diameter by one micron. The plasma etching of NbN is anisotropic and leaves almost vertical sides on the resulting stepped NbN/MgO/NbN mesas. The creation of vertical sides is strongly promoted by the dense columnar nature of the NbN films.

The over-etching of the upper NbN layer is the crucial step in reducing the junction area below that generated by standard lithography.

Sputtered SiO_2 planarization and subsequent lift-off etching to remove the chromium is still under investigation. If lift-off proves successful, further aluminium metallization will be applied and etched to serve as counter-electrode contacts.

3. RESULTS

The multiple NbN films sputtered so far in Ar/N_2 atmospheres have critical temperatures of 13 to 14k. It is expected that this result may be improved to 16K by sputtering in an atmosphere of Ar, N_2 and CH_4.

The currently produced, 3 nm thick MgO layers are of good quality. Plasma etch-through testing reveals a fault rate not exceeding 4 faults per square centimetre.

Very complex circuitry with feature sizes as small as 0.75 μm has been repeatedly and reliably etched in NbN single and compound films. The straight line edge roughness is currently 0.2 μm peak to peak. NbN film grain size improvement will reduce this to better than 0.1 $μm_{p-p}$.

The smallest round mesa junctions made to date have diameters of 2.5 μm. By taking full advantage of over-etching of 1.5 μm lithography, mesas as small as 0.25 μm effective diameter may be produced.

205 GHz SIS RECEIVER DEVELOPMENT

W. R. MCGRATH[1], C. N. BYROM[1], B. N. ELLISON[2], M. A. FRERKING[1],
H. G. LEDUC[1], R. E. MILLER[3], J. A. STERN[2]

1: Jet Propulsion Laboratory, California Institute of Technology, Pasadena, CA
2: California Institute of Technology, Pasadena, CA
3: AT&T Bell Laboratories, Murray Hill, NJ

ABSTRACT. This report describes preliminary tests of a 205 GHz SIS receiver using Pb-alloy and NbN junctions in a waveguide mixer mount. Double sideband receiver noise temperature as low as 113 K has been measured with a Pb-alloy junction operated at a physical temperature of 4.2 K.

There are many applications of remote sensing in the frequency range near 200 GHz. These include studies of the earth and other planetary atmospheres as well as the interstellar medium. Heterodyne receivers are the instrument of choice for these observations due to their high sensitivity and spectral resolving power. The nonlinear tunneling currents in a superconductor-insulator-superconductor (SIS) tunnel junction have been shown to provide the lowest noise mixing element at millimeter wavelengths in these receivers[1]. This report describes preliminary tests of a 205 GHz SIS receiver using Pb-alloy and NbN junctions in a waveguide mixer mount.

The junctions are fabricated on fused quartz substrates. The PbInAu-InO-PbAu junction has a normal state resistance, $R_n = 21\,\Omega$; a critical current density, $J_c = 1.7 \times 10^4$ A/cm^2; and a relaxation parameter, $\omega R_n C = 1$. The NbN-MgO-NbN junction has $R_n = 129\,\Omega$; $J_c = 4 \times 10^3$ A/cm^2; and $\omega R_n C = 13$. Reference [2] discusses the NbN fabrication.

The receiver is mounted in a commercial vacuum cryostat. A more complete description is given in Ref. [3]. The mixer is a full height waveguide mount employing a backshort and E-plane tuner[4] and integrated with a dual-mode conical horn. The first stage of the *if* system is a low-noise HEMT amplifier. A cooled 20 dB bidirectional coupler is used to evaluate the mixer *if* mismatch. Measurements on a 48×-scale model of the mixer block and *rf* choke filter indicate that this mount arrangement can properly match junctions with *rf* impedance 10–200 Ω and with $\omega R_n C < 4$.

Receiver noise temperature T_R was measured using the conventional Y-factor method of placing hot (300 K) and cold (77 K) loads in the signal path. The double sideband measurements are listed in Table 1. Good performance was obtained over the frequency range from 200 to 211 GHz for the Pb-alloy junction at a cryostat temperature of 4.2 K. The lowest value of $T_R = 113$ K was obtained at 205 GHz. This is comparable to previous measurements made at a reduced cryostat temperature of 2.4 K. Preliminary results using the NbN junction gave a receiver noise a factor of about 10 higher. This is probably the result of poor *rf* coupling due to the large capacitance of these junctions. The mixer mount parameters, as mentioned above, were optimized for the Pb-alloy junction which have much lower values of $\omega R_n C$. The *rf* embedding environment of the mixer is currently being modified to include integrated tuning elements which will allow for a proper impedance match. Higher current density junctions are also being fabricated.

The single sideband measurements are listed in Table 1. The SIS receiver showed slightly improved performance as a SSB receiver since the lowest values obtained were less than twice the DSB measurements. The best value, $T_R = 170$ K, was obtained at 203 GHz for the Pb-alloy junction. In all the measurements reported, the mixer was well matched at the *if* with a return loss typically of –6 dB to –10 dB. Values as high as –15 dB were also observed.

TABLE 1. Receiver Noise Measurements

Junction	Double Sideband		Single Sideband	
	LO Frequency [GHz]	T_R [K]	Signal Frequency [GHz]	T_R [K]
Pb-alloy	198.4	249±12	203.3 (LSB)	170±11
	205	113±11	206.5 (USB)	247±13
	211.2	131±38	207 (LSB)	178±11
			210 (USB)	194±22
NbN	207	1680±46	205.8 (LSB)	1770±47
	210	2012±190	208.6 (USB)	2235±60

Acknowledgements

The research described in this paper was carried out by the Jet Propulsion Laboratory, California Institute of Technology, and was sponsored by the National Science Foundation and the National Aeronautics and Space Administration.

References

[1] J. R. Tucker and M. J. Feldman (1985), *Rev. Mod. Phys.* **57**, 1055.

[2] H. G. LeDuc, J. A. Stern, S. Thakoor, S. K. Khanna (1987), *IEEE Trans. Magn.* **MAG-23**, 863.

[3] W. R. McGrath, C. N. Byrom, B. N. Ellison, M. A. Frerking, and R. E. Miller, to be presented at the 13th International Conference on Infrared and Millimeter Waves, Hawaii, Dec. 1988.

[4] B. N. Ellison and R. E. Miller (1987), *Int. J. IR and mm Waves* **8**.

NEW MEASUREMENTS OF THE SPECTRUM AND ANISOTROPY OF THE MILLIMETER WAVE BACKGROUND

G.M. Bernstein, M.L. Fischer, and P.L. Richards
Department of Physics, University of California
Berkeley, California 94720, U.S.A.

J.B. Peterson
Department of Physics, Princeton University
Princeton, New Jersey 08540, U.S.A.

T. Timusk
Department of Physics, McMaster University
Hamilton, Ontario L8S 4M1, Canada

ABSTRACT. Recent measurements of the diffuse background at millimeter wavelengths indicate no departure from a Planck spectrum near the peak of the blackbody curve. Anisotropy measurements indicate no structure, at the 2% level, in the recently detected submillimeter excess.

We report here the results of an April 1987 balloon flight of an instrument designed to measure the spectrum of the cosmic background radiation from 1 mm to 3 mm. A description of the instrument can be found in Peterson, Richards, and Timusk (1985). Modifications were made to the apparatus and experimental procedure in order to identify and reduce systematic errors. Results from the latest flight indicate that two effects hamper the interpretation of the data. These systematic effects will be described in detail in a forthcoming publications; they are probably responsible for the non-Planckian spectrum measured by Woody and Richards (1981). Attempts to remove the systematic effects from our data yield the upper limits to the CBR brightness temperature in 4 bands from 1 mm to 3 mm. There is no evidence for an excess of radiation near the 2.8 K blackbody peak.

Recent measurements have detected an excess of radiation in the submillimeter region (Matsumoto et al. 1988). If we assume that this excess has a spectrum extending into the millimeter region as fit by Hayakawa et al. (1987), then we may check its isotropy using data collected during our flight. Details of this analysis may be found in Bernstein et al. (1988). Our broad bandpass filter (0.5 to 4 mm) is used to obtain the upper limits on anisotropy shown in the figure. This level of isotropy strongly suggests that the submillimeter excess is not of local origin, but must rather originate at substantial redshift. Identification of the sources responsible for this excess is an exciting challenge for submillimeter astronomy.

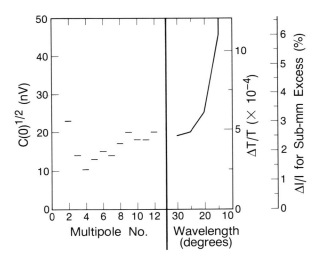

Figure 1. Anisotropy limits versus angular scale. Upper limits, at 95% CL, to the RMS sky fluctuation $\sqrt{C(0)}$ are shown for models with Gaussian fluctuations on various angular scales. Points on the left half are for models with fluctuations at multipole ℓ. The right side curve shows upper limits for models with fluctuations with angular wavelength or wavelength λ. The right hand vertical axes give the calibration in terms of changes in the 2.74 K blackbody temperature, and in terms of variations in the submillimeter excess.

This work was supported by NASA Grant #NSG-7205.

REFERENCES

    Bernstein, G.M., Fischer, M.L., Richards, P.L., Peterson, J.B., and Timusk, T. (1988) 'Anisotropy of the Diffuse Background at Millimeter Wavelengths', Ap. J. Letters (submitted).

    Hayakawa, S., Matsumoto, T., Matsuo, H., Murakami, H., Sato, S., Lange, A.E., and Richards, P.L. (1987) 'Cosmological Implication of a New Measurement of the Submillimeter Background', Publ. Astron. Soc. Japan 39, 941-948.

    Matsumoto, T., Hayakawa, S., Matsuo, H., Murakami, H., Sato, S., Lange, A.E., and Richards, P.L. (1988) 'The Submillimeter Spectrum of the Cosmic Background Radiation', Ap. J. 329, 567-571.

    Peterson, J.B., Richards, P.L., and Timusk, T. (1985) 'The Spectrum of the Cosmic Background Radiation at Millimeter Wavelengths', Phys. Rev. Letters, 55, 332.

    Woody, D.P. and Richards, P.L. (1981) 'Near-Millimeter Spectrum of the Microwave Background', Ap. J., 248, 18.

MILLIMETER AND SUBMILLIMETER INTERFEROMETRY

Wm. J. Welch
Radio Astronomy Laboratory
University of California
Berkeley, CA 94720

ABSTRACT. This review discusses the special properties of interferometers at millimeter and submillimeter wavelengths, the characeristics of existing arrays, future plans for instruments, and a few examples of current scientific results.

1. INTRODUCTION

Although an interferometeric array is technically complicated, the high angular resolution and accurate registration on the sky that it provides more than compensate for the complication. Resolution of about an arc second and registration (pointing) of 0.1-0.2 seconds have already been achieved and will be further improved. In contrast, single reflectors have much poorer resolution and pointing. Already a large amount of exciting science has come from the existing instruments.

2. SPECIAL PROBLEMS AT MILLIMETER WAVELENGTHS

2.1 *Field of View.* The basic field of view of the interferometer is given by the half power beam width of the individual elements. Even the smaller millimeter wave antennas are large compared to wavelength, so that their primary half power widths are small (e.g., 24sec. at 1mm for a 10m reflector). The majority of sources that will be observed are larger than this. Interferometer maps made in multiple adjacent antenna pointings can be combined to provide extended fields.

In addition, some special observations are needed to overcome the limitation of missing data at the center of the (u,v) plane. Since the antennas cannot be closer than one dish diameter, the visibility function is not sampled near the center of the (u,v) plane. For example, if the minimum antenna separation is s_{min}, the strength of a Gaussian source of FWHM=Θ at the center of the field will be reduced to $1/e$ if $\Theta \geq 0.5\lambda/s_{min}$, only 13 sec. for 10m dishes operating at 230 GHz. Three different schemes have been used for dealing with this problem. First, one may add map data taken with a larger antenna to the interferometer data. Second, one may add data taken with another interferometer which has smaller antennas. Third, one may scan the interferometer antennas over the extended field while synthesizing the source, keeping both cross-correlation and single antenna total power data, and combine all the data for a complete map (Ekers and Rots, 1979).

2.2 *Spectroscopy.* One of the attractions for astronomical observing at millimeter wavelengths is the rich molecular spectrum, which permits the detailed study of the chemical, physical, and dynamical states of the extended atmospheres of young and old stars and planets and the interstellar media of the milky way and other galaxies. In the 3 mm atmospheric window some

600 lines have been reported (Lovas et al., 1979), and in the Cal Tech 210-280 GHz survey of the Orion region, over 800 lines were identified (Blake et al., 1987).

Maps of molecular clouds in different molecules are often distinct, and a flexible spectrometer is required to permit observation of several lines at once in different resolutions. The hybrid correlator provides the best flexibility (Weinreb, 1985; Urry et al., 1985).

2.3 *Atmospheric Effects.* Two effects of atmospheric water vapor and oxygen are important in millimeter interferometry. The more important is the absorption, which is stronger at the shorter wavelengths and restricts observation to windows between strong lines and bands. Because the water vapor atmospheric scale height is about two kilometers, mountain top sites are driest. Submillimeter observations must be done from the highest sites, while it is possible to observe in the 3 mm and 1.5 mm windows at only 1 km above sea level.

The other effect of the atmosphere is the fluctuation that it introduces in the phase of signals at each antenna. This has the effect of blurring the image that is produced by the aperture synthesis and is like the "seeing" that limits resolution at optical wavelengths. Unlike the optical effects that are due to small scale fluctuations in the dry air, the radio fluctuations are dominated by variations in water vapor which have scale sizes up to a km or more. Radio "seeing" is a few tenths of a second in average conditions and very much better in good conditons(Armstrong and Sramek 1982;Hinder and Ryle, 1971; Bieging et al. 1984).

2.4 *Sensitivity.* The sensitivity of a millimeter or sub-millimeter array for a particular scientific problem depends principally on antenna size and number, system temperature (from the receivers and the atmospheric background), bandwidth, and the strength of the astronomical signal. Millimeter wave antennas are necessarily smaller and their system temperatures are larger than those of cm wave antennas. Thus, the existing millimeter arrays of 3-5 antennas have two to three orders of magnitude less flux sensitivity than the large cm wave arrays, putting them at a disadvantage for sources with flat or negative power spectra.

On the other hand, for the large class of objects in which the emission process is basically thermal, the small mm wave arrays have much better sensitivity than the cm wave arrays. A six element array of 6m antennas observing the CO(1-0) line has more than two orders of magnitude greater brightness temperature sensitivity than the VLA observing the HI line.

The way in which the array sensitivity scales with antenna size depends on the extent of the source to be mapped. For sources smaller than the antenna primary beam, the signal/noise is proportional to $\sqrt{t}nD^2$, where n is antenna number, D is antenna diameter, and t is observing time. For sources larger than the antenna primary beam, requiring multiple pointings, the signal/noise is proportional to $\sqrt{t}nD$. Since, as noted above, most sources are larger than the primary antenna beam, the second relation is the more appropriate.

2.5 *Calibration.* Visibility phase calibration follows standard practice at cm wavelengths. Antenna positions are determined by observation of a catalog of quasars with known positions. Slow thermal drifts are removed by the observation of a calibrator (quasar) near the source at intervals of the order of each half hour during the synthesis. This is made necessary by the thermal instabilities of the antennas on submillimeter scales and the slowly drifting large patches of atmospheric water vapor, even if the electronics are perfectly stable.

Amplitude calibration of the visibilities is more difficult at millimeter wavelengths, because all the compact sources, the quasars, may vary in amplitude by substantial amounts, 20%-30%

or more, in a week or less. The quasars must be frequently compared with stable objects, such as a planet or compact H II region. For a large planet, the comparison may require a single dish observation with a beam-switch for stability.

2.6 *Image Processing.* There have been remarkable advances in image processing, especially in radio astronomy, in the past decade, and further developments can be expected, driven in part by the special problems of millimeter wave synthesis. These are (a) the importance of multiline, multichannel mapping, (b) the need to do "self-calibration" with relatively low brightness spectral line data, and (c) the need to routinely combine data taken in many adjacent fields (pointings) to produce large field spectral line maps.

3. THE MILLIMETER AND SUBMILLIMETER ARRAYS

3.1 *The Operating Arrays.* At present writing, four arrays are in operation. These include the 5-element array of 10m dishes at Nobeyama in Japan, the 3-element array of 10m dishes at Owens Valley, California, the 3-element array of 6m dishes at Hat Creek, California, and the 3-element array of 15m dishes at the Plateau de Burre in France. Their present operating parameters are summarized in Table 1.

3.2 *Expansion and Other Plans.* Further expansion is either in progress or in planning for all the systems above. At IRAM, a fourth antenna is planned as well as operation at 1mm wavelength. Currently, the Berkeley Illinois Maryland Array at Hat Creek is adding four new 6m antennas, an increase in IF bandwidth to 830 MHz, and 1mm receivers. For the OVRO array, a digital correlator with 256 channels and 500 MHz bandwidth is under construction. In addition, three new antennas are planned for that array. At Nobeyama, receivers for 1mm operation are under construction, and there is a long range plan for a large array of perhaps 25 antennas.

There are two new arrays in the planning stages. A VLA class array is being designed by the NRAO for operation up to 360 GHz with high spatial and spectral resolution and high sensitivity. Details of this plan are discussed by R. L. Brown elsewhere in this volume.

At the Smithsonian Astrophysical Observatory in Cambridge, Massachusetts, an array is under design for submillimeter wavelengths. The current concept is 6 or more antennas of approximately 6m diameter with surfaces smooth enough to permit operation at the atmospheric limit of about 800 GHz. The choice of site will be either Mauna Kea, Hawaii, or Mount Graham in Arizona. Baselines of up to a kilometer are planned. Present activity includes the development of SIS receivers, the beginning of a design study, and the general buildup of a technical staff.

4. EXAMPLES OF RECENT RESULTS

Figures 1, 2, and 3 are recent maps made with the three presently operating arrays. Figure 1 is a map of the central 1.5 kpc region of the galaxy Maffei 2 in CO(1-0) with an angular resolution of 5 seconds (Ishiguro et al., 1989). The map includes all velocities in the V_{LSR} range of -156 to 78 km/sec. The extended linear feature appears to be a galactic bar of molecular gas. Figure 2 shows a map of M51 made at 7 sec. resolution with the Owens Valley Millimeter Interferometer in integrated CO(1-0) emission, superimposed on an Hα photograph (Vogel et al., 1988). The spiral arm molecular clouds in CO are parallel with the Hα but displaced, suggesting that there is a spiral density wave in whose frame newly formed massive stars lie slightly downstream from the molecular clouds from which they formed. Figure 3 is an 89 GHz continuum map of the core of the giant galactic molecular cloud W49 made with the BIMA at Hat Creek The angular resolution is 2.5 sec, and the squares in the figure correspond to compact HII regions in a 5 GHz VLA map (Welch et al., 1987). The HII regions at the brightest 3mm peaks have very high turnover frequencies and some dust emission.

Acknowledgement It is a pleasure to thank M. Ishiguro , C. Masson, D. Downes, P. Myers, and R. Brown for current information about instruments and plans at their observatories.

References

Armstrong, J.W., and Sramek, R.A. 1982, *Radio Sci.*, **17**, 1579.
Bieging, J., Morgan, J., Welch, W., Vogel, S., and Wright, M. 1984, *Radio Sci*, **19**, 1505.
Blake, G., Sutton, E., Masson, C., and Phillips, T. 1987, *Ap.J.*, **315**, 621.
Ekers, R. and Rots, A. 1979, in *Image Formation from Coherence Functions in Radio Astronomy*, ed. C.van Schoonveld (Dordrecht: Reidel).
Hinder, R., and Ryle, M. 1971, *M.N.R.A.S.*, **154**, 229.
Ishiguro et al., 1989, *Ap.J.*, in press.
Lovas, F., Snyder, L., and John, D. 1979, *Ap. J. Suppl.*. **41**, 451.
Urry, W., Thornton, D., and Hudson, J. 1985, *Pub.A.S.P.*, **97**, 745.
Vogel, S., Kulkarni, S., and Scoville, N. 1988, *Nature*, **334**, 402.
Weinreb, S. 1985, NRAO internal memo.
Welch, W., Dreher, J., Jackson, J., Terebey, S., and Vogel, S. 1987, *Science*, **238**, 1550.

Table 1. CHARACTERISTICS OF THE FOUR OPERATING MILLIMETER ARRAYS

Observatory:	NRO	OVRO	BIMA	IRAM
Location:	Nobeymea	Owens Valley	Hat Creek	Plateau de Burre
Altitude:	1350m	1216m	1050m	2550m
Antennas:	5 @ 10m	3 @ 10m	3 @ 6m	3 @ 15m
Baselines:	560m EW 520m NS	200m EW 220m NS	300m EW 200m NS	288m EW 160m NS
Receivers: (GHZ)	22-24(HEMT) 40-50(SIS) 80-120(SIS)	85-120(SIS) 200-300(SIS)	75-116(Shottky)	80-116(SIS)
Correlator:	250MHz cont. Spectral Line: 1024 channels BW:320MHz	180MHz cont. Spectral Line: 3 filter banks: res:.05,1,5MHz	480MHz cont. Spectral Line: 512 channels, 4 windows, BW:320MHz	500MHz cont. Spectral Line: 512 channels, 4 windows, BW:320MHz

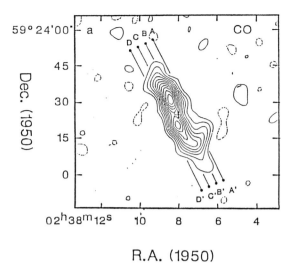

Figure 1 : An image of the galaxy Maffei 2 in integrated CO(1-0) emission made with the Nobeyama Millimeter Array (Ishiguro et al., 1989). The resolution is about 5 sec. and the map includes emission over the VLSR range -156 to 78 km/sec.

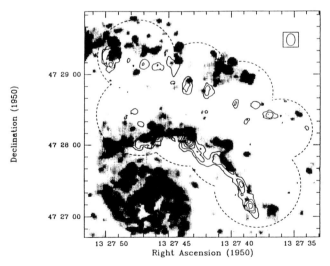

Figure 2 : The contours show an image of integrated CO(1-0) emission from M51 made with the OVRO interferometer (Vogel et al., 1988), superimposed on a grey scale image of Hα. The angular resolution is 7 sec. The dotted line surrounds the area mapped in CO.

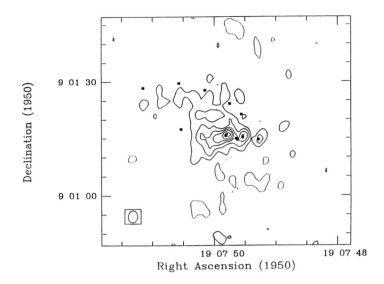

Figure 3: An 89 GHz continuum map of the core of W49 made at Hat Creek. The angular resolution is 2.5 arc seconds, and the squares denote 5 GHz continuum peaks from a VLA map (Welch et al., 1987). The relative registration of the maps is 0.3 sec.

THE MILLIMETER ARRAY

ROBERT L. BROWN
National Radio Astronomy Observatory*
Edgemont Road
Charlottesville, VA 22903-2475

1. SCIENTIFIC EMPHASIS OF THE MILLIMETER ARRAY

The need for a high resolution imaging telescope at millimeter and submillimeter wavelengths has been noted in many contexts over the last decade or so and that need is underscored in these proceedings. At the 1985 Green Bank MMA scientific workshop, the scientific emphasis of the Millimeter Array was focused on the following desired capabilities:

- Sub-arcsecond imaging at 115 GHz and higher frequencies;
- Wide-field imaging, mosaicing;
- Rapid imaging, "snapshots" of high fidelity;
- Sensitive imaging at high frequency (to 350 GHz);
- Simultaneous multi-band operation.

Together these capabilities define a unique instrument; astronomers using the MMA will explore entirely new scientific areas. Two examples are illustrative. The combination of high sensitivity and sub-arcsecond angular resolution at frequencies of 230 and 350 GHz provided by the MMA will permit the photospheric emission from hundreds of nearby stars to be detected and imaged, the stellar radii to be determined and the positions of the stars established to astrometric precision. The same combination of instrumental parameters will provide images at a resolution equal to that of the Hubble Space Telescope of the redshifted dust emission from galaxies at the epoch of formation ($z = 5-10$). Indeed, these "protogalaxies" will be the dominant source of background confusion at 1 mm at levels approaching 1 mJy.

The scientific arguments leading to definition of the MMA are outlined by Wootten and Schwab (1988).

2. THE MILLIMETER ARRAY DESIGN CONCEPT

High sensitivity implies that the total collecting area of all the individual elements in the array be made as large as possible, while fast imaging is achieved by distributing that area over many elements. The precise definition of how many elements and the size of the individual antennas is then made by minimizing the total array cost. Subarcsecond imaging places a constraint on the array dimension: 0."1 at 230 GHz, for example, requires an array of maximum extent 3 km. Finally, sensitive imaging at high frequency demands that the MMA be located on a high altitude site with excellent atmospheric transparency. Considerations such as these drive the design of the MMA; they are described in detail by Brown and Schwab (1988). However, the MMA design is still evolving in response to changing technology and in recognition of new scientific opportunity. Additional input in this regard is most welcome.

The current MMA design concept is summarized in Table 1.

TABLE 1. DESIGN CONCEPT OF THE MILLIMETER ARRAY

ARRAY
  Number of Antennas: 30-40
  Total Collecting Area: 1750 m^2
  Angular Resolution: 0."07 λ_{mm}

ANTENNAS
  Diameter: 7.5-8.5 m
  Precision: $\lambda/40$ at 1 mm
  Pointing: 1/20 beamwidth
  Transportable

CONFIGURATIONS
  Compact: < 100 m
  Intermediate: 300-1000 m
  High Resolution: 3 km

FREQUENCIES
  Emphasis on: 200-350 GHz
  Capability at: 70-150 GHz; 9 mm
  Desirable: Simultaneous Multi-band

SITE
  High Altitude--suitable for precision imaging at 1 mm

REFERENCES

Brown, R. L. and Schwab, F. R. 1988, 'Millimeter Array Design Concept', Volume II of the NRAO MMA Design Study.

Wootten, H. A. and Schwab, F. R. 1988, 'Science with a Millimeter Array', Volume I of the NRAO MMA Design Study.

*Operated by Associated Universities, Inc., under contract with the National Science Foundation.

THE NOBEYAMA MILLIMETER ARRAY: NEW DEVELOPMENTS AND RECENT OBSERVATIONAL RESULTS

M. ISHIGURO, R. KAWABE, K.-I. MORITA, T. KASUGA, Y. CHIKADA,
J. INATANI, T. KANZAWA, H. IWASHITA, K. HANDA, T. TAKAHASHI,
H. KOBAYASHI
Nobeyama Radio Observatory
Nobeyama, Minamimaki, Minamaisaku, Nagano 384-13, JAPAN

S.K. OKUMURA, Y. MURATA, S. ISHIZUKI
Department of Astronomy, Faculty of Science, Univ. of Tokyo
Bunkyo-ku, Tokyo 113, JAPAN

ABSTRACT. New developments for upgrading the Nobeyama Millimeter Array(NMA, Ishiguro et al., 1984) and the recent results from millimeter-wave spectral-line observations are presented.

1. NEW DEVELOPMENTS

Each 10-meter antenna has been equipped with three frontend receivers for 22-24 GHz, 40-50 GHz and 80-116 GHz bands. Cooled HEMT amplifiers are used for 22-24 GHz band and the noise temperature is about 100 K. The receivers for 40-50 GHz and 80-116 GHz are SIS mixers using Nb/Al-AlOx/Nb junctions developed at Nobeyama. The mixers for two frequency bands are installed into a single 4 K closed cycle refrigerator. Typical noise temperatures of the SIS receivers are 200 K(SSB) for both frequency bands.

Local oscillators for the SIS mixers were designed to compromise the flexibility in frequency tuning and the stability in phase. The LO signals for 80-116 GHz are obtained by frequency-doubling the outputs of the 40-60 GHz Gunn oscillators. IF amplifiers for the mixers are cooled 1.4 GHz FET, and 90° phase switching is used for separating upper and

TABLE 1. PERFORMANCE PARAMETERS OF THE NMA

FREQUENCY	22 GHz	49 GHz	115 GHz
FIELD OF VIEW	5.3'	2.4'	1.0'
MAXIMUM RESOLUTION	4.2"	1.9"	0.8"
VELOCITY COVERAGE	4300 km/s	2000 km/s	800 km/s
APERTURE EFFICIENCY	60 %	50 %	40 %
SENSITIVITY(1 σ, calculated value)			
Δ Srms (320 MHz, 8 hr.)	0.7 mJy	1.9 mJY	3.4 mJY
Δ Trms (1 MHz, 8 hr.)	2.4 K	6.8 K	12 K

lower sidebands. The mixers and LO's have been made remotely controllable by a local-area-network of small computers.

A 320 MHz bandwidth A/D convertor system with digital delays was completed in autumn of 1987 and has been in operation with FX correlator system(Chikada et al., 1987), for line and continuum observations. It consists of six pairs of 3-bit A/D convertors which corresponds to real and imaginary parts of complex samples. The computing time for making multi-channel maps has been greately reduced by using vectorized version of AIPS installed on the vector processor FACOM VP50. The performance parametersof the NMA are shown in Table 1.

2. RECENT OBSERVATIONAL RESULTS

From the test observations done in 1987, it turned out that the NMA has enough sensitivity to do millimeter-wave spectral-line observations. For strong sources, SELF-CALIBRATION has been applied successfully and high dynamic range($\sim$800:1 for Sgr-A continuum map) is obtained.

Aperture synthesis observations with various lines(CO, CS, SiO, etc.) started in winter of 1988. Figure 1 shows examples of ^{12}CO(J=1-0) mapping of galaxies with a single configuration(10 baselines). The resolution of the maps is about 5". The quality of the M82 map is comparable with other multi-configuration observations using Hatcreek or OVRO interferometers. From the Maffei 2 observation, we have found a central narrow ridge of molecular gas (1000 pc x 200 pc) and a ring-like feature (500 pc x 240 pc) with a large non-circular motion of 60 km/s(Ishiguro et al., submitted to Ap.J., 1988).

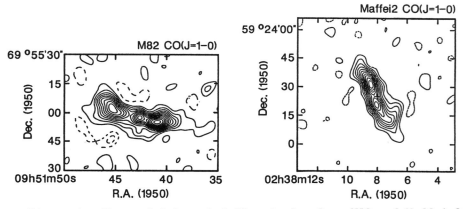

Figure 1. Maps of integrated CO emission from M82 and Maffei 2

REFERENCES

Ishiguro, M. et al.(1984) 'THE NOBEYAMA MILLIMETER-WAVE INTERFEROMETER', Proc. Int. Symp. Millimeter and submillimeter Wave Radio Astronomy, 75-87.
Chikada, Y. et al.(1987) 'A 6 x 320-MHz 1024-CHANNEL FFT CROSS-SPECTRUM ANALYZER FOR RADIO ASTRONOMY', Prc. IEEE, vol.75, 1203-1210.

MILLIMETER VLBI
Melvyn Wright, Radio Astronomy Laboratory,
University of California, Berkeley, CA 94720

ABSTRACT. We have obtained ~ 50 μarcsec resolution using global VLBI at 100 GHz. This is sufficient to resolve ly-scale structure in nearby quasars. Millimeter observations are required to probe the optically thick synchrotron components seen at cm. wavelengths. In 7 experiments since 1981 we have observed the brightest quasars which show rapid time variations at mm wavelengths. The observations imply sub-milliarcsec structure in all the sources observed. We have traced the structure of 3C84 following a flare in 1980. The changing structure can be understood in terms of the migration of high-energy electrons from a ~ 0.1 pc nucleus into a ~ 0.3 pc halo. Superluminal motion on scales as small as 1 ly are inferred at 3 epochs in 3C273.

OBSERVATIONS

We used antennas at the Hat Creek Radio observatory (HCRO), Owens Valley Radio Observatory (OVRO), Kitt Peak (KTPK), Five College Radio Observatory (QBBN), Onsala Space Observatory (ONSA), and Nobeyama (NBYM). The experiments are summarized in Table 1. We used the MKIII VLBI system with 52 MHz bandwidth. The techniques developed include simultaneous observations at 5 GHz to determine clock parameters, and phased array interferometer observations. Further technical details are discussed by Backer[1]. Atmospheric turbulence limits the coherent integration time to ~ 100 s.[2]

With VLBI arrays of 6 or more antennas we can use global fringe fitting and self-calibration techniques to obtain hybrid maps with dynamic range of ~ 100:1[3], sufficient to reliably trace the change in source structure between experiments. In the early experiments, with 3 or 4 antennas we are limited to model fitting of the observed visibility amplitudes and closure phases, and maps with dynamic range ~ 10:1

Table 1 - Summary of λ 3mm VLBI Experiments

Epoch	Stations used	Sources detected
OCT 1981	HCRO-OVRO	3C84
MAY 1982	HCRO-OVRO-KTPK	3C84, 3C273, 3C279, 3C345
APR 1983	HCRO-OVRO-KTPK	3C84, 3C273, OJ287, 1921-293
APR 1984	HCRO-OVRO-QBBN	3C84, 3C273, 3C345, 2145+06
MAR 1985	HCRO-OVRO-QBBN	3C84, OJ287, 3C273, 3C279, 1921-293, 3C345, 2145+067
MAR 1987	HCRO-OVRO-QBBN -KTPK	3C84, OJ287, 3C273, 3C279, 3C345, NRAO530
MAR 1988	HCRO-OVRO-QBBN -KTPK-NBYM-ONSA	3C84, OJ287, 3C273, 3C279, 3C345, BLLAC

RESULTS

The results on 3C84, associated with the Seyfert-like nucleus of NGC1275, show the evolution of a core-halo structure following a flare in 1980[4,5]. The unresolved core, < 0.1 pc in size, has decreased monotonically from 14 Jy in 1982 to < 0.5 Jy in 1987, corresponding to the total flux density decay since the flare in 1980. The halo, relatively constant in flux density until 1985, changed between 1985 and 1987. The changes in source structure can be interpreted as due to the decreased supply of high-energy electrons with synchrotron lifetimes of ~ 2 yr from the radio nucleus into the halo. A narrow jet, detected in 1985 has evolved into a resolved component, separated from the nucleus by 0.2 pc.

Our observations of 3C273 have followed flares within a few months at 3 epochs. From the measured size of the new components we infer superluminal motion on scales as small as 1 ly and with $\gamma \sim 8\,h^{-1}$.

The limited data on the other sources observed imply a range of sub-milliarcec structure which can be mapped with the existing array of millimeter telescopes. Addition of more telescopes in the next 1-2 years will improve the u-v coverage, especially at low declinations. The techniques and telescopes can be extended to observations at 230 GHz to provide 10 μarcsec resolution. Comparison of infrared, optical and X-ray variations, and of maps at 1.3, 3, 7, and 13 mm to compute the spatial and spectral evolution of the source structure will lead to a much clearer understanding of radio flares and jet formation in radio sources.

This research was made possible through the efforts many people including: D.C.Backer, J.E.Carlstrom, R.L.Plambeck, W.J.Welch, (U.C.Berkeley), C.R.Masson, A.T.Moffet, S.Padin, A.C.S.Readhead, D.Woody, A.Zensus, (Caltech), A.E.E.Rogers, (Haystack), J.M.Moran, (SAO), C.R.Predmore, R.L.Dickman, (U.Mass.), D.T.Emerson, J.Payne, (NRAO), L.Baath, A.Kus, B.Ronnang, R.Booth, (Onsala), H.Hirabayashi, N,Inoue, M.Morimoto, (Nobeyama).

REFERENCES

1 Backer, D.C., 1984, URSI Symposium on Millimeter and Submillimeter Wave Radio Astronomy, ed. J.Gomez Gonzalez (Grenada), P93.
2 Rogers, A.E.E, Moffet,A.T.,Backer, D.C., and Moran,J.M., 1985, Radio Science, 19, 1552.
3 Baath, L.B., this conference.
4 Backer, D.C., et al., 1987, Ap.J., 322, 74.
5 Wright,M.C.H., et al., 1988, Ap.J., 329, L61.

THE SUBMILLIMETER WAVE ASTRONOMY SATELLITE

G. J. MELNICK
Harvard-Smithsonian Center for Astrophysics
60 Garden Street
Cambridge, Massachusetts 02138
USA

ABSTRACT. As part of its Small Explorer Program, NASA recently selected the Submillimeter Wave Astronomy Satellite (SWAS) for flight, with a scheduled launch aboard an expendable Scout launch vehicle in 1993. During its baselined 2-year lifetime, SWAS's primary objective will be to conduct a large-scale, high spectral resolution survey of galactic molecular clouds in low-lying transitions of water, molecular oxygen, atomic carbon, and isotopic carbon monoxide.

The path that leads from the quiescent molecular clouds observed throughout the Galaxy to the formation of stars and planets is poorly understood. Among the main uncertainties in this process is the chemical composition of dense interstellar clouds which, in turn, determines the species available to radiate energy out of these clouds, thus facilitating their collapse. Since some atoms and molecules are substantially better coolants than others, a direct determination of the abundance of species which are predicted to be major gas coolants will bear on such crucial questions as: (1) cloud stability against collapse, (2) varying star formation efficiencies observed in different parts of the Galaxy, and (3) the exact process by which energy is removed from collapsing gas, permitting the formation of stars and planets.

Since some of the most abundant atomic and molecular constituents of dense interstellar clouds possess low-lying transitions with energy differences ($\Delta E/k$) between 15 and 30 K, and since these levels will be readily excited in most clouds, the submillimeter wavelength region is particularly useful for our understanding of the chemistry, energy balance, and structure of dense molecular clouds. Unfortunately, with the exception of a few narrow windows of modest atmospheric transmission, the submillimeter region is inaccessible from ground-based facilities. To overcome this problem, NASA recently selected SWAS as part of the Small Explorer Program. SWAS's primary goal will be to perform both pointed and survey (scanning) observations of dense ($n(H_2) \geq 10^3$ cm^{-3}) galactic molecular clouds in the lines of four important species: (1) the H_2O ($1_{10} \rightarrow 1_{01}$) 556.936 GHz ground-state ortho transition, (2) the O_2 (3, 3 $\rightarrow$ 1, 2) 487.249 GHz transition, (3) the CI ($^3P_1 \rightarrow {}^3P_0$) 492.162 GHz ground-state fine structure transition, and (4) the ^{13}CO ($J = 5 \rightarrow 4$) 550.926 GHz rotational transition. A large-scale survey in these lines is virtually impossible from any platform within the atmosphere due to telluric absorption.

SWAS will consist of a highly efficient, 55-cm diameter off-axis Cassegrain antenna, two heterodyne radiometers with Schottky Barrier diode mixers, and a single broadband (1.4 GHz) acousto-optical spectrometer (AOS). The AOS bandwidth is sufficient to permit simultaneous observation of all four proposed lines, while the spectral resolution of the AOS, 1 MHz, will give SWAS a velocity resolution of approximately 0.6 km s^{-1}. The AOS will be provided by the University of Cologne.

During its baselined two-year lifetime, SWAS will map local ($d \leq 1$ kpc) clouds, such as Orion, Taurus, ρ Ophiuchi, and Perseus, in each of the four lines as well as perform a large-scale survey of galactic molecular clouds. The nominal spatial resolution of the instrument is about 4.5 arcminutes (FWHM), though higher resolution ($\sim 2'$) studies of selected local clouds will be obtained by spatially oversampling regions of high signal-to-noise and using restoration techniques. A number of low-redshift, gas-rich extragalactic sources, such as the Magellanic Clouds, IC 342, and NGC 6946 will also be observed.

The spacecraft, supplied by NASA Goddard Spaceflight Center, will be a three-axis all-sky pointer utilizing reaction wheels. Use of a star tracker will permit a pointing accuracy of approximately 45 arcseconds. The satellite has a weight of 200 kg and is currently scheduled for launch in 1993.

The SWAS investigation team consists of Gary J. Melnick (PI, SAO), Alexander Dalgarno (SAO), Neal R. Erickson (UMass), Giovanni G. Fazio (SAO), Paul F. Goldsmith (UMass), Martin Harwit (NASM), David J. Hollenbach (NASA Ames), David G. Koch (NASA Ames), David A. Neufeld (UC Berkeley), Ronald L. Snell (UMass), Patrick Thaddeus (SAO), and Gisbert F. Winnewisser (Univ. of Cologne, FRG).

Section III

Chemistry of the Interstellar Medium

SPECTROSCOPY OF CIRCUMSTELLAR ENVELOPES WITH THE IRAM 30-M TELESCOPE

R. LUCAS and M. GUÉLIN
IRAM, Domaine Universitaire
38406 Saint-Martin-d'Hères
FRANCE

ABSTRACT. We review here recent results obtained in the study of circumstellar envelopes with the IRAM 30-m telescope.

1 Introduction

A large fraction of the material in stars is returned to interstellar space in the form of molecules and dust grains carried outward in the expanding envelopes of late-type giants. The study of these envelopes helps us to assess the effect of recently processed matter on the interstellar medium composition. The thickest envelopes, which make the biggest contribution to this process, are best studied at infrared and millimetre wavelengths. Their small angular size make them choice targets for the large new-generation millimetre instruments, such as the 30-m telescope of IRAM. We review here some of the main results obtained with this instrument during its first three years of operation. For a more general review of infrared and radio observations of circumstellar envelopes, see also Olofsson (1988).

2 The 30-m Spectral Survey of IRC+10216

A sensitive spectral survey of the carbon star IRC+10216 has been done (Cernicharo *et al.* 1988a). The searched spectrum covers over 90 GHz of frequency space, mostly in the 3-mm (80 GHz to 116 GHz) and 2-mm (130 GHz to 174 GHz) spectral windows, with some limited coverage around 1.3 mm and 4 mm. The noise level is 0.01 to 0.02 K of antenna temperature in 1 MHz channels.

About 500 lines were safely detected, most of them for the first time. The line density is very high at low frequencies, and decreases in the 2 mm and 1 mm windows. Among these transitions, 220 arise from new molecules (C_5H, C_6H, NaCl, KCl, AlCl, AlF, c-C_3H, CCS and C_3S) or new isotopic species ($C^{13}CH$, $^{29}SiCC$, $^{30}SiCC$, $^{28}Si^{13}CC$, $^{13}C^{34}S$,...). Approximately 100 lines remain unidentified to this day. The spectra contain a large number of lines from vibrationally excited molecules, such as SiS (5 lines), CS(2), C_4H(40), HC_3N(10), and others. 23 lines are detected from metal-bearing species; the long chains have also a very rich spectrum (56 lines from C_6H and isotopes).

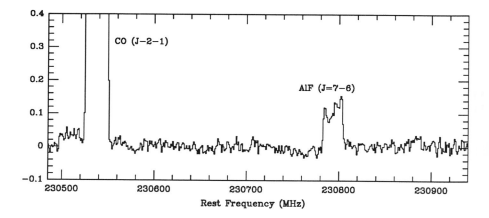

Figure 1: The J=7-6 transition of AlF at 230.8 GHz in IRC+10216

3 New Molecules

The following table gathers all molecules currently known to be present in IRC+10216 (molecules detected in the 30-m line survey are in boldface characters).

- inorganic, stable molecules:

CO	**NaCl**	**H_2S**	NH_3	SiH_4
CS	**KCl**			
SiO	**AlF**			
SiS	**AlCl**			

- organic, stable molecules: **HCN** **CH_3CN** **HC_3N** CH_4 C_2H_2 C_2H_4

- reactive molecules:

carbon chains	**HC_5N**	**HC_7N**	**HC_9N**	$HC_{11}N$	
isomers	**HNC**				
free radicals	**C_2H**	**l-C_3H**	**C_4H**	**C_5H**	**C_6H**
	CN	**C_3N**	**C_2S**	**C_3S**	C_3
cyclic	**c-C_3H**	**SiCC**	**C_3H_2**		
ion	**HCO^+**				

(The detection of HCO^+ and H_2S are based on only one transition)

Among the significant advances we may note:

- The identification of the long-chain radicals C_5H and C_6H (Cernicharo et al. 1986a, 1986b, 1987a, Guélin et al. 1986, 1987a)

- The detection of the sulfur compounds H_2S, C_2S and C_3S (Cernicharo et al. 1987b)

- The discovery of metal halides (Cernicharo and Guélin 1987). The column densities of NaCl, AlCl, KCl and AlF are in the range $10^{12} - 10^{14}$ cm^{-2}, 10^6 to 10^8 times lower than that of H_2. They agree with the chemical equilibrium abundances calculated by Tsuji (1973) and Clegg et al. (1980) for a carbon-rich stellar atmosphere. The

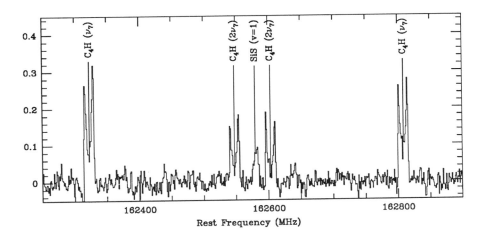

Figure 2: Transitions from the excited bending states of C_4H and the $J = 9 - 8$ transition of SiS in its v=1 vibrational state. C_4H is obviously excited in a hollow extended shell, while $v = 1$ SiS molecules are only present in the innermost regions of the envelope.

detection of AlF is now confirmed by 1.3 mm observations (Figure 1). The source of AlF emission is resolved by the 12" beam (Cernicharo et al. 1988b).

4 Vibrationally Excited Species

Numerous lines from vibrationally excited species have been observed in the spectrum of IRC+10216. These include many lines of the C_4H radical in excited bending states. (Guélin et al. 1987b, Yamamoto et al. 1987). The profile of lines from vibrationally excited species may vary substantially, indicating different radial locations in the envelopes for these species (Figure 2).

Guilloteau et al. (1987) have found strong ($\simeq 75$ Jy) $J = 1 - 0$ maser emission from HCN molecules in the $(0, 2^0, 0)$ vibrational state, in the C-rich envelope CIT 6. This is the first strong maser ever observed in a C-rich circumstellar envelope. A systematic search for HCN masers in C-rich circumstellar envelopes enabled Lucas et al. (1988) to find 6 new masers with similar luminosities and linewidths. Strong maser emission seems to be present in about 20 % of envelopes, with specific infrared colours, and with mass losses in the $10^{-6} - 10^{-5}$ M$\odot$/yr range. The CIT 6 maser was shown to be 20% linearly polarized (Goldsmith et al. 1988).

5 Isotopic Ratios

Studies of isotopic ratios in circumstellar envelopes are an important key to the understanding of the chemical evolution of the Galaxy. Several hundreds of circumstellar envelopes, both O-rich and C-rich, have now been detected in CO. Thus isotope ratios can be studied in a large number of stars. Jura et al. (1988) have found very high $^{13}CO/^{12}CO$ line intensity ratios (> 0.5) in three carbon-rich stars.

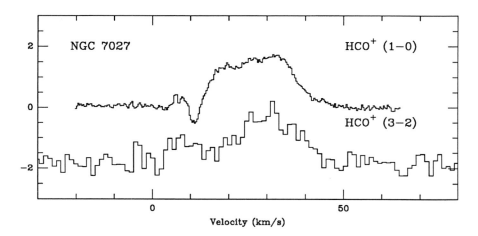

Figure 3: The 30-m spectrum of HCO$^+$ ($J = 1-0$) in NGC 7027. Absorption of the continuum is seen at negative velocities. A $J = 3-2$ spectrum obtained with the 12-m telescope at Kitt Peak is drawn at the same scale.

Reliable values of isotopic ratios in molecular species less abundant than $^{12}C^{16}O$ and $^{13}C^{16}O$ have been derived from 30-m telescope observations (Kahane et al. 1988). Optically thin lines were used, and most ratios were observed in several molecules. The silicon and sulfur isotopic ratios were found found to be very close to their solar system values. The $^{12}C/^{13}C$ ratio is smaller than the terrestrial value by almost exactly a factor of 2. ^{15}N is underabundant relative to ^{14}N by ast least a factor of 15. ^{17}O and ^{18}O isotopes have now been detected in 6 stellar envelopes (Gomez-Gonzalez et al. 1989).

6 Molecular Abundances

Up to recently, the C-rich envelope of IRC+10216 was the only object suitable for observational circumstellar chemistry. The high sensitivity of the 30-m telescope has extended this field to many stars, including several O-rich objects. New molecules such as SO_2 and SO were found (Lucas et al. 1986, Guilloteau et al. 1987).

H_2S has now been detected in 15 oxygen-rich stars in its ortho and para states (Omont et al. 1989a). The line profiles, much narrower than those of SO_2, and the high excitation temperature could indicate H_2S emission comes from the envelope innermost regions.

HCN is now found in 9 oxygen-rich envelopes with high abundances ($10^{-8} - 10^{-7}$ of H_2), which cannot be accounted for by a detailed model of nitrogen chemistry, including photodissociation by the external UV field and subsequent neutral-neutral reactions (Nercessian et al. 1988).

HCO$^+$ was expected to be detectable in circumstellar envelopes according to ion-molecule chemical models (Glassgold et al. 1986, Mamon et al.1987). After several unfruitful searches, it was finally found in several objects, both O-rich and C-rich. It is very abundant in envelopes with a central ionization source such as NGC 7027 (Figure 3), NGC 2346 (Bachiller et al. 1988), and CRL 618 (Morris et al. 1987, Bujarrabal et al. 1988, Guilloteau et al. 1988). HCO$^+$ is marginally detected in IRC+10216 (Figure 4).

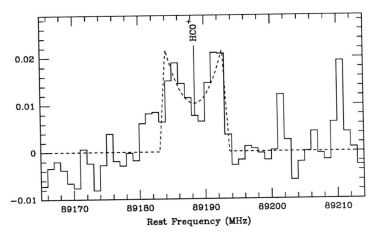

Figure 4: A deep integration spectrum on IRC+10216 at the HCO$^+$ frequency. HCO$^+$ is detected at a 5σ level, with a profile typical of a spatially resolved hollow shell.

7 Particular Objects

We review here several sources where 30-m observations have revealed new and remarkable features:

The Frosty Leo Nebula (IRAS 09371+1212) This peculiar infrared source was detected in CO by Forveille et al. (1987) and identified as a circumstellar envelope. Rouan et al. (1988) have shown that the extremely strong 60μm flux is due to a high concentration of ice on cold dust grains. A direct observation of the $43-65\mu m$ strong bands of cristalline ice was recently made by Omont et al. (1989b).

OH 0.9+1.3 An OH/IR star close to the Galactic center, which, presumably, is in a transition stage to planetary nebula. Its detection in CO (Mauersberger et al. 1988) opens new prospects for the determination of mass loss rates in this important region.

OH 231.8+4.2 This unusual object proved to be the site of an extremely rich chemistry (Morris et al. 1987). Its kinematics were studied in CO (1-0) and showed the presence of a high velocity bipolar flow aligned with the polar axis. This flow is also present in other molecular transitions such as H_2S, SO_2. A blue-shifted SO_2 jet was also found by Jackson and Rieu (1988) using the Hat Creek interferometer.

CRL 618 CO observations of post-AGB stars show high mass losses occur in the form of very high velocity winds. The CO(2-1) line observed by Cernicharo (1988) at Pico Veleta in the young planetary nebula CRL 618 shows conspicuous wings with a total extent of 400 kms^{-1}. High velocity wings are also present in the lines of HCO$^+$, HCN and HC$_3$N. 14 molecules in total have been detected in that source (Bujarrabal et al. 1988).

AKNOWLEDGEMENTS. We thank J. Cernicharo, S. Guilloteau, C. Kahane and A. Omont for the communication of unpublished results.

References

Bachiller, R., Planesas P., Martin-Pintado J., Bujarrabal, V. and Tafalla, M. 1988, preprint

Bujarrabal, V., Gomez-Gonzalez, J., Bachiller, R., and Martin-Pintado, J. 1988, *Astron. Astrophys.*, in press

Cernicharo, J. and Guélin, M. 1987, *Astron. Astrophys.* **183**, L10

Cernicharo, J., Kahane C., Gomez-Gonzalez, J. and Guélin M. 1986a, *Astron. Astrophys.* **164**, L1

Cernicharo, J., Kahane C., Gomez-Gonzalez, J. and Guélin M. 1986b, *Astron. Astrophys.* **167**, L5

Cernicharo. J., Guélin, M., Menten, K. and Walmsley, C.M. 1987a, *Astron. Astrophys.* **181**, L1

Cernicharo. J., Guélin, M., Hein, H., and Kahane, C. 1987b, *Astron. Astrophys.* **181**, L9

Cernicharo. J., 1988, Thèse d'Etat, Université de Paris VII

Cernicharo, J., Kahane, C., and Guélin, M. 1988a, in preparation

Cernicharo, J., Penalver, J., and Guélin, M. 1988b, in preparation

Clegg. R.E.S., and Wooten. H.A. 1980, *Astrophys. J.* **240**, 828

Forveille, T., Morris, M., Omont, A., and Likkel, L. 1987, *Astron. Astrophys.* **176**, L13

Glassgold, A.E., Lucas, R., and Omont, A. 1986 *Astron. Astrophys.* **157**, 35

Goldsmith, P.F., Lis, D.C., Omont, A., Guilloteau, S., and Lucas, R. 1988, *Astrophys. J.* **333**, 873

Guélin, M., Cernicharo, J., Kahane, C., and Gomez-Gonzalez, J. 1986, *Proc. Symposium "Molecules in Physics, Chemistry and Biology"*, Paris, June 1986, ed. J. Maruani, D. Reidel, Dordrecht

Guélin, M., Cernicharo J., Kahane, C. and Gomez-Gonzalez, J. 1987a, *Astron. Astrophys.* **175**, L5

Guélin, M., Cernicharo, J., Navarro, S., Woodward, D.R., Gottlieb, C.A., and Thaddeus, P. 1987b, *Astron. Astrophys.* **182**, L37

Guilloteau, S., Lucas, R., Nguyen-Q-Rieu, and Omont, A. 1986, *Astron. Astrophys.* **165**, L1

Guilloteau, S., Omont, A., and Lucas, R. 1987 *Astron. Astrophys.* **176**, L24

Guilloteau, S., Omont, A., and Lucas, R. 1988, in preparation

Gomez-Gonzales, J., Kahane, C., Cernicharo, J., and Guélin, M. 1989, in preparation

Jackson, J.M., and Nguyen-Quang-Rieu, 1988, preprint

Jura, M., Kahane, C., and Omont, A. 1988, preprint

Kahane, C., Gomez-Gonzalez J., Cernicharo, J. and Guélin M. 1988, *Astron. Astrophys.* **190**, 167

Lucas, R., Guilloteau, S., and Omont, A., 1988 *Astron. Astrophys.* **194**, 230

Lucas, R., Omont, A., Guilloteau, S., and Nguyen-Q-Rieu 1986, *Astron. Astrophys.* **154**, L12

Mamon, G.A., Glassgold, A.E., Omont, A. 1987, *Astrophys. J.* **323**, 306

Mauersberger, R., Henkel, C., Wilson, T.L., and Olano, C.A., *Astron. Astrophys.*, in press

Morris, M., Guilloteau, S., Lucas, R., and Omont, A. 1987, *Astrophys. J.* **321**, 888

Nercessian, E., Guilloteau, S, Omont, A., and Benayoun, J.J. 1988, *Astron. Astrophys.*, in press

Olofsson, H. 1988, *Space Science Reviews*, in press

Omont, A., Morris, M., Lucas, R., and Guilloteau, S. 1989a, in preparation

Omont, A., Moseley, H., Forveille, T., and Harvey, P.M. 1989b, in preparation

Rouan, D., Omont, A., Lacombe, F., and Forveille, T. 1988, *Astron. Astrophys.* **189**, L3

Tsuji, T. 1973, *Astron. Astrophys.* **23**, 411

Yamamoto, S., Saito, S., Guélin, M., Cernicharo, J., Suzuki H. and Ohishi, M. 1987, *Astrophys. J.* **323**, L149

MILLIMETER AND OPTICAL OBSERVATIONS OF TRANSLUCENT MOLECULAR CLOUDS

E.F. van Dishoeck[1], J.H. Black[2], and T.G. Phillips[3]

[1] Center for Cosmochemistry, Caltech 170–25, Pasadena CA 91125
[2] Steward Observatory, Univ. of Arizona, Tucson, AZ 85721
[3] Downs Lab. of Physics, Caltech 320–47, Pasadena CA 91125

Translucent molecular clouds with total visual extinctions $A_V^{tot} \approx 2$–5 mag are of interest because they occupy the middle ground of parameter space between the classical diffuse molecular clouds, such as the ζ Oph cloud ($A_V^{tot} \approx 1$ mag), and the dark clouds, such as TMC–1 ($A_V^{tot} \geq 10$ mag). They have the advantage that they can be studied both by millimeter emission line techniques, as well as by optical absorption line observations, provided that there is a bright star located behind the cloud. The high–latitude molecular clouds fall into the same category.

We have recently published a survey of C_2, CN and CH absorption lines toward highly reddened stars in the Southern sky (van Dishoeck and Black 1989), in which about 10 translucent clouds were identified. An excellent example is provided by the line of sight toward HD 169454 (a B1.5 I star, V=6.6 mag, $A_V^{tot} \approx 3.3$ mag) (see Figure 1). A similar survey is in progress for stars that lie behind Southern high–latitude clouds (de Vries and van Dishoeck 1988; van Dishoeck and de Vries 1988). The absorption line data indicate that the C_2 and CH column densities are strongly correlated with each other, and probably also with H_2. The observed abundances for translucent clouds can be well explained by gas–phase chemistry models which are exposed to the normal interstellar radiation field and in which carbon is slightly more depleted onto grains compared with classical diffuse clouds. In contrast, the CN abundances are found to vary greatly from cloud to cloud, and are poorly understood. The observed rotational excitation of C_2 suggests that the densities in translucent clouds are somewhat higher ($n_H \approx 1000$ cm^{-3}) than those in the classical diffuse clouds ($n_H \approx 300$–500 cm^{-3}). The measured column densities for high–latitude clouds can be reproduced in models in which the radiation field is reduced by a factor of about two compared with the normal value. The CH and CN abundances in these clouds are not abnormal compared with those in translucent clouds. The fact that the observed CH$^+$ column densities are small indicates that shock processes do not play a dominant role in the chemistry, in contrast with suggestions by Magnani et al. (1988).

Millimeter observations of translucent and high–latitude clouds are currently being obtained at several observatories. CO J=1–0 emission has been measured toward most stars, and limited maps have been made (Figure 1a, c). CS J=2–1 emission has been detected toward HD 169454 by Knapp, Drdla and van Dishoeck (1989) using the AT&T Bell Labs Telescope. CO J=3–2 emission (Figure 1d) has been observed at the Caltech Submillimeter Observatory using the Berkeley 345 GHz receiver (Sutton et al., this volume). The ratio of the CO J=1–0/J=3–2 emission is a sensitive function of density in the parameter regime of the translucent clouds, and provides an independent diagnostic of the physical conditions (van Dishoeck et al., in prep.). A preliminary analysis of the strength of the CO J=3–2 emission in translucent clouds compared with high–latitude clouds suggests that the latter have somewhat higher densities than the former. Detailed models along the lines described by van Dishoeck and Black (1988) are used to interpret the observations.

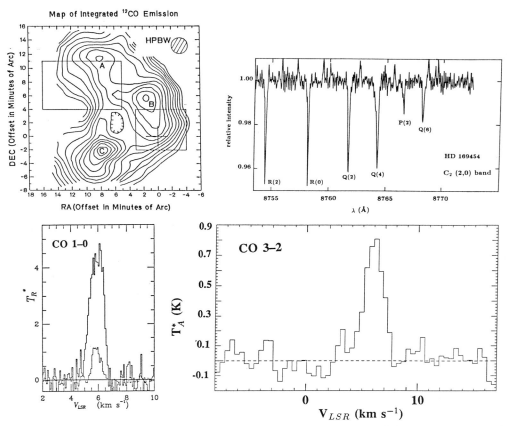

Figure 1. *Millimeter and optical observations toward HD 169454. (a): CO 1–0 map with the position of the star indicated by the cross (Jannuzi et al. 1988); (b): C_2 absorption lines (van Dishoeck and Black 1988); (c) ^{12}CO and ^{13}CO 1–0 emission (Jannuzi et al. 1988); (d) ^{12}CO 3–2 emission (van Dishoeck et al., in prep.); the antenna temperature scale has not yet been corrected for an extended source efficiency of 0.68.*

The CSO is supported by NSF grant AST 83-11849.

REFERENCES

de Vries, C.P. and van Dishoeck, E.F. 1988, *Astr. Ap.*, **203**, L23.
Knapp, G.R., Drdla, K., and van Dishoeck, E.F. 1989, *Ap. J.*, submitted.
Jannuzi, B.T., Black, J.H., Lada, C.J, and van Dishoeck, E.F. 1988, *Ap. J.*, **332**, 995.
Magnani, L., Blitz, L., and Wouterloot, J.G.A. 1988, *Ap. J.*, **326**, 909.
van Dishoeck, E.F. and Black, J.H. 1988, *Ap. J.*, **334**, in press.
van Dishoeck, E.F. and Black, J.H. 1989, *Ap. J.*, in press.
van Dishoeck, E.F and de Vries, C.P. 1988, *The Messenger*, September issue.

A SUBMILLIMETER LINE SURVEY OF SGR B2

E. C. SUTTON, P. A. JAMINET, W. C. DANCHI,
Space Sciences Laboratory, University of California, Berkeley, CA 94720 U.S.A.
C. R. MASSON, and GEOFFREY A. BLAKE
Caltech 320-47, Pasadena, CA 91125 U.S.A.

ABSTRACT. We have surveyed molecular line emission from Sgr B2 between 330 and 355 GHz. By observing two separate positions within the molecular cloud we hoped to be able to study the degree of chemical differentiation present. The two positions have quite different spectra and this seems to reflect real differences in their chemistry. Line profiles reveal additional information about the locations of chemical species within the cloud.

1. Introduction

The giant molecular cloud Sgr B2 is probably the richest and most chemically diverse region known in the Galaxy. Its microwave spectrum has been well studied, particularly in the spectral region around 100 GHz (Cummins *et al.*, 1986). Most molecular forms which have been detected in interstellar space have been seen in Sgr B2 and a number have been seen only in Sgr B2. One of the less well studied aspects of this region is the spatial dependence of the observed chemical forms and the possibility that there may be significant chemical differentiation within this cloud on spatial scales of about 1 pc. This aspect has been highlighted recently by the work of Goldsmith *et al.* (1987) who studied the spatial distribution of emission in a number of molecular transitions. They observed strong differences in the ways in which line intensities varied throughout the cloud and concluded that these variations reflected true chemical differences and not just the effects of excitation. In this work we have combined this emphasis on the study of spatial variations with an emphasis on the importance of broad-band spectral line survey work such as that of Cummins *et al.* (1986)

2. Observations

The observations described here were carried out at the 10.4m Caltech Submillimeter Observatory during the summer of 1988. The receiver, a double sideband 345 GHz SIS waveguide mixer built at Berkeley, fed a 500 MHz bandwidth AOS backend. The observations spanned the spectral region 330 - 355 GHz without gaps. The local oscillator frequency was stepped at intervals of 250 MHz, providing a highly redundant, overlapping set of observations. Most frequencies were observed four times, twice in the lower sideband and twice in the upper sideband. This provided sufficient information to identify the proper sideband for each feature which was observed.

Observations were made of two positions in the source, corresponding to the two prominent peaks in the maps of Goldsmith et al.(1987). The northernmost of the pair, designated Sgr B2(N), was seen by those authors as a peak in HNCO, OCS, and vibrationally excited HC_3N emission. It also corresponded to the northern H_2O maser source. The other position observed was the position Sgr B2(M), also an H_2O maser source and the main peak of SO emission in the map of Goldsmith et al. At our resolution of 20" these two positions (separated by about 45") are well resolved. In order to maintain the greatest consistency in our comparison of the two positions, they were always observed as a pair in rapid sequence. Typically the middle (M) source was observed first, for an integration time of about 15 minutes, followed by a similar integration on the northern (N) source.

The resulting data set consisted of 82 pairs of spectra. Between 6 and 12 such pairs could be obtained on an individual night. The entire period of observations spanned a total of about 3-4 weeks, during which not every night was devoted to this project.

3. Discussion

The data contain approximately 100 easily identifiable, resolved spectral lines. A great number of the lines are from the molecules SO_2 and CH_3OH, abundant asymmetric tops with rich spectra. Many of the species observed at lower frequencies are not prominent in these observations, largely due to the combination of the relatively low excitation present in Sgr B2 and due to the high frequencies at which this survey was conducted.

There are considerable differences between the spectrum of Sgr B2(M) and Sgr B2(N). In almost all transitions observed, the middle (M) source has the greater intensity. Many lines which appear strong in this source are very much weaker, or not even detectable, in the northern source. The differences seem to be due to real chemical differences between these two regions and not just excitation effects. Some of the molecular distributions are quite surprising. For example, Goldsmith et al. report that they saw CH_3OH only in the northern source, whereas we see it quite strongly in the middle source. Further analysis of this discrepancy must await analysis of the entire set of methanol observations, in order to separate excitation effects from those due to abundance.

There is a great variation in line profiles, both between chemical species and within transitions of a single chemical species. Some lines are seen in almost pure absorption. Others are seen partly in emission with greater or lesser amounts of superimposed absorption (self-reversal). Others are seen as virtually pure single emission peaks. Within a given species this line shape variation seems to be a rapid function of excitation energy. Otherwise between species the variation seems to be related to where in the cloud the species is located. For example, CN is seen as an absorption feature, and CN is likely to be located in regions on the cloud perimeter where it can absorb out continuum radiation from the denser cloud core.

This work has been supported by the National Science Foundation under grant AST-8715295.

4. References

Cummins, Sally E., Linke, R.A., and Thaddeus, P. 1986, *Ap. J. Suppl.*, **60**, 819.
Goldsmith, Paul F., Snell, Ronald L., Hasegawa, T., and Ukita, N. 1987, *Ap. J.*, **314**, 525.

H_3O^+ REVISITED

A. WOOTTEN[1], F. BOULANGER[2,3], S. ZHOU[4], F. COMBES[3], P. ENCRENAZ[5], M. GERIN[5], AND M. BOGEY[6]

[1] National Radio Astronomy Observatory, Charlottesville, Virginia.
[2] IPAC, Caltech, Pasadena, California.
[3] Observatoire de Meudon, F-92190 Meudon, France.
[4] Electrical Engineering Research Laboratory and Department of Astronomy, University of Texas, Austin, Texas.
[5] Ecole Normale Superieure, F-75005 Paris, France
[6] Laboratoire de Spectroscopie Hertzienne de l'Universite de Lille, F-59655 Villeneuve d'Ascq Cedex, France.

A line at 307.19241 GHz was tentatively identified with the $(J,K)= (1,1)^- \to (2,1)^+$ line (hereafter called the P(2,1) line) of para-H_3O^+ (Wootten et al. 1986, Hollis et al. 1986). New observations of this line in OMC1 with the NRAO 12 m telescope show the emission source to be unresolved by the 30" beam. Upper limits to the strength of the $(3,2)^+ \to (2,2)^-$ transition of para-H_3O^+ at 364.797427 GHz have been obtained with the MWO 4.9m telescope. Study of the excitation of this symmetric top molecule indicates that the excitation temperature of the P(2,1) line normally lies well below the rotational temperature of the molecule, rendering the line abnormally weak. Previous abundance limits for H_3O^+, based upon LTE analyses, should probably be revised upward.

The average OMC1 spectrum at points offset 30" in the cardinal directions was compared with the spectrum at the central position. Line intensities for the $4_0 \to 4_1$ A CH_3OH line in the offset spectrum have typically declined by a factor of five. There does not appear to be emission at the P(2,1) frequency in the offset spectrum at all. The decline in integrated intensity of the possible H_3O^+ line is a factor of twenty between the central position and the offset positions. Given the uncertainty in the data, this steeper decline is significant at only the 3 σ level. We conclude that the source of the 307 GHz line is more highly confined to the central position than is methanol emission, for example.

The J=30-29 line of OCS in the 364 GHz spectrum from the MWO clearly verifies the frequency tuning of the receiver. No line is present at the frequency where H_3O^+ emission is expected, within the single channel uncertainty of $T_R^*=0.5$K in the spectrum.

Limits to the abundance of H_3O^+ have assumed LTE excitation. However, the critical density for the P(2,1) line lies above 10^6 cm^{-3}, and the line may therefore be subthermally excited. Furthermore, the spectrum of H_3O^+ is complex owing to the large inversion splitting, and energy levels overlap in the K=0 and K=1 rotational ladders. Hence, the lower energy level for the 307 GHz line occurs in the J=2 state rather than the J=1 state. Radiative transitions to lower energies from the lower (2,1) level of the P(2,1) line are forbidden by dipole selection rules. Transitions into this state, then, may overpopulate it relative to expected LTE populations. This could result in an unusually low excitation temperature for the P(2,1) line and consequent underestimate of the H_3O^+ abundance. To analyze the importance of this effect, we have modelled the excitation of H_3O^+, using a large velocity gradient approximation to solve the transfer of radiation. Radiative rates were determined using molecular constants from sources described in Wootten et al. 1986.

Collisional cross-sections for H_3O^+ are not available. Like ammonia, H_3O^+ is a symmetric top, and like ammonia, the molecule is not rigid, possessing inversion transitions. For ammonia, cross sections have been calculated assuming rigidity, as the inversion timescale is much longer than the collisional duration. However, for H_3O^+, the inversion frequency, ν_{inv}=1.650 THz, is much higher and the two timescales are similar, complicating evaluation of the collisional rates. A simple hard-collision model has been used for the downward rates, with upward rates calculated according to the principle of detailed balance. Cross sections for molecular ions are typically 4 to 5 times those for neutral species (Green 1975), so we have chosen a constant upward cross section of 10^{-10} cm^3 s^{-1} at 100K.

These detailed calculations support our suspicion that excitation of the P(2,1) line can be subthermal over a large range of temperature, density and abundance conditions. Ignoring for the moment the effects of dust continuum excitation, we find that for a volume density of 1x10^6 cm^{-3} and column density of para-H_3O^+ of 1.5x10^{15} cm^{-2} at a kinetic temperature of 70K, which reproduces the observed line intensity, the excitation temperature of the P(2,1) line has not reached 10K while that characterizing most other transitions lies near 30K. As the density increases above 10^7 cm^{-3}, the transition rapidly becomes thermalized. Consequently, the line intensity increases unusually fast with increasing excitation, as observed in the 307 GHz line. Because of the subthermal excitation of the line under frequently encountered conditions, the limit placed on the column density by the observations is considerably higher than that estimated using LTE calculations. Using the above column density, and a total H_2 column density of $N(H_2) \sim 2 \times 10^{23}$, an upper limit for the H_3O^+ abundance becomes $X(H_3O^+) \lesssim 1 \times 10^{-8}$ for a ortho/para ratio of 3.

Radiative excitation by emission from dust has also been considered. Strong radiative excitation at the frequency of the inversion transitions could influence the excitation of the molecule, since $h\nu_{inv}/k \sim 79K$, and the dust continuum in OMC1 can be characterized by an 85 K temperature. However, the abundance of H_3O^+ is never high enough for this effect to be important.

Using the conservative upper limit to the H_3O^+ abundance of $X(H_3O^+) \lesssim 1 \times 10^{-8}$, the non-radiative excitation model, and correcting for beam dilution by the 60" MWO beam, we find the expected intensity of the 365 GHz line to be 0.4 K, just lower than the limit set by the spectrum obtained. Under LTE conditions, this line will be an order of magnitude weaker. To confirm the detection of H_3O^+, more sensitive spectra must be obtained of the 364, 385 and 396 GHz lines. To model its excitation and accurately measure its abundance, collisional cross-sections are needed.

REFERENCES

Green, S. 1975 *Ap. J.*, **201**, 366.
Hollis, J. M., Churchwell, E. B., Herbst, E. and DeLucia, F. C. 1986 *Nature*, **322**, 524.
Wootten, A., Boulanger, F., Bogey, M., Combes, F., Encrenaz, P. J., Gerin, M. and Ziurys, L. 1986 *Astr. Ap. (Letters)*, **166**, L15.

MILLIMETER AND SUBMILLIMETER STUDIES OF INTERSTELLAR HIGH TEMPERATURE CHEMISTRY

L.M. ZIURYS
Five College Radio Astronomy Observatory
University of Massachusetts, Amherst

ABSTRACT. Observational studies of interstellar high temperature chemistry have been carried out at mm/sub-mm wavelengths. A survey of the J=2-1 transition of SiO in warm and cold clouds has been conducted at the FCRAO 14 m telescope, with follow-up observations at the IRAM 30 m. SiO was not detected in any cold clouds, but was observed in all regions with $T_K > 30$ K, including a new outflow source 90" south of Orion-KL. Such measurements suggest that the SiO abundance correlates directly with kinetic temperature. Searches were also performed for two possible new "high temperature" molecules, MgH and SH^+.

1. Introduction

At present, it is thought that molecule formation in interstellar clouds occurs primarily via gas-phase ion-molecule reactions. Because a large fraction of these reactions are exothermic and contain negligible activation energy barriers, they can rapidly occur in the low temperature ($T_K \sim 10$ K) environment of interstellar clouds. Many clouds, however, contain energetic regions with temperatures $T_K > 100\text{-}1000$ K. In such sources, activation energy barriers in chemical reactions can readily be overcome, and endothermic reactions can take place, i.e. high temperature chemistry can occur, and ion-molecule reactions are no longer necessary. Also, high temperatures may be accompanied by the destruction of dust grains, releasing refractory elements into the gas phase. High temperature chemistry can therefore in principle alter ion-molecule abundances, but it is not clear which species have abundances influenced by such chemisry.

In order to examine this question, several observational studies have been carried out. They include a survey of SiO in warm and cold sources (Ziurys, Friberg, and Irvine 1988), with additional high resolution mapping performed at IRAM (Ziurys, Wilson, and Mauersberger 1988), as well as searches for two possible new "high temperature" species, MgH and SH^+ (Ziurys et al. 1988; Ziurys, Lis, and Hovde 1988).

2. Observations

The SiO J=2-1 measurements at 86 GHz were carried out in Jan.-June 1985, using the FCRAO 14 m telescope. Additional measurements were done using the IRAM 30 m antenna during May 1988. The search for MgH was conducted at the CSO in Nov. 1987, observing several hyperfine components of the species' N=1-0 transition at 344 GHz. The SH^+ measurements were performed March 1988, at the JCMT and NRAO 12 m telescopes, searching for the J=0-1 spin components of the N=1-0 transition at 345 GHz.

3. Results and Discussion

Previously, SiO had been observed towards various outflow sources where elevated temperatures are present (e.g. Downes et al. 1982). However, there were no reported searches for this species in cold, dark clouds. The SiO observations thus initially concentrated on such sources, including TMC-1, L134N, B335, and L1551. As shown in Table 1, SiO was not detected in any of these clouds, down to column density upper limits of 2-8 x 10^{10} cm^{-2}. SiO was observed in several new sources, including the shocked B and G clouds of SNR IC443 (Ziurys, Snell, and Dickman 1988), and at several new positions in OMC-1. Towards one position, ~ 90" south of Orion-KL, the SiO J=2-1 spectra showed evidence of line wings, suggesting the presence of a previously undetected outflow. Maps of the SiO J=2-1 line made at IRAM with 30" resolution show that this outflow is clearly distinct from that of KL, and mapping of the SiO J=5-4 transition shows that it is bipolar, and may be as small as 20" in extent (Ziurys, Wilson, and Mauersberger 1988). As Table 1 also shows, the sources with the largest SiO abundances are the ones with the highest gas kinetic temperatures; as the temperature becomes smaller, so does the SiO abundance. In fact, an almost linear relationship exists between the ln[SiO/HCN] and T_K, and suggests that SiO is formed by some process that contains an activation energy of ~ 90 K.

SiO therefore appears to have an abundance enhanced by high temperatures. This might also be the case for other molecules containing refractory elements. Since magnesium is cosmically abundant as silicon, Mg-containing molecules might be detectable in interstellar clouds, although searches for MgO in the past have been unsuccessful. Like MgO, however, MgH was not detected towards Orion-KL, down to column density upper limits of ~ 10^{14} cm^{-2}, assuming a 20" source, and T_{rot}=50 K. This corresponds to a fractional abundance of < 10^{-10}. Lack of magnesium-containing compounds in the ISM is curious. If SiO comes from silicate grains, one might expect to find molecules with Mg as well, since typical silicate formulas contain both silicon and magnesium.

SH$^+$ is an open-shell molecular ion, and highly reactive. It might therefore be present in hot, shocked gas. Unfortunately, the hyperfine components searched for lie in the line wings of other species, including CO, making a detection difficult. There is some evidence for the possible lines arising from SH$^+$ toward Orion-KL and Sgr B2, but further observations are necessary to confirm its existence.

Table 1. SiO Abundance vs. Kinetic Temperature[a]

Source	N_{tot}(SiO) cm^{-2}	SiO/HCN	T_K(K)
TMC-1	<2.4 x 10^{10}	<0.00018	10
L134N	<3.6 x 10^{10}	<0.0015	10
L1551	<8.4 x 10^{10}	<0.0070	9-12
B335	<3.6 x 10^{10}	<0.0039	10-12
Orion(3N,1E)	5.6 x 10^{11}	0.0022	14
IC443B	6.5 x 10^{11}	0.038	70
IC443G	1.68 x 10^{12}	0.063	33
Orion 1´.5 S	1.01 x 10^{14}	0.165	75-100
NGC7538	5.6 x 10^{12}	0.34	~220
Orion-KL	2.35 x 10^{17}	1.2	230

a) Taken from Ziurys, Friberg, and Irvine 1988.

References

Downes, D., Genzel, R., Hjalmarson, A., Nyman, L.A., and Ronnang, B. 1982, *Ap.J. (Lett.)*, **252**, L29.
Ziurys, L.M., Snell, R.L., and Dickman, R.L. 1988, *Ap.J.*, in press.
Ziurys, L.M., Friberg, P., and Irvine, W.M. 1988, *Ap.J.*, submitted.
Ziurys, L.M., Wilson, T.L., and Mauersberger, R. 1988, in preparation.
Ziurys, L.M., Blake, G.A., Buettgenbach, T., and Phillips, T.G. 1988, in preparation.
Ziurys, L.M., Lis, D., and Hovde, D.C., in preparation.

A SURVEY OF ORION A EMISSION LINES FROM 330 - 360 GHz

P. R. JEWELL
National Radio Astronomy Observatory, Campus Bldg. 65, 949 N. Cherry Avenue, Tucson, Arizona 85721 (USA)
J. M. HOLLIS
NASA - Goddard Space Flight Center, Space Data and Computing Division, Code 630, Greenbelt, Maryland 20771 (USA)
F. J. LOVAS
National Bureau of Standards, Molecular Spectroscopy Division, Bldg. 221, Rm. B265, Gaithersburg, Maryland 20899 (USA)
L. E. SNYDER
University of Illinois, Astronomy Department, 1011 W. Springfield Ave., Urbana, Illinois 61801 (USA)

ABSTRACT. A spectral survey of Orion A in the range 330 - 360 GHz has been performed with the NRAO 12 m telescope. About 180 spectral lines were detected. Most of these lines can be associated with one of 19 distinct chemical species, although a number of lines are yet unidentified. The spectrum is dominated by light symmetric or near-symmetric rotors such as CH_3OH and the asymmetric rotor SO_2. Heavy asymmetric rotors such as $HCOOCH_3$ that are prevalent at lower frequencies are diminished in intensity in this band.

The atmospheric window centered at 870 μm extends from frequencies as low as 330 GHz to as high as 365 GHz, under conditions of excellent transparency. This window contains a number of well-known transitions such as J=3→2 CO and J=4→3 HCN. To date, no thorough spectral surveys of this band have been published. Considering the emphasis that this band is certain to receive from the new generation of millimeter- and submillimeter-wave telescopes, we felt that a careful study of Orion A in this band would be valuable.

The motivation for this survey was (1) to chart the occurrence and strength of spectral lines for use in future investigations, (2) to provide a homogeneous, multi-transition database for the analysis of abundances and excitation of specific species; and (3) to catalog unidentified spectral lines for use in line identification in conjunction with lower-frequency surveys. Orion A was chosen as the source to survey because it has the richest line spectrum at these frequencies. Spectral surveys such as this have been discussed in these proceedings by E.C. Sutton and by G. A. Blake, who have previously conducted surveys of Orion A from 215 - 263 GHz (Sutton *et al.* 1985; Blake *et al.* 1986) and in Sgr B2 from 330 - 355 GHz (these proceedings).

This survey was conducted in 1988 February using the NRAO 12 m telescope at Kitt Peak, Arizona. The receiver used had dual polarization Schottky mixers, each with a receiver noise temperature of about 1800 K (SSB). The local oscillator signal was produced by a Gunn oscillator and a quadrupler and the system was readily tunable to any frequency in the band. The spectrometer consisted of two, 2 MHz x 256 channel filter banks. The primary reflector of the 12 m telescope has RMS surface deviations of 75 μm. During the portion of the survey covering the 330 - 345 GHz band, a standard subreflector was used. For the 345 - 360 GHz portion, an error-correcting subreflector, manufactured by J. Davis and C. Mayer (discussed in these proceedings), was used that lowered the effective RMS surface deviations to 55 μm, resulting in considerably improved beam efficiencies.

The frequency range surveyed was 330.5 to 360.1 GHz. A typical RMS noise on the T_R^* scale was 0.2 K, which is comparable to that achieved in previous high frequency surveys.

The observations were made in a double sideband mode, and the sideband ambiguity was resolved by observing at two local oscillator settings, separated by typically 10 MHz.

About 180 spectral lines (160 distinct features excluding unresolved blends), were found in the survey. Nineteen (19) chemically distinct species, excluding isotopic variants, were observed. At least 22 of these lines cannot be readily identified with any known transitions at this time. Compared to lower frequency surveys, the 330 - 360 GHz band is dominated by lines of CH_3OH, CH_3CN, and SO_2. An example of the high density of CH_3OH lines seen in some portions of the band is shown in Figure 1. Although undoubtedly a function of sensitivity, heavy asymmetric rotors such as $HCOOCH_3$ that so dominate the 1.2 mm band appear to have diminished intensity in the 870 μm band.

The data from this survey will be made available in ASCII FITS format upon request to the Astronomical Data Center, Code 633, NASA - Goddard Space Flight Center, Greenbelt, Maryland 20771 (USA).

REFERENCES

Blake, G. A., Sutton, E. C., Masson, C. R., and Phillips, T. G. (1986), *Ap. J. Suppl.*, **60**, 357.

Sutton, E. C. Blake, G. A., Masson, C. R., and Phillips, T. G. (1985), *Ap. J. Suppl.*, **58**, 341.

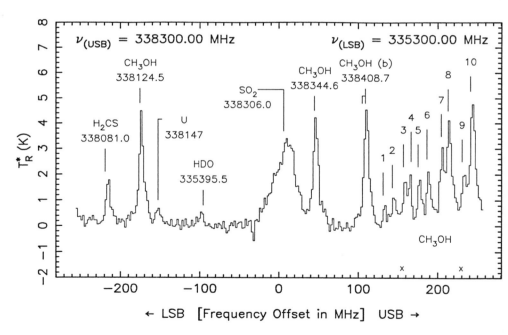

FIGURE 1. A double sideband spectrum of Orion A centered at 338.3 GHz in the upper sideband and 335.3 GHz in the lower sideband. The group of lines on the right is a manifold of $J_k = 7_k \rightarrow 6_k$ lines of CH_3OH (methanol).

Millimeter-Wave Spectral Line Survey at NRO

Masatoshi OHISHI
Department of Physics
Toyama University
3190 Gofuku
Toyama   930
Japan

Introduction.
  Molecular Spectral Line Surveys have been made in several molecular clouds : Orion KL (Johansson et al. 1984; Sutton et al. 1985; Blake et al. 1986), Sgr B2 (Cummins et al. 1986) and IRC+10216. These surveys have provided important informations on the abundance of molecules and on the excitation of molecules which are necessary to understand interstellar chemistry by comparing with the prediction from the ion-molecule reaction theory.
  The molecular spectral line survey at NRO has begun in 1982 using the 45-m radiotelescope, wide-band receivers and the acousto-optical radiospectrometers (AOS) which can observe a frequency range of 2 GHz wide simultaneously. Several new interstellar molecules have been detected by our survey : C_3O (Brown et al. 1985), C_6H (Suzuki et al. 1986), CCS (Saito et al. 1987), C_3S (Yamamoto et al. 1987),and cyclic C_3H (Yamamoto et al. 1987).
  In this paper, I summarize results of observations in 1987 winter through 1988 spring and briefly report the detection of two new interstellar molecules CH_2CN (cyanomethyl radical) and HC_2CHO (propynal) in TMC1. Details on these molecules will be found in the subsequent paper in this issue by Dr. W.M. Irvine.

Observations from 1987 winter to 1988 spring.
  In June, 1988, we have introduced an SIS receiver which covers 35-50 GHz and has a typical system temperature of $\sim$250 K. The SIS receiver enabled us to observe weaker spectral lines. We have also installed a digital Fourier transformation-type spectrometer, FX, whose frequency resolutions are variable and in a range of 10 kHz - 0.6 kHz.
  Our main target for the spectral line survey is a famous carbon-rich cold and dark cloud TMC1 whose molecular composition seems to be simple and important for the understanding of the carbon-rich interstellar chemistry. Other objects for our survey are Ori KL, Sgr B2, IRC+10216, etc..
  We have tried to detect several new interstellar molecules as follows: C_7H, C_4S, CP, HC_2CHNH, HC_3NH^+, HC_2CHO, CCO, CH_2CH, NaO and KO. We could definetly detect two new molecules (CH_2CN and HC_2CHO) and marginally detect the CCO radical. Other species were searched for without success

with a rms noise of 5-10 mK.

Figure 1 shows a spectrum of CH_2CN radical at 40.24 GHz ($N(K_a,K_c)$= 2(0,2)-1(0,1)) observed at the cyanopolyyne peak of TMC1. This radical was also observed in Sgr B2. Because the spectrum of the CH_2CN radical shows verycomplex hfs structure due to the nuclear spins of N and H, the identification was not so straightforward. Saito et al. (1988) made a laboratory experiment for the CH_2CN radical and succeeded to obtain precise molecular constants. This enabled us to identify the spectrum to the CH_2CN radical. Intensity ratios among many hfs components shown in figure 1 are very close to the theoretical values which suggests that all the spectra are optically thin. Assuming T_{ex}=5 K, we calculated the column density of the CH_2CN radical in TMC1 to be $(2-10) \times 10^{13}$ cm^{-2} which is compared to that of CH_3CN in TMC1.

HC_2CHO, which is a related molecule to C_3O, was also detected in TMC 1. The $2_{02}-1_{01}$ transition was observed at Green Bank and the $4_{04}-3_{03}$ transition was observed at Nobeyama. A column density of HC_2CHO is estimated to be $(1.5\pm0.4) \times 10^{12}$ cm^{-2}. See a subsequent paper by Dr. Irvine for further detail.

Acknowledgement.

This work was made through a collaboration among laboratory spectroscopists (S.Saito, S.Yamamoto, K.Kawaguchi, C.Yamada, H.Takeo), quantum chemists (T.Hirano, N.Nakagawa, A.Murakami) and radio astronomers (N.Kaifu, H.Suzuki, S.Ishikawa, T.Miyaji, and M.Ohishi).

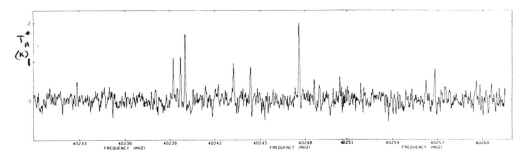

Figure 1. A spectrum of the CH_2CN radical ($2_{02}-1_{01}$) observed in TMC1.

References.
Blake, G.A. et al., 1986, Ap.J.Suppl., **60**, 357.
Brown, R.D. et al., 1985, Ap.J., **297**, 302.
Cummins, S.E. et al., 1986, Ap.J.Suppl., **60**, 819.
Johansson, L.E.B. et al., 1984, A.&Ap., **130**, 227.
Saito,S. et al., 1987, Ap.J.(Letters), **317**, L111.
Saito,S. et al., 1988, submitted to the Ap.J..
Sutton,E.C. et al., 1985, Ap.J.Suppl., **58**, 341.
Suzuki,H. et al., 1986, P.A.S.J., **38**, 911.
Yamamoto,S. et al., 1987a, Ap.J.(Letters), **317**, L119.
Yamamoto.S. et al., 1987b, Ap.J.(Letters), **322**, L55.

OBSERVATIONS OF THE CH_2CN $1_{01}-0_{00}$ AND $4_{04}-3_{03}$ TRANSITIONS

WILLIAM M. IRVINE, S.C. MADDEN, and L.M. ZIURYS
Five College Radio Astronomy Observatory
University of Massachusetts, Amherst, Massachusetts 01003, USA

P. FRIBERG and Å. HJALMARSON
Onsala Space Observatory, S-43900 Onsala, Sweden

H.E. MATTHEWS
Herzberg Institute of Astrophysics
National Research Council of Canada, Ottawa, Canada K1A 0R6

B.E. TURNER
National Radio Astronomy Observatory
Charlottesville, VA 22903, USA

The cyanomethyl radical (CH_2CN) has recently been identified for the first time astronomically, and the first laboratory measurements of its rotational spectrum have been carried out (Irvine et al. and Saito et al., both Ap. J. (Lett.), in press, 1988). This species is the heaviest non-linear molecular radical identified to date in interstellar clouds. Its abundance appears to be equal to or greater than that of the more familiar species CH_3CN (methyl cyanide), both in the cold dark cloud TMC-1 and in the giant molecular cloud Sgr B2. These initial published reports included primarily tabular material rather than figures. For this reason we include here some of the spectra.

The observations at 20 GHz were carried out with the NRAO 43m antenna during a K-band spectral survey towards the "cyanopolyyne peak" of the archetypical cold cloud TMC-1. No strong unidentified features were discovered over the entire 2 GHz survey range until the vicinity of 20.12 GHz was reached. There some 10 CH_2CN line components appear within an interval of about 25 MHz, a portion of which is shown in Figure 1a (spectral resolution 10 kHz). At high resolution (2.5 kHz) each component shows a narrow peak and a broad lower frequency shoulder, similar to that observed for C_4H (Fig. 1b; note frequency-switching artifacts and weaker satellite at +130 kHz). Corresponding spectra of Sgr B2(OH) exhibit a complicated mixture of absorption and emission, presumably due to the presence of regions of differing temperature and density along the line of sight to the Galactic Center (Fig. 1c; frequency scale centered at 20120 MHz, assuming VLSR=62 km s^{-1}, and spectral resolution 160 kHz). The highest frequency transition definitively detected in TMC-1, $4_{04}-3_{03}$ at 80480.3 MHz, is shown in Figure 1d as observed with the FCRAO 14m antenna (includes some frequency-switched data, producing the negative artifact near 80479.7 MHz).

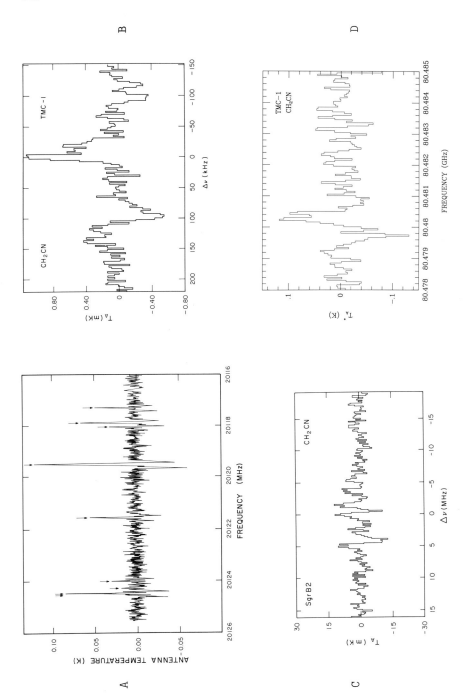

Figure 1. (a) Components of the $1_{01}-0_{00}$ transition of CH_2CN observed in TMC-1; (b) high resolution spectrum of the component at 20118.014 MHz; (c) the $1_{01}-0_{00}$ transition observed toward Sgr B2; (d) the $4_{04}-3_{03}$ transition observed in TMC-1 with 50 kHz resolution.

HETERODYNE SPECTROSCOPY OF C II IN MOLECULAR CLOUDS

A.L. BETZ, R.T. BOREIKO, and J. ZMUIDZINAS[‡]
Space Sciences Laboratory, University of California
Berkeley, CA 94720

ABSTRACT. High-resolution spectra of the 158 μm fine-structure line of ionized carbon have been obtained with a heterodyne spectrometer toward a number of galactic molecular clouds. Strongly reversed C II line profiles are seen in sources noted for similar behavior in low-J ^{12}CO. The C$^+$ producing the optically thick ($\tau \geq 1$) absorption component is cool or subthermally excited, with excitation temperatures typically less than 25-50 K. Column densities for the absorption component are $\geq 10^{18}$ cm^{-2}, similar to values predicted by UV-photodissociation models for C II emitting gas. Thus the total C$^+$ column density in the vicinity of these molecular clouds is significantly greater than previously estimated from integrated intensity measurements.

1. Introduction

Line emission from the 158 μm fine-structure line of ionized carbon has long been recognized as a primary cooling mechanism for the low density gas at the boundaries of molecular clouds (Dalgarno and McCray 1972). Russell et al. (1980) first succeeded in detecting the C II line in Orion in observations with a cooled grating spectrometer at modest spectral resolution ($\delta\lambda/\lambda \sim 0.007$). This result confirmed the theoretical prediction that the line would be strong and therefore important for cooling diffuse molecular gas. Subsequently these researchers established the existence of large C II haloes surrounding galactic H II regions (Russell et al. 1981). The first observations of C II at a spectral resolution sufficient to measure the Doppler velocity (to a few km s^{-1} accuracy), and to resolve the line in a number of extragalactic sources, were done by Crawford et al. (1985) and Lugten (1987), who used a cooled Fabry-Perot spectrometer with a velocity resolution of ~ 30 km s^{-1} ($\delta\lambda/\lambda \sim 10^{-4}$). This resolution in itself, however, is not adequate to resolve the C II line in most galactic sources.

Since these pioneering efforts with incoherent spectrometers, the technology of heterodyne spectroscopy has advanced to the point that coherent receivers are now available at frequencies as high as 2000 GHz. Such instruments offer to the far-infrared the high spectral resolution and velocity-scale accuracy long enjoyed at radio wavelengths. We are now able to compare precisely the emission observed from

[‡] Now at Astronomy Department, University of Illinois, Urbana, IL 61801

neutral (CO) and partially ionized (C$^+$) gas in UV-heated regions.

The first heterodyne observations of the C II line were reported by Boreiko et al. (1988). The spectra taken in Orion-Trapezium region at 0.8 km s^{-1} (5 MHz) resolution revealed that the C II linewidth is 5 km s^{-1} (FWHM), similar to that observed for millimeter wave CO lines. However, there were also significant variations in linewidth and velocity, and even evidence for multiple velocity components, in spectra observed from various nearby locations in Orion A.

2. C II Optical Depths

Current theoretical models of photodissociation regions generally predict low optical depths ($\tau < 1$) for C II line emission (e.g., Tielens and Hollenbach 1985). This conclusion proceeds from the argument that UV radiation will penetrate into a cloud to $\sim A_v = 4$, that this depth corresponds to a hydrogen column density of 10^{21} cm^{-2}, and therefore that the column density of ionized carbon will be $\sim 3 \times 10^{18}$ cm^{-2}. This column density will yield an optically thin line ($\tau < 1$) whenever the linewidth is > 6.9 km s^{-1} for an excitation temperature $T_{ex} = 100$ K, or > 3.4 km s^{-1} for $T_{ex} = 200$ K.

The question of the ^{12}C II optical depth is important for evaluating the effectiveness of the line in cooling diffuse gas and also because many authors have suggested that the 158 μm C II line should be good for measuring the ^{13}C/^{12}C isotopic abundance ratio without introducing the type of systematic errors implicit in multi-isotope CO comparisons. Given that we don't know the optical depths of C II lines *a priori*, the derivation of the ^{12}C/^{13}C isotopic ratio would require that all the C II lines are optically thin.

The J = 3/2 upper level of the ^{12}C II fine structure line is the lowest excited state in ionized carbon, and lies only 63 cm^{-1} (hν/k = 91 K) above the J = 1/2 ground state. The next highest state requires UV-excitation to a level 61,000 K above ground. Consequently, in the relatively cool molecular regions of interest here, ionized carbon is for all practical considerations just a two-level system. This makes calculations easy but interpretation hard. The reason is that with a two level system there are not enough observable quantities to separate questions of excitation from those of abundance in explaining a single observed emission line. Even if we assume LTE conditions, we still have two unknowns but only one observable. We need additional independent information on either the excitation or the abundance of the species. For example, observations of the isotopic ^{13}C II lines can in principle aid us in interpreting the ^{12}C II optical depth, *provided we already know the isotopic ratio accurately*. In the C II spectra taken by Boreiko et al. (1988), a weak feature was detected at the correct frequency for the strongest component of the ^{13}C II triplet. If this feature is indeed ^{13}C II, then from the apparent ^{12}C II/^{13}C II intensity ratio, the ^{12}C II emission must be optically thick ($\tau > 1$) if the ^{12}C/^{13}C isotopic ratio is >12. For an *adopted* ratio of 60, the authors concluded that the ^{12}C II line has an optical depth of $\tau \sim 5$ for two observed positions in Orion. It remains to be seen, however, if the weaker ^{13}C II subcomponents can be detected for verification, and also what isotopic ratio is correct. For example, the lower ratio of 40 suggested by Blake et al. (1987) would correspondingly reduce the estimated ^{12}C II optical depth to $\tau = 3 \pm 1$.

Generally speaking, if a spectral line can be observed in absorption, a lower limit to the optical depth can be determined without concern for the excitation. This gives us a tractable problem: one equation in one unknown. Of course, some knowledge of the excitation is needed before any estimates of abundance can be made. With these points in mind we observed a number of molecular clouds with known strong reversals in the emission spectra of millimeter wave CO lines. The object was to look for "self"-absorption in C II, to compare the observed profiles with those of CO, and thereby to measure the column density of ionized carbon in such regions more accurately.

3. Receiver System

Because a detailed description of the spectrometer has been presented elsewhere (Betz and Zmuidzinas 1984), only a few highlights are repeated here. The receiver uses a cooled (77 K) Schottky diode (Univ. VA type 1I7) in a corner-reflector mixer mount (Zmuidzinas et al. 1988) and an optically pumped far-infrared (FIR) laser as the local oscillator. The laser oscillates on the 1891.2743 GHz line of CH_2F_2 (Petersen et al. 1980), which lies 9.3 GHz below the C II rest frequency of 1900.5369 GHz (Cooksy et al. 1986). The IF amplifier is centered at this offset frequency. The system noise temperature of the receiver mounted on the telescope is 26,000 K (SSB). Measured in units of $h\nu/k$, this system noise level is equivalent to that of a 115 GHz CO receiver with $T_{sys} = 1600$ K. Given that the frequency width of a 0.8 km s^{-1} channel corresponds to 5 MHz, one can integrate down to a few Kelvin post-detection noise level in a matter of minutes. Because the earth's atmosphere is completely opaque at the C II line frequency even at high mountain sites, observations are conducted from the Kuiper Airborne Observatory operated by NASA. The 91 cm telescope aboard the aircraft produces a 43 arcsec beam at $\lambda = 158$ μm. Calibrations of line intensities are relative to the brightness temperature of the Moon, which is reasonably well known. Most observations are done in sky-chopping mode with beam separations between 8 and 10 arcmin.

4. Observations and Analysis

Observations of 5 sources known for strong CO line reversals were done at various times in 1987 and 1988 as part of a larger survey of C II in galactic clouds. Figures 1 and 2 show observed spectra for particular positions in two of these sources: NGC 2024 and W51. For W51 we have also plotted the profile of the 230 GHz J=2-1 ^{12}CO line taken by Stacey (1987) on the 12-m Kitt Peak telescope. The benefit of the higher resolution possible with heterodyne spectroscopy is immediately obvious. The linewidth (FWHM) of the apparently reversed component in the NGC 2024 spectrum is only 1.6 km s^{-1}. For all 5 of the observed sources (including W3, W49, and Mon R2 which are not shown here), the profiles of the ^{12}C II lines follow the shapes of the optically thick ^{12}CO lines remarkably well and are quite dissimilar to the shapes of the thinner isotopic CO lines. From this we conclude that the apparent reversals in the C II lines are real absorptions from ionized carbon in cooler or subthermally excited foreground gas, and are not just an artifact produced by multiple velocity components seen only in emission. While quantitatively we cannot do more than speculate on the optical depth of the emission (for reasons mentioned above), we are in a good position to place limits on the depth of the absorption component.

A lower limit for the optical depth of the absorption can be derived from inspection of the reversed profile. We have in this case:

$$\min \tau_{abs} = -\ln\left[\frac{T_A^*(\text{dip})}{T_A^*(\text{peak})}\right],$$

where we take the excitation temperature of the absorbing gas to be 3 K (or equivalently zero). This gives us $\tau_{abs} \geq 1.0$ for NGC 2024 and $\tau_{abs} \geq 2.2$ for W51-IRS2. With a more refined model using Gaussian absorption and emission components, we get $\tau_{abs} \geq 2.2 \pm 0.6$ for NGC 2024 and $\tau_{abs} \geq 4.4 \pm 2.7$ for W51-IRS2. The corresponding lower limits to the C^+ column densities for the absorption component alone are $1 \times 10^{18}\,\text{cm}^{-2}$ and $3 \times 10^{18}\,\text{cm}^{-2}$ for NGC 2024 and W51-IRS2, respectively.

We see that in some sources there is likely as much ionized carbon in absorption as predicted by model calculations for the C^+ emission component. It is possible some of this absorption is produced by gas previously identified from its spatially extended emission as the C II halo component in galactic H II regions (Russell et al. 1981). On the other hand, we see C II line reversals only in sources for which strong CO line reversals are noted, which suggests that the C^+ seen in absorption is kinematically and probably energetically associated with molecular material rather than the extended atomic gas believed to give rise to the C II halo emission. These points and others are discussed in more detail in an upcoming manuscript on reversed C II line emission in galactic sources (Betz et al. 1988).

5. Conclusions

We have basically three simple conclusions we can draw from the analysis of all the reversed-line sources:

(1) In dense molecular clouds the C II line profiles are generally similar to those of optically thick ^{12}CO (and not ^{13}CO), which suggests a common physical origin.
(2) The optical depth of C II in absorption is typically ≥ 1 in reversed-line sources. The column density of this colder or subthermally excited C II typically exceeds $\sim 1 \times 10^{18}\,\text{cm}^{-2}$; hence, the total abundance of ionized carbon is significantly higher than that estimated from observations of unresolved emission lines.
(3) Excitation temperatures for photodissociation regions derived from the 158 μm C II/63 μm O I integrated intensity ratio should be viewed cautiously. Not only may the C II profiles be optically thick in emission, but they frequently show optically thick absorption reversals. Moreover, both these effects are likely to be even more significant in the 63 μm O I line.

This work was supported in part by NASA under grant NAG2-254 for research in airborne astronomy.

References

Betz, A.L., and Zmuidzinas, J. (1984), 'A 150 to 500 μm Heterodyne Spectrometer for Airborne Astronomy', Airborne Astronomy Symposium, NASA Conference Publication 2353, 320-329.

Betz, A.L., Boreiko, R.T., and Zmuidzinas, J. (1988), 'Reversed emission from ionized carbon in molecular clouds', (manuscript in preparation).

Blake, G.A., Sutton, E.C., Masson, C.R., and Phillips, T.G. (1987), 'Molecular abundances in OMC-1: the chemical composition of interstellar molecular clouds and the influence of massive star formation', *Ap.J.*, **315**, 621-645.

Boreiko, R.T., Betz, A.L., and Zmuidzinas, J. (1988), 'Heterodyne Spectroscopy of the 158 Micron C II Line in M42', *Ap. J. (Lett.)*, **325**, L47-L51.

Cooksy, A.L., Blake, G.A., and Saykally, R.J. (1986), 'Direct measurement of the fine-structure interval and g_J factors of singly ionized atomic carbon by laser magnetic resonance', *Ap. J. (Letters)*, **305**, L89-L92.

Crawford, M.K., Genzel, R., Townes, C.H., and Watson, D.M. (1985), 'Far-infrared spectroscopy of galaxies: the 158 micron C^+ line and the energy balance of molecular clouds', *Ap.J.*, **291**, 755-771.

Dalgarno. A., and McCray, R. (1972), 'Heating and ionization of HI regions', *Ann. Rev. Astron. Astrophys.*, **10**, 375-426.

Lugten, J.B. (1987), 'Velocity resolved far-infrared spectroscopy of galactic sources', Ph.D. thesis, Univ. CA, Berkeley.

Petersen, F.R., Scalabrin, A., and Evenson, K.M. (1980), 'Frequencies of cw FIR laser lines from optically pumped CH_2F_2', *Int. J. Infrared Millimeter Waves*, **1**, 111-115.

Russell, R.W., Melnick, G., Gull, G.E., and Harwit, M. (1980), 'Detection of the 157 micron (1910 GHz) [C II] emission line from the interstellar gas complexes NGC 2024 and M42', *Ap.J. (Letters)*, **240**, L99-L103.

Russell, R.W., Melnick, G., Smyers, S.D., Kurtz, N.T., Gosnell, T.R., Harwit, M., and Werner, M.W. (1981), 'Giant [C II] halos around H II regions', *Ap.J. (Letters)*, **250**, L35-L38.

Stacey, G.J. (1987), private communication.

Tielens, A.G.G.M., and Hollenbach, D. (1985), 'Photodissociation regions. I. Basic model', *Ap.J.*, **291**, 722-746.

Zmuidzinas, J., Betz, A.L., and Boreiko, R.T. (1988), 'A Corner-Reflector Mixer Mount for Far-Infrared Wavelengths', *Infrared Physics*, (to be published).

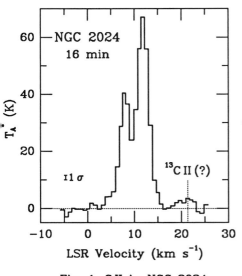

Fig. 1. C II in NGC 2024.

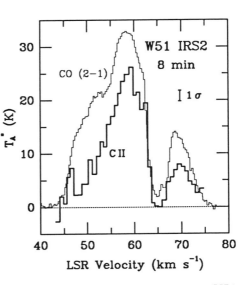

Fig. 2. C II and CO J=2-1 in W51.

ION-MOLECULE CHEMISTRY OF CARBON IN SHIELDED REGIONS

W. D. LANGER
Princeton University
PO Box 451
Princeton, NJ 08543

ABSTRACT. The abundances of neutral carbon and carbon molecules are calculated using the updated chemical model of Graedel and Langer. The inclusion of new reactions and rate coefficients increases substantially the CI abundances in shielded regions even for C/O<1.

1. INTRODUCTION

The observed CI abundance in interstellar clouds ranges from 1 to 20 percent (Keene et al. 1985; Genzel et al. 1988). In shielded regions this range is much larger than had been predicted by models of the ion-molecule chemistry (cf. Langer et al. 1985) because the total gas phase C/O ratio is expected to be less than one and almost all the carbon is locked up in CO. A number of explanations, such as nonequilibrium chemistry, dynamical processes, or a high C/O ratio >1, were proposed to reconcile this problem. Recent high spatial observations of CI and C+ near the interface of molecular clouds and OB associations suggest another possibility, that UV photodissociation in a clumpy cloud structure, where the UV can penetrate deeply, produces CI (Genzel et al. 1988). In other regions, for example B5, this solution may not prevail because the evidence from CO and IR emission suggests that the UV does not penetrate to the core (Langer et al. 1989). However, the model of the ion-molecule chemistry has changed in recent years, largely as a result of new laboratory measurements, and here we re-examine the CI abundance in shielded regions of clouds.

2. CHEMICAL MODEL

The approach and detailed information used in modeling the chemistry of dense clouds is given in Langer et al. (1984), but updated to include changes in the chemistry since 1982 (Langer and Graedel 1989). The most important changes are: 1) the dissociative recombination of H_3^+ with electrons is set to zero; 2) the replacement of Langevin ion-molecule reaction rates for molecules with large permanent dipole moments by the much larger values appropriate at low temperatures (Adams, Smith, and Clary 1985); 3), photodissociation of CO proceeds by line rather than continuum absorption, and, 4) the inclusion of UV produced by cosmic ray interactions with H_2. The model was run for 15 cases where the following parameters were varied: density: $10(3)$ to $5 \times 10(4)$ cm^{-3}; Temperature: 10 to 40 K; C/O ratio: 0.7 to 1.3; and cosmic ray ionization rate: (1 to 5)x $10(-17)$/sec. In addition various changes in some reaction rates were considered for the nitrogen chemistry.

3. RESULTS and DISCUSSION

The updates in the chemistry of interstellar clouds changes considerably the abundances for carbon, especially for the important cases where C/O < 1. The key results for the latter condition are that the fraction CI/CO, is in the range 3 to 5% (corresponding to a fractional abundance of $(2-3) \times 10(-6)$). In the current models where C/O<1 the increase in carbon is due to the larger reaction rates of ions with polar molecules which produce CI. For example, He+ reacting with CN, CS, and C2 and C+ with CH, CH2, and C2H produce C at a rate about 10 times larger than earlier models using the room temperature Langevin rates extrapolated to low cloud temperatures (<50 K). Cosmic ray-produced UV photodissociation of CO contributes about 10% of the CI in our model, but this is likely to be an underestimate by a factor of three. Other carbon bearing molecular abundances are generally in good agreement with observations, except for CN which is 10 to 20 times too large. Eliminating CN would only lower the calculated CI abundance by a factor of two. A combination of low densities (~$10(3)$/cm(3)), low temperatures (~10K), and moderate cosmic ray ionization rate ($5 \times 10(-17)$/s) yield CI/CO of 5%.

Finally, there are other areas of uncertainty in the chemical models that could lead to a larger CI abundance. The primary destruction mechanism for CI is a reaction with O2, whose rate is measured at room temperature. The reaction rate is likely to be smaller at low temperatures analagous to the cases for CI with OH and CH (cf. Graff 1989). If either the rate is smaller or the O2 abundance lower a further factor of a few increase in CI is likely. In conclusion, the ion-molecule chemistry, updated to include the changes discussed here, is capable of explaining the observed CI in shielded regions.

4. References

Adams, N, Smith, D, and Clary, D 1985, Ap.J.Lett., **296**, 31-35.
Genzel , R. et al. 1988, Ap. J., **332**, 1049-1057.
Graff, M. 1989, Ap. J., in press.
Keene, J. et al., 1985, Ap. J., **210**, 679-689.
Langer, W., et al., 1984, Ap. J., **277**, 581-604..
Langer, W. and Graedel, T. 1989, Ap. J. Suppl., in press.
Langer, W, et al., 1989, Ap. J. in press.

158 μm [CII] Line Emission from Galaxies

G. J. Stacey[1], R. Genzel[2], J. B. Lugten[3], and C. H. Townes[1]

[1] Dept. of Physics, University of California, Berkeley, CA USA
[2] Max-Planck-Inst. for Extraterrestrial Physics, Garching, FRG
[3] Inst. for Astronomy, Univ. of Hawaii, Honolulu, HI USA

ABSTRACT. We have measured 158 μm [CII] line emission from a sample of thirteen gas rich galaxies. The [CII] line emission arises in warm (T ~ 300 K), dense (N_H ~ $10^3 \rightarrow 10^4 cm^{-3}$) photodissociated gas at the interface regions between giant molecular clouds and ionized gas regions and is an important gas coolant for these regions. We find the integrated [CII] to $^{12}CO(1 \rightarrow 0)$ line intensity ratio to be constant for starburst galaxies and star formation regions in our own galaxy. This suggest that much of the observed ^{12}CO line emission from these starburst galaxies arises in the warm molecular gas immediately interior to these photodissociation regions (PDR's), and not from the cold disk molecular cloud component as is commonly assumed.

1. Introduction and Observations

[CII] line emision has previously been reported from a sample of six starburst galaxies (1). We have expanded this sample to include a wide variety of galactic spectral types and luminosity classes. We made our observations using the UC Berkeley Tandem Fabry-Perot spectrometer (55" beam) onboard the Kuiper Aiborne Observatory. We detected eleven of the thirteen new galaxies investigated, five of which we mapped with our new three element spatial array (2). Figure 1 shows representative spectra obtained on the starburst galaxy NGC 2146 and the normal Sc galaxy NGC 5907. For comparision, we have superposed the ^{12}CO (J = 1 $\rightarrow$ 0) spectra observed at the [CII] positions with a similar sized beam (2,3). It is clear that there is excellent spatial and spectral correlation between the two lines indicating that they arise from in or about the same gas.

2. Results and Discussion

The CO(1 $\rightarrow$ 0) line emission in galaxies is presumed to be dominated by emission from the cold molecular disks. This cold component will produce very little [CII] line radiation due to the lack of nearby ionizing radiation (4). The integrated [CII]/CO line intensity ratio should therefore be substantially less in galaxies than in galactic star formation regions, and vary depending on just how the UV energy density

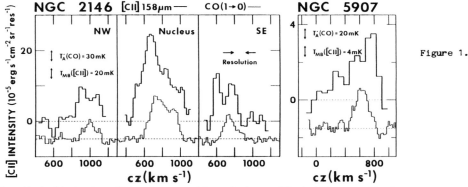

Figure 1.

is distributed. This is exactly what we find (Figure 2) for the normal spiral galaxies (●) and non-starforming clouds (○) in our own galaxy. However, for the starburst galaxies (★), the integrated line intensity ratio is the same as that for star formation regions (☆) in the Galaxy. This suggests that the CO line emission from these galaxies arises in the warm molecular gas associated with star formation regions and not from the cold disk component. Thus, the CO line intensity to mass conversion ratio derived for the cold disk clouds in our galaxy may be totally inappropriate for these starburst galaxies.

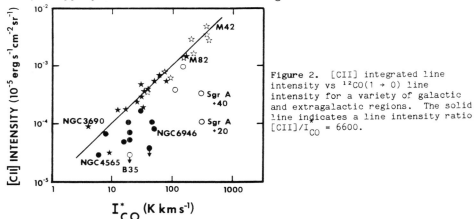

Figure 2. [CII] integrated line intensity vs $^{12}CO(1 \to 0)$ line intensity for a variety of galactic and extragalactic regions. The solid line indicates a line intensity ratio $[CII]/I^*_{CO} = 6600$.

Acknowledgements We thank Gregory Engargiola for data provided in advance of publication and the staff of the KAO for their enthusiastic support. This research was supported by NASA grant NAG2-208.

References

(1) Crawford, M.K., Genzel, R., Townes, C.H., and Watson, D.M. 1985, **Ap.J.**, 291, 755.
(2) Stacey, G., Genzel, R., Lugten, J., and Townes, C. 1988 in prep.
(3) Young, J.S., 1988, **Ap.J. (Letters)**, 331, L81.
(4) Wolfire, M.G.; Hollenbach, D. and Tielens, A.G.G.M. sub. **Ap.J.**

HOT QUIESCENT GAS IN PHOTODISSOCIATION REGIONS
CO AND C+ OBSERVATIONS OF NGC 2023

D. T. JAFFE, J. E. HOWE
Department of Astronomy, University of Texas at Austin
R. GENZEL, A. I. HARRIS, J. STUTZKI
Max-Planck-Institute for Extraterrestrial Physics
and G. J. STACEY
Department of Physics, University of California, Berkeley

ABSTRACT: We present observations of the reflection nebula NGC 2023 in the millimeter and submillimeter lines of CO and the C+ 158 µm fine structure line. The CO J = 7 → 6 emission coincides with a shell of fluorescently excited H_2. This H_2 shell lies between a partially ionized region and the molecular shell seen in optically thin millimeter CO lines. The warm CO layer is at least 75 K. Both the morphology and the high temperature of the gas argue for UV heating in the inner part of the molecular shell.

1. INTRODUCTION AND OBSERVATIONS

Recent observations of the J = 7 → 6 transition of CO have led to the discovery of hot (150-500 K), dense (n_{H_2} = 2 x 10^4 - 3 x 10^5 cm^{-3}), quiescent gas in luminous star formation regions (e.g., Harris et al. 1987). The association of the hot molecular gas with sources of intense UV radiation (10^4-10^5 times the general interstellar radiation field) and the narrow line widths suggest that UV radiation heats this gas which lies in the neutral interface zone between the H II region and the cooler bulk of the molecular cloud. In order to study this suggestion, we began a study of NGC 2023 which surrounds the B1.5 star HD 37903. This reflection nebula offers a number of advantages over the regions studied previously: (1) The characteristics of the UV source are known. (2) Fluorescent molecular hydrogen emission in this source demonstrates the influence of UV radiation (Hasegawa et al., 1987). (3) The UV field is about 10^3 times the ISRF, providing some variety in the sample. (4) Molecular lines in the cloud surrounding the reflection nebula are narrow (Δv ~ 2 km s^{-1}), eliminating the possibility of shock heating. (5) The source is nearby (D = 450-500 pc), making details of its structure more accessible. We mapped NGC 2023 in CO J = 2 → 1 and in $C^{18}O$ 2 → 1 with the NRAO 12m telescope (θ_{BEAM} = 32 arcsec), the CO 3 → 2 transition at the UTexas Millimeter Wave Observatory (θ_{BEAM} = 46 arcsec), the CO 7 → 6 transition at the NASA-IRTF (θ_{BEAM} = 32 arcsec), and the C+ $^2P_{3/2}$ → $^2P_{1/2}$ transition with the Kuiper Airborne Observatory (θ_{BEAM} = 55 arcsec) to produce the cross-scans in Figure 1.

2. RESULTS AND DISCUSSION

The morphology of the molecular and neutral atomic emission in NGC 2023 support a picture in which UV radiation from the exciting star heats the gas at the inner edge of the molecular cloud. Maps of the H_2 S(1) and CO J = 1 → 0 lines (Gatley et al., 1987) show an incomplete shell of molecular material surrounding HD 37903. This shell has a maximum in the southeast quadrant, about 1.5 arcmin from the star (Figure 1). C+ 158 μm emission peaks closer to the star than the fluorescent H$_2$ and drops rapidly to the southeast. The presence of the C^{18}O J = 2 → 1 peak farther from the star than the H$_2$ demonstrates that the molecular hydrogen emission originates on the UV illuminated side of the molecular cloud. The optically thin C^{18}O 2 → 1 emission implies a mass for the southeast quadrant of the molecular shell of 50 M$_\odot$. There is also some C^{18}O emission northwest of the star which indicates that the cool molecular cloud does extend around behind the star. Figure 1 shows that the CO J = 7 → 6 emission peaks at the position of the H$_2$ S(1) line maximum. This peak would be much more pronounced if allowance were made for beam dilution (32" for CO 7 → 6 vs. 20" for the H$_2$ line) and undersampling. Overall, these cross-cuts and our mapping results show a clear trend from C+ emission closer to the star to fluorescent H$_2$ to cooler CO farther from the star.

The spectroscopic evidence also supports UV photons as the ultimate source of heating for the high temperature molecular gas. The widths of the CO 7 → 6 lines range from < 1 km s^{-1} to 1.4 km s^{-1}, implying that any turbulence in the cloud is subsonic. The CO 7 → 6 lines are bright (Planck brightness temperature T$_{pl}$ = 75 K). This T$_{pl}$ implies a kinetic temperature ≥ 75. The gas emitting the 7 → 6 line could be considerably hotter than this if beam dilution of the bright shell or the high opacity of the cooler parts of the shell mask some of its strength. In any case, the temperature of this gas is high for a moderate luminosity source and is substantially greater than the dust temperature of 40-50 K measured at the south and east ends of the shell (Harvey, Thronson, and Gatley, 1980). This work was supported in part by NASA Grant NAG 2-419 and NSF grant AST 8611784 to the University of Texas at Austin.

REFERENCES

Gatley, I., et al. 1987, *Ap. J. (Letters)*, **318**, L73.
Harris, A. I., Stutzki, J., Genzel, R., Lugten, J. B., Stacey, G. J., and Jaffe, D. T. 1987, *Ap. J. (Letters)*, **322**, L49.
Harvey, P. M., Thronson, H. A., and Gatley, I. 1980, *Ap. J.*, **235**, 894.
Hasegawa, T., Gatley, I., Garden, R. P., Brand, P. W. J. L., Ohishi, M., Hayashi, M., and Kaifu, N. 1987, *Ap. J. (Letters)*, **318**, L77.

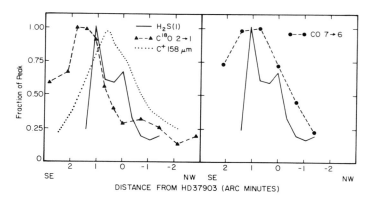

Figure 1: Scans across NGC2023. All scans show line flux except CO 7 → 6 which shows brightness temperature.

DCO+ IN NEARBY DENSE CORES

Harold M. Butner
Dept. of Astronomy, University of Texas at Austin
Austin, Tx 78712 USA

1. Introduction

Myers and Benson (1983) conducted a survey of nearby molecular cores for NH_3. They identified 27 such cores, with an average $T_K \sim 10$ K and an average density of 10^4. This sample allows one to investigate physical conditions in low mass cores, with a variety of techniques. In addition, it is possible to test chemical models of processes such as deuterium fractionation. We observed the NH_3 cores using the DCO^+ and $H^{13}CO^+$ (J=1-0) lines with the NRAO 12 meter, and the DCO^+ (J=2-1) line with the MWO 5 meter. The line intensities were corrected to T_R, where a source size of 2' was assumed for η_c. We compare the observed DCO^+/HCO^+ ratio with current models of deuterium fractionation, and we compare the derived properties of the DCO^+ core with those derived from NH_3 and CS.

2. Deuterium Fractionation

The isotope exchange $\quad H_3^+ + HD <=> H_2D^+ + H_2 + \Delta E$

is the ancestral reaction in the formation of DCO^+. The energy difference, ΔE, causes the abundance of H_2D^+ to be temperature dependent. Any molecule, such as DCO^+, that forms from H_2D^+ will also have a temperature dependent abundance. Chemical models, such as those of Millar, Bennett and Herbst 1988, suggest that at low temperatures there is a substantial enhancement of DCO^+ over the cosmic deuterium abundance. For the conditions found in the NH_3 cores, with $T_K \sim 10$ K, a ratio $R(DCO^+/HCO^+) \sim 0.02$ to 0.05 is predicted.

In order to determine the DCO^+/HCO^+ ratio, we calculated the total column density of the DCO^+ and $H^{13}CO^+$ molecules from the observed J=1-0 lines. $H^{13}CO^+$ was chosen because the main isotopic species, HCO^+, is optically thick in the J=1-0 line. An excitation temperature of 10 K was assumed, and a $^{12}C/^{13}C$ ratio of 75 (Wilson *et al.*, 1981). The average value of the DCO^+/HCO^+ ratio, ~0.034, agrees well with the Millar *et al.*, 1988 models.

3. Comparison of DCO^+, NH_3 and CS

From the DCO^+ (J=1-0) and (J=2-1) lines, it is possible to estimate the densities of the emission regions. Using LVG models, and assuming the gas kinetic temperature was 10 K, we derive an average density of 10^5 cm^{-3}. This is higher than the estimates derived from NH_3, suggesting that the DCO^+ emission is originating from a higher density regime than the NH_3. The DCO^+ density estimates are in better agreement with those of CS (Zhou *et al.*, 1988).

Inspection of the DCO^+ linewidths reveals several trends. First, the DCO^+ linewidths are broader in those cores which have associated IRAS sources, than in those which do not. A similar pattern was pointed by Beichman *et al.*, 1986, who noted that in the Myers and Benson sample, the $C^{18}O$ and NH_3 linewidths are substantially broader in cores with embedded sources. Zhou *et al.*, 1988 find that CS also shows this broadening. Secondly, while DCO^+, CS, and $C^{18}O$ show similar linewidths, NH_3 linewidths are consistently narrower. A comparison of the DCO^+ and NH_3 linewidths (see Figure 1) shows that this is true for both cores with and cores without associated IRAS sources. There are several possible explanations for this difference, including linewidth broadening due to large optical depths.

In the case of DCO^+, the optical depths of the J=1-0 line found by our column density estimates and the LVG models are similar, with typical optical depths of order 0.5 in the DCO^+ (J=1-0) line. This is insufficient to explain the broadening relative to NH_3. A minimum optical depth of ~4 would be required to explain the broadening as purely an optical depth effect. A comparison between a number of molecular probes will be necessary to sort this problem out and identify whether optical

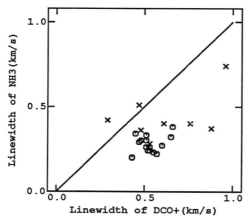

Figure 1: Comparison of DCO^+ and NH_3 linewidths. Cores without associated IRAS sources (o), cores with associated IRAS sources (x). The straight line shows where the linewidths would be equal.

depth, turbulence effects, or possibly chemical differentiation are responsible for the large differences in linewidths.

References

Beichman, C. A., Myers, P. C., Emerson, J. P., Harris, S, Mathieu, R., Benson, P. J., and Jennings, R. E. 1986, *Ap. J.* **307**, 337.
Millar, T. J., Bennett, A., Herbst, E. 1988, submitted to *Ap. J.*
Myers, P. C., and Benson, P. J. 1983, *Ap. J.* **266**, 309.
Wilson, R. W., Langer, W. D., and Goldsmith, P. F. 1981, *Ap. J.* **243**, L47.
Zhou, S., Wu, Y., Evans, N. J. II, Fuller, G., and Myers, P. C. 1988, in preparation.

CIRRUS CLOUD CORES: A DEFICIENT CHEMISTRY?

B. E. TURNER[1], LEE-J. RICKARD[2], AND XU LAN-PING[3]
[1] National Radio Astronomy Observatory, Charlottesville, Virginia.
[2] E.O. Hulbert Center for Space Research, Naval Research Laboratory.
[3] Department of Geophysics, Beijing University.

ABSTRACT

A survey of 28 core regions in 16 different high-latitude cirrus clouds has been made in the 2 cm lines of H_2CO, the 1.5 cm line of C_3H_2, and the J=1-0 and 2-1 lines of HC_3N. The H_2CO results imply typical core densities of 4(4) cm^{-3}, and extinctions of 2 to 20 magn. Fractional abundances are well determined for 7 cores and are found to be less than those of galactic plane clouds by factors of 10, 50, and >50 for H_2CO, C_3H_2, and HC_3N respectively. Possible explanations are discussed. IRAS emissitivities indicate "normal" grains in the cirrus cores, not a small-sized stochastically-heated population. The cirrus cores are probably gravitationally bound.

1. PHYSICAL CONDITIONS IN CIRRUS CORES

The 28 cirrus cores surveyed in the 2 cm line of H_2CO and in C_3H_2 and HC_3N were selected on the basis of previous detections of the 6 cm H_2CO line (at resolution 2.6') and the 2.6 mm line of CO, by Magnani et al (1988, Ap.J. 326, 909). With the 140 ft telescope we detected 2 cm H_2CO in 12 cores (resolution 2.1') and mapped it at 5 points with 2' spacing in 7 cores, for which core sizes $\geq$ 5' are found. The 2 and 6 cm lines have identical widths in all cases, and are optically thin, so they sample the same (core) gas. An LVG analysis yields densities ranging from 1.2(4) to 7.2(4) cm^{-3}, with a median of 4(4) cm^{-3}, corresponding to a typical extinction of $\sim$ 9 magn if a core size of 5' is adopted. Lower limits for core masses range from 0.3 to 10 $M_\odot$, and are typically within a factor of a few of the virial masses, so that most of the cores are likely gravitationally bound. If such cores are not to produce a star formation rate greatly in excess of that observed, they must be virialized, just as galactic plane clouds cores appear to be.

2. FRACTIONAL ABUNDANCES IN THE CIRRUS CORES

2.1. H_2CO

The LVG analysis and assumed core radius of 5' gives a median fractional abundance $X = \int dv/dR$ of 4(-10) for a gas temperature T_K= 15 K as indicated by grain color temperatures derived from IRAS data. These values of X are upper limits, since core sizes may be larger than 5'. By contrast, $X(H_2CO)$ has been reliably derived for only 4 cold galactic plane clouds and has a median value 2(-8) if TMC-1 is included, or 5(-9) if it is excluded. We conclude that $X(H_2CO)$ is at least 10 times larger in galactic plane clouds than in the 7 well studied cirrus cores.

2.2 C_3H_2

The 1_{10}-1_{01} line at 18.3 GHz was detected in 16 of the 18 cirrus cores searched. We use an LTE analysis identical to that of Bell et al (1988, Ap.J. 326, 924) for 13 galactic plane cold clouds, which have similar temperatures and densities. We find cirrus column densities $N(C_3H_2)$ average a factor 34 less than the average found by Bell et al for 13 galactic plane clouds. Assuming C_3H_2 and H_2CO are coextensive, we find $X(C_3H_2) = X(H_2CO)[N(C_3H_2)/N(H_2CO)]$ has median value 1(-10), again a lower limit because of incomplete mapping of the cores. The galactic plane clouds average $X(C_3H_2) = 5(-9)$, a factor of 50 larger.

2.3 HC_3N

HC_3N was detected in neither J=1-0 nor J=2-1 transitions in 17 cirrus cores searched. An LVG analysis yields $X(HC_3N) < 2.4(-11)$ for all objects searched. By contrast, the average value for 4 well studied galactic plane clouds is 1.1(-9) if TMC-1 is excluded (Snell et al 1981, Ap.J. 244, 45). Thus HC_3N is at least 50 times larger in galactic plane clouds than in cirrus cores.

3. THE CIRRUS CLOUD GRAINS

Accurate IRAS radiances $I(\lambda)$ have been derived at all 4 IRAS wavelengths for all positions where C_3H_2 or H_2CO was detected. From these, color temperatures and continuum opacities have been derived. The observational results may be summarized as follows: (i) grain color temperatures T_{gr} (from $100\mu m/60\mu m$ ratio) fall in the range $14 \leq T_{gr} \leq 18$ K (ii) T_{gr} does not correlate with A_v for the ensemble of cores studied; (iii) no IRAS flux is detected at 12 or 25 μm for any core; (iv) $I(100\mu m)/A_v$ correlates with A_v, being largest at low A_v and decreasing to a constant value of ~ 1 MJy/sr/magn for $A_v \geq 5$ magn; (v) $\tau(100\mu m)/A_v$ has an average value 5(-4). The interpretation of these results is as follows: point (iii) implies there is no stochastically heated small-grain population. Therefore, T_{gr} is reliably determined. Point (v) is consistent with the model of Draine and Lee (1985, Ap.J. 285, 89) for normal-sized graphite + silicate grains. Point (i) implies at least "skin" heating of the cores by external UV, since $T_{gr} > 10$ K. Points (iv) and (ii) imply that the heating is ONLY a skin effect.

4. A DEFICIENT CHEMISTRY FOR THE CIRRUS CORES?

The factors by which $X(H_2CO)$, $X(C_3H_2)$, $X(HC_3N)$ in cirrus cores fall below those for galactic plane clouds are $\sim$10, 50, and > 50 respectively. These cannot be explained by: (i) non-dissociating shocks, which should INCREASE these abundances by factors of 10-1000, 10(2)-10(4), and $\sim$10(3) respectively (Mitchell 1984, Ap.J. (Suppl) 54, 81); (ii) dissociating shocks, which deplete molecules for only $\sim$10(2) years after the shock has passed; (iii) early time chemistry (Herbst and Leung 1988, Ap.J. in press) which INCREASES the abundance by factors of 10(2), 10(2)-10(3), and 10(3)-10(4) respectively; (iv) UV dissociation, which is negligible for the high extinctions we find.

Possible explanations for the relative deficiency of cirrus abundances are: (i) a lower metallicity. This is consistent with lack of massive star formation (the core masses are too small), but it is not clear why the galactic plane and cirrus gas would not be reasonably well mixed; (ii) a NORMAL ion-molecule chemistry for the cirrus clouds, together with an ENHANCED chemistry for galactic plane clouds. It is well known that $X(C_3H_2)$ and $X(HC_3N)$ are much greater in galactic plane clouds than are predicted by standard models (e.g. Herbst and Leung 1988). By contrast, our cirrus values of $X(C_3H_2)$ and $X(HC_3N)$ agree well with the predicted steady-state values. Perhaps shocks are much more effective in galactic plane clouds, or their dynamical timescales are much shorter, effectively creating early-time chemistry.

Section IV

Star Formation

STAR FORMATION IN ACCRETION DISKS

Ralph E. Pudritz
Dept. of Physics, McMaster University, Hamilton, ON L8S 4M1

1. Introduction

The route from the formation of a molecular cloud to the appearance of an association of young stellar objects with a well defined IMF is a formidable challenge for modern theories of star formation. No theory of star formation can be complete without answering two fundamental questions: 1. what determines the mass reservoir for a star, and 2. what determines the accretion rate out of that reservoir.

Star formation proceeds in cores and fortunately, strong clues about the controlling agent for core formation have recently been unearthed. Since neither rotation nor thermal energy densities are comparable with gravity on supra-core scales, only gravity and magnetism are left. Myers and Goodman (1988) compiled many workers' magnetic observations of molecular clouds and showed that for a wide variety of cloud types, the magnetic energy density is comparable with gravity. Hydromagnetic effects therefore are likely to be a controlling factor in core formation.

Accretion rates out of mass reservoirs depend upon whether accretion is spherically or axially symmetric. Spherical accretion supposes that the angular momentum problem has largely been resolved at the moment of collapse, and that magnetic fields are not important. Collapse into disks (axial symmetry) must ensue if these conditions are violated. The most important indication that the latter has happened is that dynamically significant magnetic fields in clouds (eg. Taurus) are actually well ordered on scales of several pcs. This implies that sheets or filaments must form and observations show that molecular clouds are not just clumpy, but STRINGY. The second clue is that molecular disks abound around young stellar objects. They have sizes and masses which strongly suggest that the collapse occurred early on before much angular momentum or magnetic field were lost (see eg. Pudritz 1986). The only recourse for star formation in the aftermath of collapse into a disk, is to develop by accretion flow through the disk itself.

2. Core Formation

The magnetic data give strong support to the idea of Arons and Max (1975) that hydromagnetic waves can support molecular clouds against collapse. When magnetic energy densities are comparable to gravity, they showed that the amplitude of Alfven waves in the molecular gas would be of order the observed supersonic line widths. While Alfven waves are transverse modes that create purely velocity

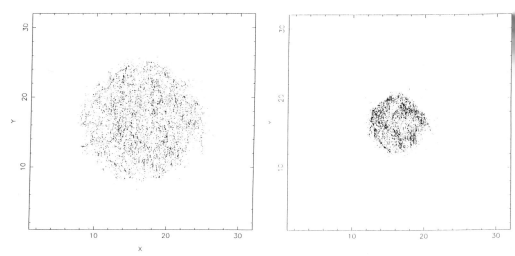

FIGURE 1. N-body simulation (20,000 gas particles) of a uniform gas being perturbed by spectrum of fast magnetosonic modes (from Carlberg and Pudritz 1988). Frame (a) is a slice through the cloud at 3 free-fall times, (b) at 4 free-fall times. We observe structures with overdensities as high as 100; the simulation is not self-consistent at higher values. Note the presence of smaller scale fragments and the later appearance of filamentary structure.

fluctuations there is another hydromagnetic wave, the so-called fast magnetosonic mode, that must also occur in a magnetized molecular cloud. In cold molecular gas, these modes obey virtually the same dispersion relation as Alfven waves, ie

$$\omega = kV_A^o \qquad (1.)$$

where V_A^o is the Alfven speed in the undisturbed cloud, ω is the wave frequency, and $k = 2\pi\lambda^{-1}$ is the wave number. The important difference is that fast modes are compressive and produce density fluctuations $\delta\rho$ that are directly proportional to velocity fluctuation $\delta\mathbf{v}$, or

$$\delta\rho/\rho_o = -\mathbf{k}.\delta\mathbf{v}/kV_A^o \qquad (2.)$$

This implies that not only should a wave field support a molecular cloud against global collapse, but it MUST also create a spectrum of density fluctuations in the cloud. Because a molecular cloud is strongly self-gravitating, density fluctuations will grow by drawing in gas from the surrounding underdense regions. Ambipolar diffusion in overdense regions will lead to a slight weakening of magnetic support there, which will lead to even larger density fluctuations. Thus Ray Carlberg and I (1988) have argued that such wave spectra can be the prime movers in starting the formation of structure in molecular clouds.

What fragment spectrum would be produced by this process? Suppose that density peaks are created at velocity peaks in a wave propagating through a cloud (as predicted by 2.). For a 1-D wave, the number of density fluctuations created by the waves clearly scales as $N \propto \omega$. If the wave field is a fairly disordered, 3-D magnetic field then clearly $N \propto \omega^3$. But since these are Alfvenic type waves (at least as far as their velocity dispersion is concerned), then from (1.) the spectrum of clumps is $N \propto \lambda^{-3}$. We need only convert scales to masses using one of Larson's relations, $m(\lambda) \propto \lambda^2$ to find that $N \propto m^{-3/2}$. Thus the differential mass spectrum predicted by this simple wave picture is

$$\frac{dN}{dm} \propto m^{-5/2} \qquad (3.)$$

While there are many simplifing assumptions here (eg. for a D-dimensional magnetic field the spectral index is $-(D/2) - 1$, a spectrum rather than a single mode is probably involved, etc), the result warrants deeper examination.

Ray Carlberg and I have used a 3-D, N-body code (1988) that mimics isothermal gas dynamics in order to test our idea of producing structure in clouds by wave perturbations. We introduce a spectrum of five modes which range in wavelength from 1/3 down to 1/12 of the size of the cloud. This range ensures that gravitational back reaction is much slower than the wave frequency on the large wave-length end, and that the smallest scale wave is just larger than that scale which would be damped by friction (ie ambipolar diffusion) effects. The equation of motion of each of the neutral "gas" particles (20,000) is modified by friction between the neutrals and ions, supposed to be "glued" to the hydromagnetic mode. Thus the effects of ambipolar diffusion are correctly incorporated.

Some of our results shown in Figure 1a and 1b. These are cuts through our self gravitating clouds (initially a smooth, homogenous sphere) at 3 and 4 free-fall times into the run. The clouds are so cold that in 1 free-fall time they would have collapsed into a cell 2 units in size without the hydromagnetic wave support. The simulations clearly demonstrate that wave support against gravity works (here, the energy density in the waves is 25 percent that of gravity). The effects of ambipolar diffusion are also apparent in the overall contraction of the cloud. A large variety of structure is also seen. The clouds always fragment on a small scale near the cut-off scale first. It takes a longer time for the larger scale filament structure to emerge because the longer scale waves have a longer wave period (see 1.). Our simulations show that the fragments are not transitory structures, but grow with time by accreting nearby material and neighbouring smaller fragments. Their motion can be considered simply as oscillating back and forth around a "home" position. Our recent experiments have varied the power law spectral index for the wave field. The figures are for a wave spectrum $\delta v \propto k^{-\beta}$ where $\beta = 3/2$. For wave spectra $\beta \leq 1.0$, we see a less strong fragmentation on small scales because the smaller scales have more support by wave excitations. For input spectra $\beta > 1.0$ the input wave power available is being used to stabilize the larger scales. Our experiments show that the small scales then break out far more strongly.

3. Accretion Disk Geometry

If star formation proceeds by accretion through disks, then theoretical models can be constrained using observations of the density, temperature, and mass distributions through all parts of a disk. While mm observations provide constraints on 3-5" scales, a good IR camera on a high resolution telescope such as CFHT can get down to sub 1" scales. Recently we (Monin, Pudritz, Lacombe, and Rouan 1988)

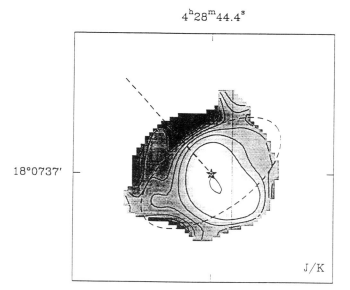

Figure 2. Ratio of J and K images of HL Tau taken on CIRCUS IR camera (from Monin et al 1988). Dotted circle represents circle inclined by 20-30°, the inclination angle of the disk. Axis shows direction of $H\alpha$ jet.

obtained high resolution IR images of the protostellar disk around HL Tau using the INSU CIRCUS camera. This camera consists of a 32 × 32 CID array, equipped with the usual J, H, K, L, and M filters. The focal plate scale was 0.5" per pixel so that the total field of view was 16 × 16 "

We found that the IR images are slightly flattened, but not enough to warrant the idea that an edge on disk is being seen. However, there is a systematic trend in the distortions in each image, with the short ("bluer") wavelength images being more extended towards the N- E, and the longer near IR wavelength images being more extended towards the S-W. This colour gradient is shown in Figure 2 which clearly indicates an alignment of the colour *gradient* across HL Tau with the H_α jet. We interpret this structure as strong evidence for a tilted disk (perhaps by 20-30°) extended out to 500 AU in radius. An object of mass M_d rotating at v_ϕ, has a corotation radius of

$$R_{cor} = 500(M_d/0.6M_\odot)(v_\phi/\mathrm{kms}^{-1})^{-2} \quad \mathrm{AU} \qquad (4.)$$

At near IR wavelengths $\leq 2\mu m$, scattering and absorption of optical radiation dominates over thermal re-emission. Thus, these images can be used to derive the surface of last scattering of an accretion disk around HL Tau. In order to have reasonable scattering, the optical depth should be near unity. In fact, we estimate that it is $\tau \geq 0.15$. Depending upon the grain composition, Monin et al deduce that the total disk mass is $\geq 0.7 \times 10^{-2} M_\odot$ (graphite-silicates) up to $0.7 \times 10^{-1} M_\odot$ (pure silicates). Our images suggest that accretion disks have concave surfaces rather than convex because the star always appear to lie in the centre of the scattered light maximum.

4. Accretion Disk Structure and Evolution

How is angular momentum transported through disks? One possibility is magnetic braking by hydromagnetic disk winds. Would such a braking law have observable consequences that are different than predicted by standard viscous disk theory? Remember that as long as disks undergo *steady state* evolution, the temperature distribution across the disk must follow the $T_d \propto r^{-3/4}$ (see eg Pringle 1981). In the case of purely viscous angular momentum transport through disks, the gravitational binding energy that must be released at each radius in a Keplerian disk, is radiated away. Since protostellar disks are present at the same time that bipolar outflows are active, it is likely that winds originate from disks (eg Pudritz 1986). The angular momentum and gravitational binding energy are then largely carried off from each radius of the disk by the wind.

Consider the angular momentum equation for steady state disk structure

$$\dot{M}_a dj/dr = dg/dr \qquad (5.)$$

where the specific angular momentum $j \equiv rv_\phi$, the torque is g, and $\dot{M}_a$ is the constant accretion rate. Now for viscous disks, the torque is $g_{vis} = (1.5\nu\Sigma)\Omega r^2$ (see Pringle 1981). As is well known, Equation (5.) constrains the product of the viscosity and the surface density in order that the angular momentum of a Keplerian disk be transported away at a rate determined by $\dot{M}_a$. If a disk wind carries off the angular momentum, it is possible to cast the disk angular momentum equation in similar form as (5) except now the torque is $g_{wind} = (f_g \dot{M}_w (r_A/r)^2)\Omega r^2$. Comparing this with the purely viscous disks, it is clear that the wind mass loss rate $\dot{M}_w$ and the lever arm (ratio of Alfven radius to r squared) play the same role physically as the combination of viscosity and surface density. Thus, *winds adjust their mass loss rate and geometry* in such a way to carry off the disk angular momentum. The requirement of steady state drastically simplifies models for disk winds, as it does for purely viscous disk evolution. If the wind carries off (as mechanical energy) a constant fraction of the gravitational binding energy at each disk radius, then the standard temperature law will again be predicted. Deviations from such a law for steady state disks would imply that for a wind loss mechanism the wind would be relatively more efficient at energy extraction at different disk radii.

It is a pleasure to thank the organizers of this stimulating conference for the invitation to speak and the Herzberg Institute for its financial support. I have learned a great deal from my collaborators Ray Carlberg, Francois Lacombe, Jean-Louis Monin, and Daniel Rouan. This work was supported by NSERC of Canada.

REFERENCES

Arons, J. and Max, C.E. 1975, *Ap. J. Letters*, **196**, L77.

Carlberg, R.G. and Pudritz, R.E. 1988, *M.N.R.A.S.*, submitted.

Monin, J-L, Pudritz, R.E., Lacombe, F., and Rouan, D. 1988,
    Astron. Ap. Letters, in press.

Myers, P.C., and Goodman, A. 1988, *Ap. J. Letters*, **326**, L27.

Pudritz, R.E. 1986, *P.A.S.P.*, **98**, 709.

Molecular Clouds and Bipolar Flows

Göran Sandell
Joint Astronomy Centre
665 Komohana Street, Hilo, HI 96720

ABSTRACT. This paper is devoted to observations of dust emission in the mm and submm continuum carried out during the first year of JCMT. I present recent high resolution work done on molecular clouds and HII regions, bipolar outflows and visible Pre-Main-Sequence stars (T Tauri, FU Orionis, and Herbig Ae/Be stars). With the improved resolution now available, the clouds start to resolve into individual cores, signalling the presence of embedded young stars. Many of these cores are also surrounded by asymmetric dust distribution, suggesting that disks or strong density gradient regions surround young stars. This is even more obvious when bipolar outflows are considered. The massive ones are surrounded by extended relatively hot disks, while the low luminosity sources generally show an unresolved dense core and an extended circumstellar disk, perpendicular to the the direction of the outflow. The PMS stars are generally unresolved. For T Tauri stars we derive a dust-emissivity index, $\beta \leq 1$, which suggests that density and temperature gradients cannot be neglected in the analysis, or that the stars are surrounded by unusually large dust grains.

1. INTRODUCTION

This talk presents results in the mm/sub-mm continuum obtained with JCMT[1] and its common user bolometer receiver, UKT14. UKT14 is discussed in detail by Duncan et al. (1988). The improved resolution that can be achieved by the new generation of sub-mm telescopes make continuum observations attractive. The interpretion and analysis of dust emission is easier than that of spectral lines, because in the sub-mm regime one can normally assume that the dust is optically thin, while molecular lines are typically affected by radiative transfer, excitation and chemical effects. However, it should be pointed out that even though the

---

[1] The James Clerk Maxwell Telescope is operated on a joint basis between the United Kingdom Science and Engineering Research Council (SERC), the Netherlands Organisation for the Advancement of Pure Research (ZWO), the Canadian National Research Council (NRC), and the University of Hawaii (UH).

interpretion of dust emission appears to be relatively straight forward, there are still many unknowns, like the size and composition of dust grains and the dust-to-gas ratio. These topics are discussed e.g. in the reviews by Hildebrand (1983) and Emerson (1988).

2. MOLECULAR CLOUDS AND COMPACT HII REGIONS

Giant molecular clouds and HII regions are usually associated with massive star-formation, and are very bright sub-mm sources. Most of them are rather distant, and previous continuum work at 1mm or below has had rather course spatial resolution, HPBW $\geq$ 20" (see e.g. Westbrook et al., 1976; Gordon and Jewell, 1987; Keene et al., 1982). Some recent work on the IRAM 30m (Chini, 1988) has been carried out at 1.3mm with 11" resolution showing structure previously not seen. Otherwise the field has been dominated by millimeter aperture synthesis telescopes like the Berkeley array and the Owens Valley Radio Interferometer (OVRO) typically operating at wavelengths about 3mm, although OVRO has recently also started observations at 1.4mm (Mundy, 1988). These interferometers have superior resolution compared to any single dish telescope, but are only sensitive to compact structures. With telescopes like JCMT capable of working down to 350μm, 6" resolution can now be achieved in the sub-mm region. This is similar to resolutions typically used in the mid-infrared, and better than what can currently be achieved in the far-infrared.

A sample of JCMT observations of GMC:s are presented below. Richardson et al. (1988a) mapped W3 at 1.1mm, 800μm and 350μm at intermediate resolutions, HPBW ~ 19"-16". They find that IRS5 becomes the dominating source at 350μm, and that the western core appears to be elongated North-South, possibly indicating two separate sources. Richardson et al. (1988b) mapped the DR21 & DR21(OH) complex at 1.1mm and 800μm. DR21(OH) is found to be very deeply embedded and appears to be optically thick at 350μm. With a resolution of 19" DR21(OH) breaks up into four distinct components: DR21(OH), DR21(OH)SW, DR21(OH)S, and DR21(OH)N, suggesting that a cluster of massive young stars have recently formed in this region. The first three sources have also been detected by Mangum et al. (1988) using the OVRO interferometer in the 2.7mm continuum and in the J=1-0 $C^{18}O$ emission. Mangum et al. cite additional OVRO observations that resolve DR21(OH) into a binary, with the two components separated by ~ 9".

The whole OMC-1 complex has been mapped by Gear et al. (1988) on JCMT at 800μm with a 13.5" beam and a 60" region around IRc2 at 450μm with a 7.5" beam, while Lis et at. (1988) have mapped the SgrB2 region in several sub-mm bands. These studies demonstrate what one can easily do with a telescope like JCMT.

3. BIPOLAR FLOWS

There is not much of a difference between studying molecular clouds or bipolar flows, because mass outflow is an essential part in the early evolution of young stars. The bipolar outflows discussed below are with one exception nearby low-luminosity stars. These have been selected by rather stringent criteria: 1) clear bipolar structure, 2) the outflow close to the plane of the sky, and 3) evidence for disks from molecular line observations or polarization studies. Since one of the main objectives is to search for disks, we maximize our chances by looking at nearby objects and edge-on disks. This strategy has proven to be very effective, because most of the objects we studied so far have revealed clear disks or disk like structures (see e.g. Dent et al., 1988; Sandell et al. 1988a, Chandler et al., 1988).

The mapping of sub-mm emission done at JCMT of nearby bipolar flows (16293-2224, SSV13, B335, NGC2071, LkHα234 etc) all give a quite clear consistent view. The star driving the flow is generally unresolved and surrounded by extended emission perpendicular to the flow direction. In several cases one may also see faint emission in the flow direction. This emission is at low levels and still needs to be confirmed. If present, the likely cause is either hot dust associated with the flow itself or emission from the cavity walls. This means that continuum mapping may become a powerful tool to study the morphology of outflows. Although the central source is still unresolved at 13.5" resolution (i.e. the achievable resolution at 800μm), increased resolution seems to reveal the same symmetry as seen on large scales, i.e. "interstellar" disks. The core, here assumed to be an accretion and/or collimating disk, is found to be quite massive, one to a few solar masses.

Again, only a few of the results will be discussed in more detail: 16293-2422, SSV13 and B335.

16293-2422: This is a low-luminosity (L ~ 27 $L_\odot$), not very prominent bipolar outflow in Ophiuchus (d = 160pc). It has, however, some unique features, which warrant a further study. Mundy et al. (1986), using the OVRO interferometer found evidence for a compact rotating disk, while Mundy et al.(1988) using the VLA found an even more extended ammonia disk. Sandell et al. (1988b) mapped this source at 1.1mm, 800 and 450μm. At 1.1mm they find an elliptical, but still unresolved core with a total mass of ~ 1 $M_\odot$, very similar to the rotating ^{13}CO disk by Mundy et al. (1986). This disk is very likely a circumstellar accretion disk. At 800 and 450μm the basic shape is still elliptical, but with the increased resolution it splits up into two components, a cross , most clearly seen towards NW in the 450μm image (at p.a. 0º and 310º). In addition the 1.1mm image also reveals a more extended faint N-S ridge (interstellar disk ?), which, provided it has the same temperature as the core (T_d ~ 30K), has a mass of .1 -.2 $M_\odot$. However, if it is substantially cooler, it may be more massive.

SSV13 This is a well studied low luminosity (L ~ 48 $L_\odot$) outflow source in NGC1333 (d ~ 350pc) associated with the chain of Herbig-Haro objects 7-11 and a spectacular high velocity CO flow. Sandell et al. (1988a) have mapped the source at 1.1mm and 800μm and find a an unresolved source coinciding with SSV13 as well as a ridge perpendicular to the outflow. Some of this ridge could be due contribution from other sources. $H_2O(B)$, the maser source and compact HII region (Snell and Bally, 1986) lies on this ridge and so does the "protostellar" source SV13b, detected 15" SW of SSV13 by Grossman et al. (1987) using the OVRO interferometer at 3mm. However, the failure to detect this source at 1.4mm (Mundy, 1988), makes its existence more doubtful. However, there is definitely a disklike structure surrounding SSV13. In addition to the compact emission surrounding SSV13 and the more extended circumstellar disk (D ~ 40" - 60"), the 800µm map also show an arc of emission to the SE curving towards HH7. This emission coincides quite well with the blueshifted arc of CO emission seen by Grossman et al., and is most probably dense material associated with the outflow cavity. The mass associated with the unresolved SSV13 emission is 0.2 $M_\odot$.

B335: This is a large well collimated outflow in the plane of the sky emanating from the core of the Bok globule B335. Due to very high extinction in the core region, the star driving the flow is not seen even in the near-IR. The luminosity of the star is very low, ~2.9 $L_\odot$, if the distance is assumed to be 250pc. Chandler et al. (1988) has mapped the core at 800µm (HPBW ~ 17", total map extent 2'x2'), and at 450µm with a diffraction limited beam (HPBW ~ 7.9", map size 20"x20"). The 800µm map shows an unresolved core surrounded by faint emission extending N-S, i.e. perpendicular to the direction of the outflow. There appears to be some emission in the outflow direction, but this needs to be verified. At 450 m the source size is $\leq$ 7", but the core is clearly surrounded with fainter extended emission. There is a narrow N-S ridge (the collimating disk?) as well as a tongue of emission to the West. The latter feature may be associated with the redshifted flow or the funnel surrounding it. A least squares fit to the whole mm- and sub-mm spectrum indicates that the dust emission may be marginally optically thick at 350µm (τ ~ 1) and cold (T_k ~ 15 K). The total core mass is 1.9 $M_\odot$, while the mass of the extended 800µm "interstellar" disk is ~ 10 $M_\odot$.

4. PRE-MAIN-SEQUUENCE STARS.

The presence of dusty disks or envelopes around PMS-stars has been predicted or confirmed by many different observational techniques: IR excesses, near-IR absorption features, near-IR speckle, polarization studies, highly collimated jets, and interferometry observations of dust continuum at 2.7mm (see e.g. Weintraub et al., 1988 and references therein). The 2.7mm continuum work at OVRO have both high spatial resolution and high sensitivity, but has so far only covered a few a the brightest stars. While JCMT cannot really match the spatial resolution of OVRO, the large spectral coverage: 2mm - 350 µm with filters in all the major atmospheric windows, makes it ideally suited for detailed studies of the protoplanetary disks around PMS objects. Below are briefly summarized the results of two such studies.

T Tauri and FU Orionis stars: Weintraub et al. (1988) report millimeter and submillimeter detections of 12 low mass PMS stars: 9 T Tauri stars, 1 SU Auriga star, and 2 FU Orionis stars. Many of these stars are first detections at mm/submm wavelengths. The submm spectra of T Tauri stars are found to be less steep than what would be expected from thermal dust emission. The dust emissivity index, ß, is found to range between 0 - 1.5 for the T Tauri stars, while ß is ~1 for the FU Ori stars, i.e more similar to the ß-index range 1-2 found for dust associated with molecular clouds and compact HII regions. The low values for β indicate that density and temperature gradients cannot be neglected for the T Tauri stars or that the grains are unusually large in the disks surrounding T Tauri stars. Three stars (T Tau, HL Tau, and DG Tau) were found to be extended on the 1000-2000 AU level. The cloud masses are $\sim 10^{-2} M_\odot$ for the T Tauri stars and $\geq 10^{-1} M_\odot$ for the FU Ori stars.

Herbig Ae/Be and peculiar PMS-stars: Herbig Ae/Be stars are the more massive equivalents to T Tauri stars. In the sample of 11 stars observed by Sandell and Duncan (1988), most of the stars were found to be associated with extended emission. If one only takes the stars where the circumstellar emission can be accurately determined (LkHα198, LkHα234, PV Cep, AB Aur, and R Mon), a preliminary analysis of the data indicate that β ≥ 1.0. The possible exception is LkHα198, for which β = 0.6 ± 0.3. This star was also mapped by Sandell and Duncan, who found it to be elongated at 800μm. The masses of the circumstellar material associated with these stars span a much larger range than the T Tauri stars, i.e. from $\sim 0.01\ M_\odot$ (AB Aur) to $\sim 10\ M_\odot$ (LkHα234), with perhaps the majority in the range of $0.1 - 1.0\ M_\odot$.

References:
Chandler, C.J., Gear, W.K., Sandell, G., Duncan, W.D., Hayashi, S., Griffin, M.J., Hazell, A.S.: 1988, MNRAS, submitted
Chini, R.: 1988, this volume
Dent, W.R.F., Sandell, G., Duncan, W.D., Robson, E.I.: 1988, MNRAS,subm.
Duncan, W.D., Robson, E.I., Ade, P.A.R., Griffin, M.J., Sandell, G.: 1988, MNRAS, to be submitted
Emerson, J.: 1988 in "Formation and Evolution of Low Mass Stars", Proc. of the NATO Adv. Study Inst., Eds. A.K. Dupree and M.T.V.T. Lago
Gear, W.K., Robson, E.I., Sandell, G., Duncan, W.D.: 1988, in prep.
Gordon, M.A., Jewell, P.R. : 1987, Ap. J. **323**, 766
Grossman, E.N., Masson, C.R., Sargent, A.I., Scoville, N.Z., Scott, S., Woody, P.G.: 1987, Ap. J. **320**, 356
Hildebrand, R.H.: 1983, Quart. J. R. Astr. Soc. **14**, 267
Keene, J., Hildebrand, R.H., Whitcomb, S.E.: 1982, Ap.J. **252**, L11
Lis, D.C., Goldsmith, P.F., Hills, R.E.: 1988, this volume
Mangum, J.G., Wootten, A., Mundy, L.G.: 1988, this volume
Mundy, L.G.: 1988, this volume
Mundy, L.G., Wilking, B.A., Myers, S.T.: 1986, Astrophys. J. **311**, L75
Mundy, L.G., Wootten, H.A., Wilking, B.A.: 1988, in preparation
Richardson, K.J., Sandell. G., White, G.J., Duncan, W.D., Krisciunas, K. 1988a, A&A., in press
Richardson, K.J., Sandell, G., Krisciunas, K.: 1988b, A&A, submitted
Sandell, G., Aspin, C., Duncan, W.D., Robson, E.I., Dent, W.R.F.: 1988a, A&A., to be submitted
Sandell, G., et al.: 1988b, in preparation
Sandell, G., Duncan, W.D.: 1988, in preparation
Snell, R. L., Bally, J.: 1986, Ap. J. **303**, 683
Weintraub, D.A., Sandell, G., Duncan, W.D.: 1988, Ap. J. (Letters),subm.
Westbrook, W.E., Werner, M.W., Elias, J.H., Gezari, D.Y., Hauser, M.G., Lo, K.Y., and Neugebauer, G.: 1976, Ap. J. **209**, 94

SUBMILLIMETER EMISSION FROM SMALL DUST GRAINS ORBITING NEARBY STARS

E. E. BECKLIN
Institute for Astronomy, University of Hawaii
2680 Woodlawn Drive
Honolulu, Hawaii 96822 USA

B. ZUCKERMAN
Department of Astronomy
University of California, Los Angeles
Los Angeles, California 90024-1562 USA

The most exciting discovery of the Infrared Astronomical Satellite (IRAS) was intense far-infrared radiation from various nearby main-sequence stars. This radiation, which is produced by numerous dust particles, is suggestive of solar-systemlike objects in orbit about the stars. We have detected and measured, at submillimeter wavelengths, properties of the particulate clouds around Vega, Fomalhaut, and Beta Pictoris. These measurements yield accurate estimates of particle size, spatial distribution, and minimum total mass for these three dusty clouds. In particular, the particles appear to be sufficiently small that they are probably being supplied continually by a very substantial population of much larger objects, perhaps comets or asteroids. The widespread occurrence of excess far-infrared emission from nearby stars (Backman and Gillett 1987; Aumann 1988) suggests, therefore, that a majority of single main-sequence stars are surrounded by vast clouds of comets and/or asteroids. A limited map of the submillimeter emission toward Fomalhaut is consistent with the possibility that the radiating particles are substantially depleted near the star, which could be due either to destruction of the grains or accumulation into larger objects such as planets.

The IRAS measurements at 12, 25, 60, and 100 μm were hampered by poor spatial resolution so that deduced properties of the particulate clouds were rather model-dependent (Gillett 1985; Backman and Gillett 1987). In the case of β Pic, additional information at much better spatial resolution is available from ground-based optical (Smith and Terrile 1984; Gradie et al. 1987; Artymowicz et al. 1989; Lagrange-Henri et al. 1988) and mid-infrared (Backman et al. 1989; Telesco et al. 1988) observations so that detailed modeling has been attempted (Artymowicz et al. 1989; Backman et al. 1989). Vega has been detected at far-infrared wavelengths from NASA's Kuiper Airborne Observatory (Harvey et al. 1984; Harper et al. 1984).

In the present paper we report observations, primarily at 800 μm wavelength, obtained with the 15 m James Clerk Maxwell Telescope (JCMT) at Mauna Kea Observatory in Hawaii. The JCMT was equipped with a cooled liquid ^{3}He bolometer and filters with effective wavelengths of 800 and 450 μm. The focal plane aperture was 65 mm at f/35, which provided a beam size, measured using the planet Uranus, of 16''.5 and 20'', full width at half-maximum power (FWHM), at 800 and 450 μm, respectively. Data were obtained by chopping the telescope secondary mirror at 7.8 Hz between a star and a nearby reference position 40'' away in azimuth, while the telescope was nodded every ten seconds, so that the star appeared first in one beam and then in the other. Uranus and Mars were used as flux calibration sources. The atmosphere was measured to be stable during each night of our August 1988 run so that rather reliable estimates of flux density could be derived for each of the six program stars in Table 1. Therefore, we believe that calibration errors are less than the statistical errors. The listed errors are entirely statistical and represent one standard error of the mean. We are confident that major systematic errors are not large in comparison with statistical errors, since the three stars in our sample that have the weakest IRAS fluxes (ϵ Eri, τ Eri, and HR 451) are not detected at 800 μm, whereas the three strongest IRAS sources are all detected.

TABLE 1
OBSERVATIONAL DATA

Star	D (pc)	λ (μm)	Offsets Relative to Star α	δ	F_ν (mJy)	$M_d(M_\leftmoon)$*
Vega (α Lyr)	8.1	800	0''	0''	21.5 ± 5.4	0.1
β Pic	16.6	800	0''	0''	80 ± 14	0.5†
						2‡
		800	7''.4 E	15'' N	11.7 ± 17	
		800	7''.4 W	15'' S	40 ± 26	
Fomalhaut (α PsA)	6.7	800	0''	0''	35 ± 6.5	0.1
		450	0''	0''	511 ± 178	
		800	8''.2 W	14''.2 N	40 ± 11	0.2
		800	14''.2 E	8''.2 N	23.6 ± 12	0.1
ε Eri	3.3	800	0''	0''	7.7 ± 7.7	≲0.01
τ_1 Eri	13.7	800	0''	0''	−7.7 ± 7.3	
HR 451	51	800	0''	0''	0 ± 12	

*Minimum mass of dust particles (in Moon masses, $M_\leftmoon = 7 \times 10^{25}$ g) necessary to account for the flux observed at 800 μm.
†Mass in small grains ($a \lesssim 10$ μm), see text.
‡Mass in large grains ($a \sim 1$ mm), see text.

As indicated in Table 1, we acquired 800 μm data at a few positions offset by one full beam width (16'') from β Pic and from Fomalhaut. The two positions near β Pic lie on the disk detected in scattered optical light (Smith and Terrile 1984). In neither position did we detect statistically significant flux, although if we average the fluxes from the two positions, the result is positive at nearly the 2 σ level. Substantial flux was detected offset from Fomalhaut. We will discuss these data in more detail.

In Figure 1 we present the far-infrared to submillimeter spectra for four of the six program stars over the region whence the 60 μm IRAS flux originates. For both Vega and Fomalhaut, the upper of the two plotted 800 μm points is the 800 μm flux that we estimate would be measured were we to integrate over the region of 60 μm emission. This estimate is based on the measured IRAS source sizes (Gillett 1985; Backman 1988), and in the case of Fomalhaut, our limited mapping data at 800 μm. Because information on particle size and composition can be deduced from the shape of the low-frequency tail of the Planck curve, we have plotted in Figure 1 blackbody fits to the IRAS fluxes. In each of the four stars the long wavelength extrapolations of these blackbodies pass above our best estimate of the total 800 μm flux. This is particularly convincing for β Pic and ϵ Eri, where the geometry and source size are not in question. Also, for Vega, our 800 μm results are consistent with extrapolation of the 193 μm flux measured by Harper et al. (1984).

The decline of the measured flux below the blackbody curve at long wavelengths is most naturally explained as decreased grain emissivity resulting from the emission wavelength being much larger than the size of the dust particles. We therefore conclude that the grains observed around stars typically have radii $a \lesssim \lambda/2\pi$ where $\lambda \sim 200$ μm. Compared with previous models for Vega, this upper limit size is comparable to that estimated by Harper et al. (1984) but is much smaller than the millimeter-size particles preferred by Gillett and Backman (1987) and Gillett (1985). For Fomalhaut, our grain-size estimate is comparable to their more recent estimate (Backman and Gillett 1987). For β Pic, previous investigators have deduced that grain sizes are much smaller than 1 mm, based on the spatial extent and temperature of the grains that produce the observed infrared emission.

One millimeter is a particularly significant size since larger grains will survive the Poynting-Robertson drag force for the estimated ages of the stars (about 10^9 yr) at their current distances, approximately 100 AU, from the stars. However, since the 800 μm fluxes imply that, typically, $a \ll 1$ mm, a different model appears to be required. The most plausible model is one in which the observed small particles are continually supplied by much larger bodies, perhaps comets or asteroids (Harper et al. 1984; Weissman 1984), and then lost as they spiral into the star under the influence of the Poynting-Robertson effect. An alternative model, whereby the observed grains are primordial, i.e., coeval with the stars, were formed at distances about 1000 AU, and have spiraled into 100 AU during the past 10^9 yr, appears less plausible. For any reasonable initial distribution of particle size and separation from a central star, one would, at the present time, expect a wide range of particle temperatures leading to an infrared emission distribution that would be significantly wider, not narrower, than a blackbody spectrum. Also, evidence exists (Backman et al. 1989; Telesco et al. 1989) for substantial numbers of very small, $a \sim 1$ μm, particles near β Pic. Such particles must have been produced in the recent past. Therefore, although the case is not yet ironclad, the most likely source of the infrared radiating particles near Vega-like stars appears to be large parent objects similar to comets and/or asteroids.

A lower limit to the size of the dominant radiating particles may be estimated by considering the temperature and the extent of the region that produces the 60 μm IRAS emission. Typical values for Vega and Fomalhaut are 75 K and 20″ diameter (FWHM) (Gillett 1985; Harper et al. 1984), which are consistent with each other only if the particles, like blackbodies, radiate at 60 μm (i.e., $a \gtrsim \lambda/2\pi = 10$ μm). If the particles were much smaller than this, they would then have difficulty reradiating the stellar energy that they absorb at optical and ultraviolet wavelengths and, 10″ from the stars, they would be much hotter than 75 K. Therefore, in agreement with the estimate of Harper et al. (1984)

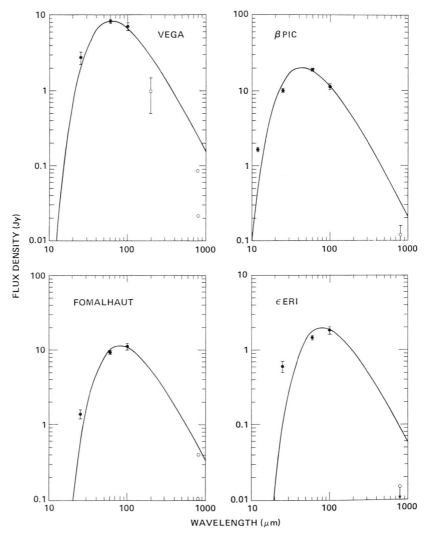

Fig. 1. Infrared spectra of four Vega-like stars. The ordinate is flux density measured in Janskys; the abscissa is wavelength in microns. The filled data points are from IRAS (Gillett 1985) and depict excess emission above that expected from the stellar photospheres. The solid curves are blackbodies that roughly fit the IRAS data. The open circles at 800 μm are based on data from the JCMT reported in Table 1. At 800 μm, emission from the stellar photospheres is expected to be a few orders of magnitude smaller than the plotted flux densities. For β Pic the plotted point is our estimate of the total 800 μm flux within the region producing 60 μm emission and is based on three measurements along the disk. For Vega and Fomalhaut, we plot two different values for the 800 μm flux. The lower point for each star is the minimum 800 μm flux observed. The upper points are our best estimates of the 800 μm flux that is likely to originate from the region that produces the 60 μm radiation measured by IRAS (Gillett 1985; Backman 1988). The open circle data point for Vega near 200 μm is from Harper et al. (1984).

or Vega, the dominant radiating particles probably have sizes that lie within the range $3 \lesssim a \lesssim 100$ µm for Vega and Fomalhaut. These particles are much larger than interstellar grains, somewhat larger than particles ($a \sim 1$ µm) in the comas of solar system comets that dominate the observed scattered sunlight and thermal emission (Hanner 1980), and comparable in size to the grains that dominate the zodiacal emission in our solar system (Reach 1988). There appears to be general, though perhaps not universal, agreement that comets are the primary source of the dust grains in the zodiacal cloud.

For β Pic the situation appears somewhat different. The measured IRAS size (Gillett 1985) is consistent with the infrared color temperatures only if the grains are inefficient radiators at $\lambda \gtrsim 60$ µm. This suggests dominant particle radii of order a few microns or smaller, consistent with the best size estimates ($a \sim 1$ µm) from recent ground-based mid-infrared observations (Backman et al. 1989; Telesco et al. 1988).

In a steady-state model where particles with $a \ll 1$ mm spiral in due to Poynting-Robertson drag, the radial density distribution $\rho(r)$ will be inversely proportional to r, the distance between the grain and the star. If the main region of particle production is fairly close to the star, as would be likely if sublimation and/or parent body collisions are important, then outside this region $\rho(r)$ would drop much more rapidly than r^{-1}. Previous models (Smith and Terrile 1984; Artymowicz et al. 1989) of β Pic suggest that $\rho(r) \propto r^{-3}$ beyond 100 AU. The two off-center points in Table 1 each lie 265 AU from β Pic along the direction of the optical disk (Smith and Terrile 1984; Gradie et al. 1987; Artymowicz et al. 1989). The apparent rapid decline in the submillimeter flux at this offset from β Pic is more consistent with the model of $\rho(r) \propto r^{-3}$ than $\rho(r) \propto r^{-1}$.

Because of our limited mapping data, we cannot yet precisely define a value of the dust emissivity index, p, where the dust emissivity at $\lambda \gg a \propto \lambda^{-p}$. For Vega and Fomalhaut, the data in Figure 1 and Table 1 suggest that $p \lesssim 1$ between 100 and 800 µm. For ϵ Eri, $p \gtrsim 1$ between 100 and 800 µm. The situation near β Pic may be more complicated. As we mentioned above, the IRAS data are explained by small grains such that p is already $\gtrsim 1$ near 60 µm. If so, then there appears to be some emission at 800 µm at β Pic that is in excess of that expected from these small grains. Therefore, the 800 µm flux may be produced, in part, by a population of large grains ($a \sim 1$ mm).

The submillimeter is a favorable spectral region for measuring the total mass of dust in a cloud (Hildebrand 1983; Sopka et al. 1985) because optical depths are small and, for dust temperatures of interest, one is on the Rayleigh-Jeans portion of the blackbody emission curve. Therefore, derived dust masses, M_d, depend only inversely on the uncertain dust temperature, T, as may be seen from the following equation from Hildebrand (1983):

$$M_d = \frac{K F_\nu D^2}{B(\nu, T)} (\lambda/0.025)^p \qquad (1)$$

Here F_ν and D, given in Table 1, are the measured submillimeter flux from and distance to the star in question, B is the blackbody intensity equal to $2kT/\lambda^2$ in the Rayleigh-Jeans limit, and all quantities are in cgs units. The numerical coefficient, $K = 0.05$, in units of g cm^{-2}, is our best estimate of one over the opacity per gram of dust grains at a wavelength of 250 µm based on Hildebrand (1983), Sopka et al. (1985), and Jura (1988). Inserting appropriate values for the various quantities, we derive the minimum masses of dust grains, given in the right-hand column of Table 1, that are required to account for the measured 800 µm fluxes. Equation (1) is valid in the limit that $a \ll \lambda$. Because Gillett (1985) assumed larger particle sizes, he derived dust masses that were typically one order of magnitude larger than those given in Table 1, although contributions to the 800 μm flux from positions off Fomalhaut and possibly Vega that we did not have time to observe would somewhat diminish this discrepancy. The estimate of minimum dust mass for β Pic listed in the right-hand column of Table 1 assumes a simple model where one-half the 80 mJy detected at the (0, 0) position is due to small grains ($a \lesssim 10$ μm) and one-half is due to large grains ($a \sim 1$ mm). In any event, the total minimum mass of dust particles required to account for the observed fluxes is, in each case, less than or of order the mass of the Moon. But, as we have argued before, a much larger mass of much larger parent objects is likely also to be present.

Without additional mapping data, we cannot yet address the question of whether there is a substantial population of large dust grains far from Fomalhaut that are too cold to radiate effectively at $\lambda \lesssim 100$ μm. Such particles would not have been detected by IRAS. The two offset positions in Table 1 lie along (8.''2 W, 14.''2 N) and perpendicular to (14.''2 E, 8.''2 N), the major axis of the dust emission as determined by IRAS (Gillett 1985). The orientation suggested by the 800 μm data is, therefore, in agreement with that determined by IRAS, but it would be premature to overinterpret this result because only two offset positions were measured. The strength of the emission at (8.''2 W, 14.''2 N) relative to (0'', 0'') indicates that the column density of dust grains along lines of sight passing near the star is significantly less than along some lines of sight 16'' from Fomalhaut. This lends qualitative support to the contention (Backman and Gillett 1987) that the dust is substantially depleted within a radius of 4'' of Fomalhaut due either to destruction of the grains or accumulation into larger objects such as planets.

Ending on a somewhat speculative note, we remark on a few potential implications of these results. Marochnik et al. (1988) have recently argued that the Sun is surrounded by an enormous retinue of comets that contains a total mass and angular momentum that is an order of magnitude greater than that carried by the nine planets in the solar system. Based on the infrared luminosity of the known zodiacal dust cloud in the inner solar system and the infrared luminosity of the Vega-like stars (Backman and Gillett 1987), the total mass of dust particles near a Vega-like star is likely to be one to four orders of magnitude greater than the mass of dust near the Sun. Therefore, even allowing for the possible existence (Backman and Gillett 1987) of an as-yet-undiscovered particulate cloud beyond Pluto, it appears entirely reasonable to suppose that the Vega-like phenomenon implies a mass of comets at least equal to that in our solar system. It has been argued (Backman and Gillett 1987; Aumann 1988) that Vega-like particulate clouds are "the rule, rather than the exception" near single stars. If so, then comet clouds could be a significant reservoir of angular momentum that would help to alleviate the classical angular momentum problem in the formation of single stars from the collapse of rotating interstellar molecular clouds. In addition, models for the formation of planets often proceed in a two-step fashion. First, tiny interstellar grains clump together to form planetesimals of order the size, about 10 km, of comets and asteroids. Then the planetesimals accumulate to form planets. The present results, in conjunction with those of Backman and Gillett (1987), Aumann (1988), and Marochnik et al. (1988) suggest that at least the first step proceeds in the vicinity of a substantial fraction of all solarlike stars.

We thank Drs. Saeko Hayashi and Goeran Sandell of the JCMT staff for assistance, Mr. Robert L. O'Daniel and Louise Good for editing and typing the manuscript, and Drs. Peter Goldreich and Dana Backman for comments. This research was partially supported by grants from NASA and the NSF to the University of Hawaii and from the NSF (AST 87-17872) to UCLA.

References

Artymowicz, P., Burrows, C., and Paresce, F.: 1989, *Astrophys. J.*, in press
Aumann, H. H.: 1985, *Astron. J.* **96**, 1415–1419.
Backman, D. E.: 1988, private communication.
Backman, D. E., and Gillett, F. C.: 1987, in J. L. Linsky, and R. E. Stencel (eds.), *Cool Stars, Stellar Systems, and the Sun*, Springer-Verlag, Berlin, pp. 340–349.
Backman, D. E., Gillett, F. C., and Witteborn, F. C.: 1989, *Astrophys. J.*, in press.
Gillett, F. C.: 1985, in F. P. Israel (ed.), *Light in Dark Matter*, Reidel, Dordrecht, pp. 61–69.
Gradie, J., Hayashi, J., Zuckerman, B., Epps, H., and Howell, R.: 1989, in *Proceedings of the 18th Lunar and Planetary Conference*, Part 1, Lunar and Planetary Institute, Houston, pp. 351–352.
Hanner, M. S.: 1988, in I. Halliday and B. A. McIntosh, (eds.), *Solid Particles in the Solar*, Vol. 90 in the series IAU Symposiums, Reidel, Dordrecht, pp. 223–235.
Harper, D. A., Loewenstein, R. F., and Davidson, J. A.: 1984, *Astrophys. J.* **285**, 808–812.
Harvey, P. M., Wilking, B. A., and Joy, M.: 1984, *Nature* **307**, 441-442.
Hildebrand, R. H.: 1983, *Quar. J. Roy. Astron. Soc.* **24**, 267–282.
Jura, M. 1988, private communication.
Lagrange-Henri, A. M., Vidal-Madjar, A., and Ferlet, R.: 1988, *Astron. Astrophys.* **190**, 275–282.
Marochnik, L. S., Mukhin, L. M., and Sagdeev, R. Z.: 1988, *Science* **242**, 547–550
Reach, W. T.: 1988, *Astrophys. J.* **335**, 468–485.
Smith, B. A., and Terrile, R. J.: 1984, *Science* **226**, 1421–1424.
Sopka, R. J., et al.: 1985, *Astrophys. J.* **294**, 242-255.
Telesco, C. M., Becklin, E. E., Wolstencroft, R. D., and Decker, R.: 1988, *Nature* **335**, 51–53.
Weissman, P.R.: 1984, *Science* **224**, 987–988.

RESULTS FROM 1.4 MILLIMETER WAVELENGTH INTERFEROMETRY AT THE OWENS VALLEY

L. G. Mundy (Astronomy Program, Univ. of Maryland, College Park, MD), A. I. Sargent, N. Z. Scoville, S. Padin, and D. P. Woody (OVRO, Caltech, Pasadena, CA)

1. Introduction

This paper summarizes the first interferometric results in the 1.4 mm band from the Owens Valley Radio Observatory (OVRO) Millimeter Wave Array. Relative to 2.7 mm, observations of thermal dust and molecular emission at 1.4 mm benefit from increased emissivity and brightness temperature sensitivity. For optically thin emission in the Rayleigh-Jeans limit, dust emission increases as λ^{-3} to λ^{-4} assuming a λ^{-1} to λ^{-2} wavelength dependence for the dust emissivity, while molecular emission increases faster than λ^{-2} due to the Planck function and the larger statistical weights of higher J levels. In the optically thick limit, both increase as λ^{-2}. This λ^{-2} advantage can also be expressed in terms of the brightness temperature sensitivity; for a given beam size, the rms brightness temperature corresponding to a specific rms flux level is a factor of 4 lower for the 1.4 mm band than for the 2.7 mm band.

2. Instrument and Observations

The OVRO Array consists of three 10.4 m antennas (34" primary beam at 220 GHz) equipped with 1.4/2.7 mm dual channel SIS receivers (100-200 K DSB at 220 GHz). Data presented here were acquired in March and April 1988 at λ = 1.366mm, the $C^{18}O$ J=2-1 transition frequency; atmospheric optical depths during the observations varied from 0.2 to 0.5. Maps with 3-6" resolution were made of six sources (HH 7-11, L1551 IRS5, DR 21(OH), NGC 7538 IRS1, Arp 220, and NGC 3690) and "snapshot" flux measurements were made of an additional 15 sources. These results mark the commissioning of the OVRO array at 1.3 mm.

3. Results

Continuum flux at the $\geq$ 400 mJy level was detected from all sources, except NGC 3690 ($\leq$ 120 mJy). In the mapped objects, the emission was found to arise from unresolved sources ($\leq$ 4"). Based on the rate of increase in flux from centimeter radio wavelengths, emission from thermal dust is dominant in more than 70% of the 21 objects surveyed. In two galatic sources L1551 IRS5 and DR 21(OH), $C^{18}O$ J=2-1 emission was detected easily and mapped. In the DR 21(OH) region, two compact sources of 1.4 mm line and continuum emission separated by 7" (0.1 pc) were found. These coincide with two H_2O maser centers (Gensel and Downes 1977), one of which is also coincident with the OH maser position itself (Norris et al. 1982) The virial mass estimate for the region (based on the ~ 4 km s^{-1} velocity difference between the sources) and those based on $C^{18}O$ integrated intensity and dust continuum flux are all in the range 100-130 $M_\odot$. These compact millimeter wavelength sources are clearly at the center of activity in the DR 21(OH) region; a forming binary or multiple-star system with two 15$M_\odot$ stars, or many $\geq$ 6$M_\odot$ stars could produce the observed luminosity.

An unresolved ($\leq$ 3.5") continuum source was also detected in Arp 220 coincident with the radio and near-IR peak (Neugebauer et al. 1987. Our 1.4 mm flux, 140 ± 20 mJy, is ~ 50% of that found in an 11" beam (Chini, this meeting). Using the standard dust to gas mass conversion (Hildebrand 1983), the mass inside 3.5" (650 pc) radius is 3×10^9 $M_\odot$. This implies a mean volume density of 50-100 H_2 cm^{-3}, similar to that within giant molecular clouds in the Galaxy.

For the 10 objects with little or no centimeter wavelength radio emission and compact millimeter emission, the ratio of 1.4 to 2.7 mm flux averages 7.8 ± 1.1, yielding a spectral index ($\lambda^{-\alpha}$) α = 3.0 ± 0.2 for the observed dust emission. Assuming this emission is optically thin and $h\nu \gg kT_{dust}$, the resulting dust emissivity is proportional $\lambda^{-1.0 \pm 0.2}$, rather than λ^{-2} as generally expected for both crystalline and amorphous carbon materials. Grains with two dimensional structure, such as layered silicates or amorphous carbon (Tielens and Almondola 1986) or fluffy grains (Wright 1987) can exhibit this shallower wavelength dependence. Alternatively, it is possible that the emission is partially optically thick at λ =1.4 mm; in this case the average source size for our sample must be $\leq$ 0.7" and mass estimates based on the continuum flux would increase by a factor of ~ 10, invalidating the rough agreement between the various mass estimates for DR 21(OH).

More complete presentation of these 1.4 mm line and continuum results are given by Woody et al. (1988) and Padin et al. (1988).

4. References

Genzel, R., and Downes, D. 1977, Astr. Ap. Suppl., **30**, 145.
Hildebrand, R. H. 1983, Quart.J.R.A.S., **24**, 267.
Norris, R. P., Booth, R. S., Diamond, P. J., and Porter, N. D. 1982, M.N.R.A.S., **201**, 191.
Neugebauer et al., 1987, A.J., **93**, 1057.
Tielens, A.G.G.M., and Allamandola, L. J. 1987, in Interstellar Processes, ed. D. J. Hollenbach and H. A. Thornson (Riedel: Dordrecht, Holland), p. 397.
Padin et al. 1988, Ap.J., accepted.
Wright, E. L. 1987, Ap.J., **320**, 818.
Woody et al. 1988, Ap.J., accepted.

A SURVEY OF GALACTIC SOURCES FOR SUBMILLIMETER EMISSION FROM HIGH EXCITATION MOLECULES

Neal J. Evans II, Harold M. Butner, Shudong Zhou, Charles E. Mayer
John E. Howe, William F. Wall, and Steven C. Laycock
Department of Astronomy and Electrical Engineering Research
Laboratory, The University of Texas at Austin

ABSTRACT: We report the results of a survey of galactic sources for submillimeter emission from the $J = 4 \rightarrow 3$ lines of HCN and HCO$^+$ and the $J = 7 \rightarrow 6$ line of CS. The purpose of the survey was to determine how common these lines would be in regions of star formation and, by extension, how useful submillimeter spectroscopy can be in determining the physical conditions in molecular clouds. For this purpose, we selected sources known to be regions of active star formation.

The survey was done in January 1988 with the 5-m antenna of the Millimeter Wave Observatory (MWO) and a cooled Schottky receiver. The telescope was equipped with an error-correcting secondary, which greatly improved its performance at these wavelengths. The aperture efficiency improved from 0.08 to 0.31; the beam efficiency improved from 0.22 to 0.46. In addition to the dramatic improvements to the efficiencies, the beam became narrower and the near sidelobes, which had been substantial, became essentially undetectable.

The 3 lines were readily detected in most of the sources (11/13 for HCN, 11/12 for HCO, and 11/14 for CS). Typical T_A^* were about 2 K. We conclude that these lines can be detected in many sources. Only Orion was mapped, but several sources were observed at more than one position and seen to be extended. The three lines have similar central velocities, suggesting that the emission comes from similar regions.

Since we have only one line of each molecule in most sources, we cannot yet apply detailed excitation analyses to determine the densities. For 3 sources which we have previously studied in detail (M17, S140, and NGC2024), we can examine the consequences of the new measurements. Our previous work on these transitions (Evans et al. 1987) was done without the error-corrector and we were forced to make rather large corrections for the source coupling to the beam; because of the substantial near sidelobes the corrections were quite dependent on assumptions about the source structure. With the cleaner, narrower beam now available, we find substantially stronger lines in some sources, most notably S140 and NGC2024. Whereas the old data were generally consistent with a single density fitting all the transitions, the new data suggest that a higher density component is present. Determination of the physical conditions in the higher density component will require new measurements with matched resolution, both in lower J transitions and in still higher J transitions, such as the $J = 10 \rightarrow 9$ line of CS or the $J = 9 \rightarrow 8$ lines of HCN and HCO$^+$; the last two have already been detected in Orion (Stutzki et al. 1988; Jaffe et al. 1988).

TABLE 1. RESULTS

Source	HCN J = 4 → 3		HCO+ J = 4 → 3		CS J = 7 → 6	
	T_A^* (K)	ΔV (km/s)	T_A^* (K)	ΔV (km/s)	T_A^* (K)	ΔV (km/s)
W3 (OH)	1.9	4.2	1.8	6.2	1.2	4.1
GL490	---	---	1.5	2.9	---	---
Orion A	2.6	12.8	8.8	6.3	6.1	3.8
	6.0	37.4	5.2	63.0	3.6	15.6
S235	---	---	---	---	---	---
NGC 2024	3.1	2.6	3.7	8.3	4.1	1.8
Mon R2	2.3	4.5	---	---	0.8?	2.9
S255	2.1	6.3	---	---	2.0	3.2
M17SW (1E,1N)	2.8	7.7	4.1	5.3	2.7	4.3
W51 (1W)	2.8	8.7	5.1	8.7	2.7	6.5
S88 (0.5E)	0.9	4.4	1.7	5.0	---	---
DR21 (OH)	2.8	5.2	4.3	3.6	2.0	5.3
W75N	---	---	4.8	5.8	2.3	4.9
S140	2.9	4.2	5.3	4.7	1.6	2.9
Cep A	2.8	6.1	4.4	6.1	1.1	6.6

This research was supported by NSF Grant AST86-11784 to the University of Texas.

REFERENCES

Evans, N. J., II, Mundy, L. G., Davis, J. H., and Vanden Bout, P. A. 1987, *Ap. J.*, **312**, 344.
Jaffe, D. T., Genzel, R., Harris, A. I., and Stutzki, J. 1988, *Ap. J.*, in preparation.
Stutzki, J., Genzel, R., Harris, A. I., Herman, J., and Jaffe, D. T. 1988, *Ap. J.*, in press.

MM AND SUB-MM OBSERVATIONS OF THE PROTOSTELLAR DISKS

Saeko S. Hayashi
Joint Astronomy Centre
665 Komohana Street
Hilo, Hawaii 96720, USA

ABSTRACT. Millimeter and sub-millimeter transitions of molecular lines demonstrated that the protostellar disks have hierarchical structures. The compact, highly excited regions are embedded in the cold, massive envelopes. The excitation and the distribution of those lines suggest vigourous interaction between the disks and the outflows.

1. OBSERVATIONAL APPROACH

Between the placental molecular clouds and stars, the protostellar cores/disks are intermediate in the spatial scale (1 to less than 1/100 pc) and in the time scale. To catch an actual site of the star formation, a series of radio observations have been made toward the active star forming regions. The radio telescopes used are Nobeyama 45 m (Japan) and 15 m JCMT (Hawaii). They give similar spatial resolution, which corresponds to a linear scale of 0.05 pc at Orion's distance. In this paper, we summarize CS observations at 49/98/343 GHz.

2. RESULTS AND INTERPRETATION

A prototype of the disk image was composed by millimeter observations of CS 1-0 (49 GHz) and 2-1 (98 GHz). These sub-pc strucutres are quite massive (up to 10^3 M☉) and sometimes show a hint of rotation, though the rotational velocity tends to be smaller than the equivalent velocity of Keplerian (e.g. Kaifu 1987, 1988). Typically those mm-transitions delineate the regions of $n(H_2) = 10^4 - 10^5$ cm^{-3}.

Current sub-mm experiments show, first, that the CS 7-6 emission (342 GHz) is quite strong. Toward massive star forming regions typically of L = 10^4 L☉, the CS intensities are as following : Tr (J=1-0) = 2 - 4 K, Tr(J=2-1) = 4 - 9 K, and Tr (J=7-6) = 2 - 5 K, respectively. To excite this high J level, high volume density of $n(H_2) > 10^6$ cm^{-3} and high temperature are required (e.g. Snell et al. 1984).

Second, the distribution of CS 7-6 is quite compact -- a few tenth of a pc -- compared to the extended envelope of the mm-transitions. Third character is that the morphological relation between the CS 7-6 condensations and the molceular outflows imply strong coupling between them.

Cepheus A is an instructive case of an onion-like distribution. Within an envelope of about 3'x2' shown in CS J=1-0/2-1, the 7-6 emission has a strong contrast in 1'x0.5' scale (e.g. Hayashi et al. 1988). Suggeted in this source is an elongation of CS 7-6 along the outflow. This is also obvious in the cases of NGC2071 and S140. Particularly toward S140, the blueshifted CS emission is displaced to the south, similar to the CO blueshifted lobe (Hayashi, M. et al. 1985, 1987). Further implication of the activity around the outflow is shown in NGC2024 hot spots. The individual clumps in its northern outflow lobe are more enhanced in CS 7-6, and the intensity varies a lot more than low-J lines.

The excitation of high-J CS can be attributed to the density enhancement or the high temperature caused by the outflow activity. The dense clumps may be pre-existing but also snow-plowed by the outflow pressure. The boundary of the wind and the ambient material may be shock heated to have the high temperature or the anomalous CS abundance. The CO J=3-2 (and higher J) data suggest that the kinetic temperature is 50 K or more in the vicinity of those protostars. Then which mechanism is dominant, density or excitation ? Both are responsible for the dissipation process of the protostellar disks. More studies to come.

This work is a joint project by the astronomers from Hawaii, Japan, UK, Canada, and more.

References.

Hayashi, M., Omodaka, T., Hasegawa, T., and Suzuki, S.
    (1985), Astrophys. J., 288, 170-174.
Hayashi, M., Hasegawa, T., Omodaka, T., Hayashi, S. S., and
    Miyawaki, R. (1987), Astrophys. J., 312, 327-336.
Hayashi, S. S., Hasegawa, T., and Kaifu, N. (1988),
    Astrophys. J., 332, 354-363.
Kaifu, N. (1987), in M. Peimbert and J. Jugaku (eds.),
    Star Forming Regions, 275-285.
Kaifu, N. (1988), in IAU 20-th General Assembly.
Snell, R. L., Mundy, L. G., Goldsmith, P. F., Evans, N. J.,
    II, and Erickson, N. R. (1984),
    Astrophys. J., 276, 625-645.

STUDIES OF STAR FORMING REGIONS IN M33

Christine D. Wilson
Dept. of Astronomy, 105-24
California Institute of Technology
Pasadena, CA 91125

ABSTRACT. Molecular clouds similar to Galactic Giant Molecular Clouds have been detected in six of seven fields mapped with the Owens Valley interferometer. Comparison of single dish CO data with B and V photometry is used to study the star formation efficiencies in gas-rich and gas-poor regions.

1. Introduction

M33 is an Sc-type galaxy in the Local Group at a distance of ~ 0.8 Mpc. Its proximity and favorable inclination ($\sim 50^\circ$) make it a prime candidate for studies requiring high spatial resolution. Previous single dish CO line observations suggest that M33 has a relatively low molecular gas content (Young and Scoville 1982), yet optical studies reveal numerous OB associations (Humphreys and Sandage 1980). Molecular clouds similar to Galactic Giant Molecular Clouds (GMCs) have recently been detected in the giant HII region NGC 604 and in the nucleus with the Owens Valley Millimeter Interferometer (Boulanger et al. 1987; Wilson et al. 1988).

2. Properties of Individual Clouds

We have mapped seven fields in the inner 1 kpc of M33 and detected molecular clouds in six of the fields. The diameters, velocity widths, brightness temperatures, and masses of the clouds are very similar to Galactic GMCs (Sanders, Solomon, and Scoville 1985). Figure 1 shows the interferometer map of the largest cloud overlaid on a blue CCD frame. The cloud shows a good spatial correlation with a region of strong optical extinction.

In general, the interferometer maps detect less than 50% of the single dish flux. There are three possible explanations for this missing flux. First, the flux may be from large clouds on the edges of the interferometer map which are included in the single dish measurements but not in the interferometer flux (which includes only emission out to the half power point). Second, the missing flux may be due to the missing spatial frequencies in the observations (the zero spacing and baselines longer than 60 m). Finally, the flux may originate in small clouds below the detection limit of the interferometer, i.e. clouds with diameters smaller than about 20 pc. Since in our Galaxy $\sim 85\%$ of the molecular flux is from structures larger than 20 pc (Sanders, Scoville, and Solomon 1985), this would imply that the number of clouds as a function of size in M33 is quite different from the Galactic distribution. This might indicate that cloud formation mechanisms are less efficient in M33 or that the clouds are destroyed faster.

3. Star Formation Efficiencies

CO data from the NRAO 12 m antenna have been combined with B and V photometry from the Canada-France-Hawaii telescope to compare the star formation efficiency (SFE) in regions of the galaxy with and without detected CO lines. The field studied was 2'x3' and the gas-rich regions have on average at least 2.5 times more molecular gas than the gas-poor regions. The color-magnitude diagrams of the two regions are very similar, in that the average reddening and the mass of the most massive star are the same in both fields.

Both regions contain the same number of high mass stars per unit area, which may indicate that regions with abundant molecular gas have a *lower* SFE. However, the effects of stellar drift and stellar obscuration may hide any differences between the two regions. Stars less massive than 15 $M_\odot$ can drift from one region to the other if they are born with a random velocity of 10 km s^{-1}, but the presence of more massive stars in the gas-poor regions cannot be explained unless the random velocities are higher by a factor of two. If the SFEs in the two regions are comparable, the clouds in the gas-rich regions must block the stars for >50% of the stellar lifetimes. This situation could occur if the clouds in the gas-poor regions have much smaller average diameters.

4. References

Boulanger, F., Vogel, S. N., Viallefond, F., and Ball, R. in *Molecular Clouds in the Milky Way and External Galaxies* (eds. Dickman, R., Snell, R., and Young, J.), in press.
Humphreys, R. M. and Sandage, A. R. 1980, *Ap. J. Suppl.* **44**, 319.
Sanders, D. B., Scoville, N. Z. and Solomon, P. M. 1985, *Ap. J.* **289**, 373.
Wilson, C. D., Scoville, N., Freedman, W. L., Madore, B. F., and Sanders, D. B. 1988, *Ap. J.* **333**, 611.
Young, J. S. and Scoville, N. 1982, *Ap. J.* **260**, L11.

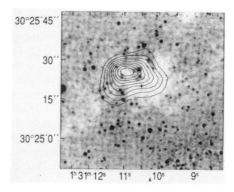

Figure 1. The interferometer map of the largest cloud is overlaid on a blue CCD frame. The peak flux in the map is 19 (Jy beam^{-1}) (km s^{-1}).

CIRCUMSTELLAR DENSE GAS IN B5 IRS1

Gary A. Fuller,
Astronomy Department,
University of California,
Berkeley, CA 94720, U.S.A.

ABSTRACT. High resolution maps showing dense clumps of gas within $\sim 5\times 10^{16}$ cm of the low luminosity young star B5 IRS1 are presented.

B5 is a dense core embedded within an extended region of extinction in Perseus, north east of the NGC1333 nebula. IRAS detected four low luminosity infrared sources associated with B5 (Beichman et al., 1984). The most luminous and most deeply embedded of these sources, 03445+3242, (also called IRS1) has been mapped with the Hat Creek Millimeter Interferometer and the FCRAO 14m telescope, in the J= $1 \rightarrow 0$ line of HCN. This line was chosen as it should trace high density gas, its critical density being around 10^6 cm^{-3}.

Figure 1 shows the emission averaged over two different velocity ranges in the average of the two outer, and optical thinner, hyperfine components (F = 1 − 1 and F = 0 − 1) of the HCN. These maps have an angular resolution of 12 by 12.8 arcsecs and are the result of combining the interferometer data with a Nyquist sampled map made with the FCRAO 14m telescope (see Fuller 1989 for full details). Three major clumps are seen in figure 1. They are labelled I, II and III.

In the maps of the main hyperfine component integrated over the same velocity ranges, clump III appears very similar and clumps I and II appear to blend together to form a single clump elongated in the same direction but offset slightly northeastwards from the positions of clumps I and II. This change of position is probably the result of optical depth and/or temperature effects in the line as the main hyperfine component is very optically thick (see below).

Observations of H^{13}CN with the FCRAO telescope give an estimate of between 5 and 17 for the optical depth in the main hyperfine component of the main isotope. This high optical depth is supported by the near unity intensity ratios for the hyperfine components in the maps, so the clump masses were derived from the measurements of the F = 0 − 1 hyperfine which has the lowest optical depth of the components. Table 1 gives the clump measured masses (assuming a fractional abundance of HCN of 1.4×10^{-9}, Fuller 1989) along with their sizes measured at the half peak contour and their virial masses. From these values it can be seen that all the clumps are consistent with being in virial equilibrium.

Clump III lies in the region of the blue shifted wing of the CO outflow from IRS1 and has a velocity that is slightly blue shifted with respect to the other two clumps (10.0 km s^{-1} for clump III compared to 10.4 km s^{-1} for clumps I and II). In addition this clump is coincident with the brightest region in the optical arc-nebula imaged by Campbell et al. (in prep.) which delineates the edge of the high velocity CO emission.

From figure 1 it can be seen that clumps I and II form an elongated structure with axial ratio of about 2:1, close to the infrared source IRS1. The direction of the elongation is nearly perpendicular to the axis of the CO outflow from IRS1. In addition the elongation is also perpendicular to the axis of the velocity gradient seen in larger scale maps of CS 2 − 1 and C^{18}O 1 − 0 (Fuller 1989), although there is no detectable velocity difference between these two clumps down to the velocity resolution of the data (0.13 km s^{-1}). However the limit on the rotational velocity difference between the clumps is a factor of several times this value. Taking a value of 0.4 km s^{-1} for the rotational velocity leads to an interior mass of 0.6M$_\odot$/sin(i) where i is the angle between the axis of rotation and the line of sight of the observer.

Table 1

Clump	Radius (10^{16} cm)	N(H$_2$) (10^{22} cm^{-2})	Mass (M$_\odot$)	Virial Mass (M$_\odot$)
I	6.3	1.0	0.24	0.43
II	3.7	2.0	0.16	0.23
III	9.0	1.2	0.60	0.60

What is the origin of this flattened structure? One interpretation is that the material may be part of a dense disk around the young star. However there are several problems with this interpretation. First no evidence for systematic rotational motion was found. Secondly, IRS1 is not symmetrically located with respect to the two clumps. Finally, and probably more troublesome for a disk hypothesis, are the possible effects of the outflow from the star. Energetically it is possible that the clumps seen near the star were originally part of a more uniform, spherical distribution of dense gas, some of which has now been removed by the action of the outflow, leaving only the flattened looking structure observed.

References

Beichman, C. A., et al., 1984, Ap.J.(Letters), **278**, L45

Fuller, G.A., 1989, Ph.D. Thesis, Astronomy Department, University of California, Berkeley

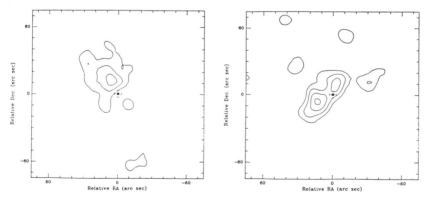

Figure 1 The figure shows the CLEANed maps of the average for the F = 0 − 1 and F = 1 − 1 emission. The left panel shows the emission centered at 10.0 km s^{-1} averaged over 0.4 km s^{-1}. The right panel shows the emission centered at 10.4 km s^{-1} averaged over 0.3 km s^{-1}. The contour levels are 1, 1.5, 2 and 2.5 K. The solid box with error bars indicates the 2μm position of B5 IRS1.

$^{12}CO(2-1)$ OBSERVATIONS OF A COMPLETE SAMPLE OF DARK CLOUDS

P.F. SCOTT, N.D. PARKER & R. PADMAN
Cavendish Laboratory
Madingley Rd.
Cambridge CB3 0HE
England

ABSTRACT. A complete sample of dark clouds is being observed in order to study the processes of star formation. Bipolar outflows have been found in 7 out of the 14 clouds studied to date while 2 further clouds show high-velocity features. It is concluded that a period of mass-loss is a widespread feature of star formation in dark clouds and could have a significant role in cloud evolution.

The James Clerk Maxwell Telescope (JCMT) has been used to map the $^{12}CO(2-1)$ emission in the vicinity of the active cores in a sample of clouds with 20 arcsec (HPBW) resolution. In order to provide a well-defined sample, suitable for statistical analysis, the following procedure was used for selection of the dark cloud sample:

i) The sample was selected from clouds in opacity class 6 ($m_v \gtrsim 15^m$) in the Lynds Catalogue;
ii) positions and extents of these 147 clouds were redetermined from the Palomar Sky Survey prints (Parker 1988);
iii) the IRAS point source catalogue was used to produce a secondary list of 72 clouds having one or more IRAS sources within their boundary. Approximately 90% of these coincidences are believed to represent true associations;
iv) a final list was selected of those clouds containing IRAS sources having good quality fluxes and spectra characteristic of "embedded cores" (Emerson 1987).

This final list, of 19 clouds, provides a homogeneous sample of clouds containing active regions, allowing a comparatively unbiassed study of the star formation process.

Enhanced CO emission was detected from most of the regions; more suprisingly, however, 7 of the 14 regions have associated (mostly bipolar) molecular outflows and two others have CO spectra with high velocity emission. A period of mass-outflow is clearly a common feature of the formation process not only of massive stars but also intermediate and lower mass objects. Several of the outflows show evidence of interactions with the parent cloud; an energy estimate for one of the outflows (L1262 - see below) suggests that such outflows are likely to be significant in the overall evolution of the clouds. The figure shows results for four of the clouds. The CO emission in L43 (i) shows well-resolved red and blue wings. The emission towards the south follows the boundary of a cavity, also identifiable in an optical

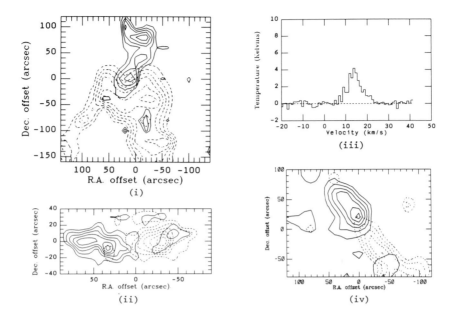

Figure 1: *Results for four of the clouds. Contour maps are of the red-shifted (solid) and blue-shifted (dashed) emission.*

image, produced as the flow expands into a lower-density region of the original cloud. A new compact, bipolar outflow has been found in L483 (ii). The red-shifted emission superposed on the 'blue' lobe may indicate outflow in a cone, aligned fairly close to the plane of the sky. The width of the spectrum centred of the core in L588 (iii) is consistent with an outflow. This interpretation is supported by the presence of pronounced absorption apparent in the 'blue' wing of many of the spectra and attributable to intervening colder ambient material swept up by the outflow. A compact bipolar outflow is seen in L1262 (iv). A position-velocity plot of the low-level emission shows that the outflowing material is progressively decelerated away from the central core, in contrast to other, higher-luminosity sources, in which acceleration is commonly seen. $^{13}CO(2-1)$ measurements for this region allow rough estimates of the flow parameters; a mass of $0.1 M_\odot$ and energy of $10^{35} - 10^{36}$ J are obtained, the latter being about one fifth of the estimated binding energy of the overall cloud.

References

Emerson J.P., 1987. Proc. IAU Symp. 115: Star Forming Regions, eds. Peimbert, M. & Jugaku, J (Reidel, Dordrecht).
Parker, N.D. 1988. Mon. Not. Roy. astr. Soc. in press.

PLATEAU EMISSION FROM ORION IN THE CO J=17-16 LINE

R.T. BOREIKO, A.L. BETZ
Space Sciences Laboratory, University of California
Berkeley, CA 94720
J. ZMUIDZINAS
Astronomy Department, University of Illinois
Urbana, IL 61801

ABSTRACT. High-resolution spectra of the J=17-16 CO line (1956 GHz) in the BN-KL region of Orion have been obtained with a heterodyne spectrometer. The profiles show a broad component with 30 km s^{-1} (FWHM) linewidth and a narrower 8 km s^{-1} component. The broader plateau emission detected over a range of transitions from J=1-0 to J=17-16 is analyzed under the assumptions of thermal equilibrium and optically thin wings to deduce an excitation temperature of 180 ± 50 K and minimum column density of 1×10^{18} cm^{-2} for CO in this component.

1. Procedure

The observation described here of the J=17-16 transition of CO at a resolution of 0.8 km s^{-1} extends the range of excitation covered by high-resolution CO data fivefold, allowing a more accurate determination of excitation temperature and column density. Comparison of CO spectra over a range of J shows that the emission in the wings is very likely to arise from the same source in all cases, and is optically thin. Available high-resolution spectra (J=1-0 and J=2-1 from Plambeck, Snell, and Loren 1983; J=3-2 from Richardson *et al.* 1985; J=7-6 from Schmid-Burgk *et al.* 1988; and J=17-16 from Boreiko, Betz, and Zmuidzinas 1989) were individually fitted to a linear combination of two Gaussians to isolate the plateau emission. The data and fit for the J=17-16 line are shown in Fig. 1. The plateau component with FWHM~30 km s^{-1} contributes 80% of the integrated intensity of this line, while the remaining 20% arises from a component with FWHM~8 km s^{-1}. The fitted peak antenna temperatures for the plateau component were then used to determine the excitation temperature T_{ex} and column density, as discussed in detail by Boreiko, Betz, and Zmuidzinas (1989). The assumptions implicit in the procedure are that the line wings are optically thin, the levels are in LTE at a common excitation temperature, and source coupling corrections are similar for all data.

2. Results and Conclusions

The excitation temperature derived from the fitting procedure is 180 ± 50 K, where the uncertainty reflects both statistical and systematic effects. The column density derived for the CO giving rise to the plateau emission is $\geq 1 \times 10^{18}$ cm^{-2}. Fig. 2

shows the data and the fit. The error bars represent uncertainties from calibration of the data and the Gaussian fitting. In cases where the errors are not quoted by the authors, a calibration uncertainty of 25% was assumed. As can be seen from Fig. 2, the low-J lines (J<5) best define the CO column density in the plateau emission, since their intensities are relatively insensitive to T_{ex} in the range of interest. In contrast, lines with J>13 are the most sensitive for determining T_{ex} once the column density is fixed. Lines with intermediate J are the strongest, and therefore the best indicators for optical depths effects at line center.

This work was performed as part of the KAO Airborne Astronomy Program under NASA Grant NAG 2-254.

References

Boreiko, R.T., Betz, A.L., and Zmuidzinas, J. (1989) 'Heterodyne spectroscopy of the J=17-16 CO line in Orion', *Ap.J.* (to be published).

Plambeck, R.L., Snell, R.L., and Loren, R.B. (1983) 'J=2-1 CO observations of molecular clouds with high-velocity gas: evidence for clumpy outflows', *Ap.J.* **266**, 321-330.

Richardson, K.J., White, G.J., Avery, L.W., Lesurf, J.C.G., and Harten, R.H. (1985) 'CO J=3-2 observations of molecular line sources having high-velocity wings', *Ap.J.* **290**, 637-652.

Schmid-Burgk, J. *et al.* (1988) 'Extended CO (J=7-6) emission from Orion Molecular Cloud 1: hot ambient gas, two hot-outflow sources', *Astron. Astrophys.* (to be published).

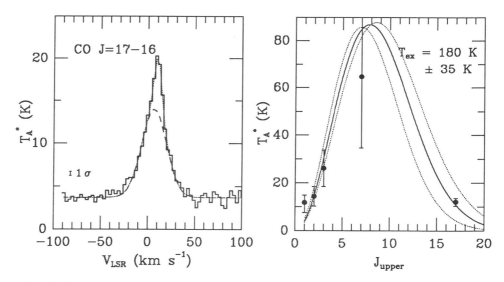

Fig. 1. J=17−16 line of CO. The dotted line shows the fit, and the dashed line shows only the plateau component of the fit.

Fig 2. Fitted antenna temperatures and predicted values at line center for the plateau emission.

STRUCTURE OF DUST DISCS AROUND YOUNG STARS

W.R.F. DENT
Royal Observatory
Blackford Hill
Edinburgh EH9 3HJ
Scotland

ABSTRACT. The JCMT has been used to map the thermal continuum emission from two young luminous stars associated with massive bipolar outflow. The results have been compared with simple models of centrally heated symmetric discs, from which it is possible to estimate the density structure of the dust.

1. G35.2N

The central star in this source lies at the centre of a high velocity CO ridge, with an orthogonal rotating disc observed in NH_3 and CS (Dent et al 1985; Little et al, 1985). It is thus an ideal candidate for elongated sub-mm dust emission, as the proposed disc would be viewed almost edge-on. Fig 1 shows the 1100 and 450μm maps of this object. The results indicate a mean kinetic temperature of 40 ± 10K, which gives a dust mass of 0.5$M_\odot$ $(Q_1/10^{-4})^{-1}$, where Q_1 is the dust emissivity at 1100μm. A v^1 emissivity law, and canonical grain sizes and bulk densities have been assumed. This agrees with dynamical mass estimates (from NH_3) if $Q_1=0.2\times10^{-4}$ which corresponds with the dust models of Draine & Lee (1984).

The 1100μm morphology (Fig. 1) generally agrees with the NH_3 disc, however, the detailed morphology of NH_3 (1,1) and (2,2) lines indicated a depletion of this gas within the central 0.1pc (Little et al. 1985). We have made model fits to a cross-section cut along the major axis of the sub-mm disc to search for a similar depletion in the dust density. The models assumed a central heating source and optically thin emission on the scale of the smallest beam. The best fit to both wavelengths was achieved with a single power law in density as a function of radius, with $n_d \propto r^{-0.5}$. This would imply a relative enhancement of the dust/gas ratio or depletion of [NH_3] within 0.1pc of the core.

An attempt was also made to model the structure perpendicular to the disc plane, however, it was found to be impossible to simultaneously fit both wavelengths. Fig 1 shows that the apparent thickness of the disc is similar at both wavelengths and the deconvolved dimensions are 16 ± 3" and 20 ± 5" at 1100 and 450μm respectively. The simplest explanation is that there is an additional energy source heating the dust out of the disc plane. This is thought to be collisional heating in the bipolar high velocity gas.

The "excess" 450μm emission has an estimated luminosity of 1-8 x 10^3 $L_\odot$ (for T_D = 40K); this compares with the flow mechanical luminosity from CO of 250 $L_\odot$. The wind luminosity is thus significantly underestimated by the CO data, and may be as much as 5-40% of the stellar bolometric luminosity. Edwards et al. (1986) reached a similar conclusion from extended FIR observations of L1551.

2. LkHα 234

This source contains an optical He-Be star, with a bipolar CO outflow (Sandell & Liseau, 1985). Figure 2 shows the 1100μm map and a cross-section along the major axis. Symmetrical "shoulders" are detected around the central star, and a fit to these data is also shown. This structure can most easily be modelled by an unresolved central source, and a ring of dust at r = 0.17pc. The ring mass is ~10 times that of the central condensation, but is less prominent because of the lower temperature. Two distinct dust temperatures are also indicated by IR spectra of Bechis et al. (1978) and the present data suggest that their 42K component arises from the extended ring. Further observation of this ring are required to determine its evolutionary state.

3. References
Bechis, K.P. et al., (1978) Astrophys J., 226, 439-454.
Dent, W.R.F. et al., (1985) Astr. Astrophys., 146, 375-380.
Draine, B.T. & Lee, H.M. (1984) Astrophys J., 285, 89-108.
Edwards, S. et al., (1986) Astrophys J., 307, L65-68.
Little, L.T. et al., (1985) M.N.R.A.S., 217, 227-238
Sandell, G. & Liseau, R. (1985) in G. Serra (ed.), Nearby Molecular Clouds, Springer, Berlin, p. 227-233.

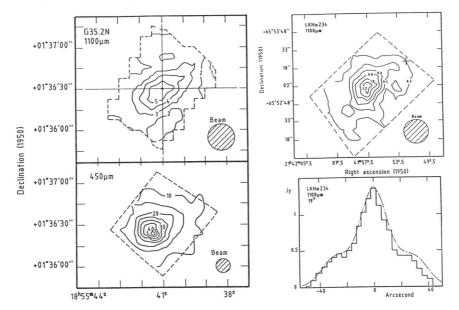

Fig 1   Continuum emission maps of G35.2N Contours are Jy/beam.

Fig 2   1100μm map of LkHα234, with cross section along major axis and model fit.

Probing the Lower Main Sequence with Molecular Clouds

R.L. Dickman[1], T.H. Jarrett[1], and W. Herbst[2]
[1] FCRAO, University of Massachusetts, Amherst MA, 01003 USA
[2] Wesleyan University, Middletown, CT 06457 USA

Apart from their inherent interest, molecular clouds can function as unique backdrops against which two basic aspects of the star formation problem can be studied[1]:

How does the faint end of the luminosity function, $\Phi(M_v)$, behave? In particular, does $\Phi(M_v)$ continue its apparent decline below $M_v \sim 13$[2], or does it rise again? Even a slightly rising luminosity function extending from $M_v \sim 16$ to the limit of the hydrogen-burning main sequence would resolve the missing mass problem in the Solar neighborhood[3].

What are the characteristics of low mass ($M < 0.5\ M_\odot$) star formation in the Milky Way at the current epoch? Are such stars being actively formed in giant molecular clouds? In dark clouds?

These questions can be addressed by using molecular clouds as opaque screens. Luminosity function studies face the difficulty of obtaining complete, distance-limited samples; this stems from the selectivity of proper motion surveys and from the necessity of culling magnitude-limited samples in order to separate luminous, distant late-type stars from nearby, intrinsically faint objects of the same color. The second problem also affects attempts to identify low-mass stars in the vicinity of (and possibly embedded in) molecular clouds.

By making deep, photometrically-calibrated multicolor observations of stars lying toward the most opaque regions of molecular clouds — regions which have visual extinctions in excess of 10 magnitudes — one can avoid most of these difficulties. The vast majority of background stars toward such areas are, in principle invisible, and embedded stars are easily distinguished by their color. Moreover, by determining the optical extinction of the molecular material, one can utilize data even for the less-obscured portions of the cloud.

Because one is concerned with identifying foreground and embedded stars in a program of this kind, one must rely on indirect methods rather than star counts[4] to estimate the extinctions due to the molecular cloud. This can be done by using either infrared or molecular data to trace the extinction. Naive IR optical depths computed from IRAS intensity data at 60 and 100μm have been found to trace molecular cloud material well up to $A_v \sim 4$ magnitudes[5,6] (IRAS is an especially attractive data resource in this connection because of its all-sky coverage and uniform quality). Recently, we demonstrated[7] that in the ρ Ophiuchi molecular cloud, 60μm optical depths correlate reasonably well with visual extinctions even *above* 10 magnitudes, provided that one avoids regions in the vicinity of outflow sources (Fig. 1).

Isotopic carbon monoxide can also be used as a dust tracer, and ^{13}CO LTE column density has long been known[8] to correlate well with visual extinction up to moderate levels ($A_v \sim 6$ mag). At still higher obscurations, the situation is more ambiguous: in Ophiuchus, the two quantities remain fairly well correlated in a reasonably linear fashion beyond $A_v = 10$ mag[9] (Fig. 2), whereas in Taurus there is evidence[10] that the molecular column density begins to saturate at $A_v \sim 5$ mag.

Once one can estimate extinctions for the nearest molecular clouds, deep photography or CCD observations provide the necessary stellar data. Given present instrumental capabilities, the luminosity function can be probed to $M_v > 18$ (though only within small sampling volumes), and one can study low-mass dwarf populations in the vicinities of nearby molecular clouds.

We have now obtained deep photographic data for two very different fields using the 4m telescopes at Kitt Peak and CTIO. One field lies toward the ρ Oph molecular cloud, the other in the direction of Heiles' Cloud 2 (TMC-1). Analysis[9] of the Ophiuchus field reveals that $\Phi(M_v)$ does not rise between $M_v \sim 14$ and $M_v \sim 18$, and we are presently engaged in a similar analysis for our Taurus data. Studies of the dwarf population problem await new, deeper CCD data.

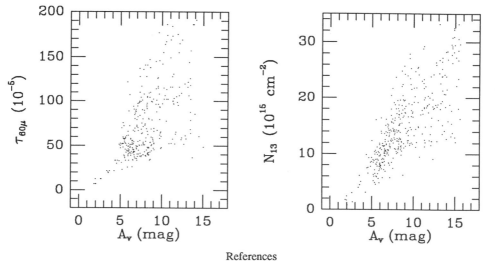

References

1. Herbst, W. and Dickman, R.L. 1983 in IAU Colloquium No. 76, *The Nearby Stars and the Stellar Luminosity Function*, A.G.W. Philip and A.R. Upgren eds. (Schenectady: L. Davis Press), p. 187.

2. Wielen, R., Jahriess, H., and Krüger, R. 1983 in IAU Colloquium No. 76, *The Nearby Stars and the Stellar Luminosity Function*, A.G.D. Philip and A.R. Upgren, eds. (Schenectady: L. Davis Press), p. 163.

3. Bahcall, J.N. 1984, *Ap. J.*, **276**, 169.

4. Dickman, R. L. 1978, *A.J.*, **83**, 363.

5. Snell, R.L., Heyer, M.H., and Schloerb, F.P. 1988, *Ap. J.*, submitted.

6. Langer, W.D., Wilson, R.W., Goldsmith, P.F., and Beichmann, C.A. 1988, preprint.

7. Jarrett, T.H., Dickman, R.L., and Herbst, W. 1988, *Ap.J.*, submitted.

8. Dickman, R.L. 1978, *Ap.J. (Suppl.)*, **37**, 407.

9. Dickman, R.L. and Herbst, W. 1988, in preparation.

10. Frerking, M.A., Langer, W.D., and Wilson, R.W. 1982, *Ap. J.*, **262**, 590.

1.0-mm CONTINUUM, ^{12}CO, ^{13}CO AND C^{18}O MAPPING OF THE NGC 6334 COMPLEX AND COMPARISON WITH IRAS OBSERVATIONS

DAN GEZARI
NASA/Goddard Space Flight Center, Greenbelt, MD

LEO BLITZ
Univ. of Maryland, College Park, MD

We present high spatial resolution well sampled observations of the NGC 6334 cloud complex in the ^{12}CO, ^{13}CO and C^{18}O isotopic species, and 1.0-mm continuum emission from cool dust, covering the far infrared emission Peaks I-V, a new Peak IV, and the cold (T=20K) sub-mm Peak I(North) cloud (Cheung et al. 1978, Gezari 1982). The ^{12}CO and ^{13}CO emission was mapped previously by Scoville and Wannier (1976) with 2 - 4 arcmin sampling. These 1.0-mm continuum maps were obtained at the CTIO 4-m telescope with 65 arcsec FWHM resolution in June 1981, and the CO line observations were made at the NRAO 36-foot telescope with 1 arcmin FWHM resolution in April 1982. Several new observational results are presented here, including the large-scale detailed structure of the 1.0-mm continuum and ^{13}CO line emission a ~0.7 pc scale (Figure 1a-c), and half beam width (0.5 arcmin) sampled ^{12}CO, ^{13}CO and C^{18}O maps of the dense Peak I(North) region (Figure 2). A comparison of the 1.0-mm continuum (Figure 1a) and CO isotopes (Figure 2) shows that the highest $T_A^*(^{12}CO) = 16K$ occurs at the position of sub-mm Peak I and compact HII region "E" (Rodriguez, Canto and Moran 1982), consistent with the distribution expected of optically thick gas heated by warm dust associated with the compact HII region. In contrast, the map of integrated $T_A^*(^{13}CO)$ shows two distinct peaks similar to the 400 μm continuum structure mapped by Gezari (1982). The highest $T_A^*(^{13}CO) = 20K$ at Peak I(North), and the ^{13}CO/^{12}CO ratio is greater there than at Peak I, indicating that the cold

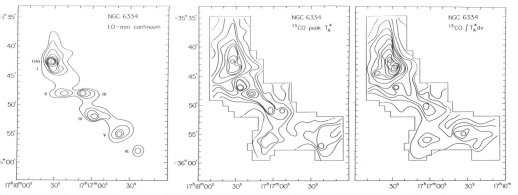

Figure 1: a) 1.0-mm continuum map of the NGC 6334 complex with 65 arcsec (FWHM) beam. The region was sampled on a 30 arcsec grid near the strong peaks, and 1 arcmin elsewhere. The contour interval in 1.0-mm brightness is 15 Janskys/beam (peak = 135); since the dust is optically thin this is equivalent to an interval of 1 x 10^{23} cm^{-2} in derived molecular hydrogen column density. (Note: Peaks labled I-IV by Cheung et al. (1978) are actually I/North, I, II and III in current nomenclature). b) Peak $T_A^*(^{13}CO)$, contour interval is 2K (peak = 20K), full beam width sampled (1 arcmin) over the region and half beam sampled near Peak I. c) Integrated $T_A^*(^{13}CO)$ line intensity, contour interval is 6 Jy km s^{-1}arcmin^{-2} (peak = 131), same sampling as in Figure 1b.

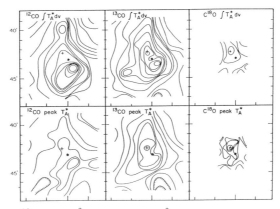

Figure 2: ^{12}CO, ^{13}CO and C^{18}O peak T_A^* and integrated T_A^* intensity maps near far infrared NGC 6334 peak I, half beam width sampled (30 arcsec). The cross (+) marks the cold sum-mm peak I(North) object, an asterisk (*) marks compact HII region "E". Peak T_A^*(^{12}CO) = 16K, peak integrated T_A^*(^{12}CO) line intensity = 166 K km s^{-1}arcmin^{-2}; peak T_A^*(^{13}CO) = 20K, peak integrated T_A^*(^{13}CO) = 131 K km s^{-1}arcmin^{-2}; peak T_A^*(C^{18}O) = 4.5K, peak integrated T_A^*(C^{18}O) line = 20 K km s^{-1}arcmin^{-2}.

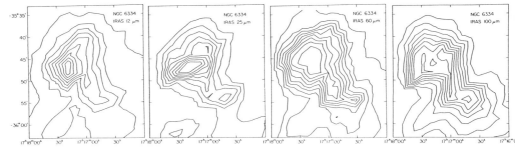

Figure 3: IRAS maps of the NGC 6334 cloud complex at 12, 25, 60 and 100 μm.

Peak I(North) source can not be optically thin in ^{13}CO and is massive, in spite of relativly weak ^{12}CO emission. The optically thin C^{18}O source (Figure 3) is very compact and also brightest at Peak I(North), the position of the highest 1.0-mm dust column density (N_{H_2} = 1 x 10^{24} cm^{-2}). The C^{18}O distribution confirms that Peak I(North) apparently has among the highest derived visual extinction (m_v > 100 mag). The ^{13}CO lines at I(North) are narrow (6 km s^{-1} FWHM) and may arise from the envelope of a protostellar cloud still in the collapse phase. The four color IRAS obvservations (Figure 3), while considerably lower in spatial resolution, show that the warmest (12 - 25 μm) dust emission is concentrated about 5 arcmin SW of the strongest 1.0-mm and CO emission, and is associated with the prominent, extended visible HII region NGC 6334. This work was partially supported by NSF grant ASF-86-18763, and by NASA RTOP 188-41-55.

Cheung, L., Frogel, J. A., Gezari, D. Y., and Hauser, M. G. 1978, L149.
Gezari, D. Y. 1982, Ap. J. (Letters), 259, L29.
Rodriguez, L. M., Canto, J., and Moran, J. M. 1982, Ap. J., 255, 103.
Scoville, N. Z. and Wannier, P. 1976 (private communication; see Scoville and Kwan 1977, "Infrared and Submillimeter Astronomy", ed. G. Fazio, D. Reidel Pub. Co.

Observations of HCO⁺ in B335

T.I.Hasegawa	St.Mary's University	Halifax
C. Rogers	University of Toronto	Toronto
S.S.Hayashi	Joint Astronomy Centre	Hilo

ABSTRACT. We present HCO⁺ observations of B335 by JCMT. They show that there is warm gas in the central region and a steep excitation gradient.

1. INTRODUCTION

B335 is a samll, isolated Bok globule believed to harbour a protostellar object. Far-infrared continuum observations reveal a region of enhanced emission with a size of 30". This region is also the centre of a bipolar flow observed in CO(1-0). Both the large optical depth of dust ($A_V>100$mag) and short dynamical time of the flow (10^4yrs) point to a very early stage of star formation.

The high column density and low temperatures throughout most of the cloud make molecular lines at millimetre wave lengths optically thick, hiding from view the dense inner core of B335. A good probe of the physical conditions in this core are the submillimetre transitions of HCO⁺, which are excited at high densities but have lower optical depth in low-density regions.

We observed B335 in the J=3-2 (268GHz) and 4-3 (357GHz) transitions of HCO⁺ and the J=3-2 (260GHz) transition of H^{13}CO⁺, using the James Clerk Maxwell Telescope in April and September, 1988.

2. RESULTS

An area of 40"x40" was mapped in the HCO⁺(3-2) line. The emitting region has a half-maximum size of 25" (10^{17}cm) and has four arms which extend along the boundaries of the east-west bipolar flow (Fig.1).

We also observed the HCO⁺(4-3) and H^{13}CO⁺(3-2) lines at a few positions (Fig.2). The (4-3) emission is confined to a smaller region (<20") and is of comparable intensity to (3-2). It is difficult to explain such a line ratio with LVG models, where emission in both lines originates from a common volume. Indeed, the (3-2) line is optically thick and may not be in LTE, as indicated by our detection of relatively strong H^{13}CO⁺(3-2) emission at the map centre. The evidence strongly suggests a warm, dense region ($T_k>20$K, $n(H_2)>10^5$cm^{-3}) surrounded by a cold region ($T_k=10$K, $n(H_2)=10^5$

cm^{-3}) with a strong excitation gradient across the interface of these regions. The densities in the core may be high enough to thermally couple the gas to the warm dust of the FIR source.

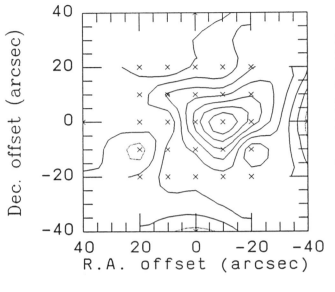

Figure 1. An integrated intensity map of HCO$^+$ (3-2). Contour interval is 0.5Kkms^{-1}. Maximum contour is 3.5Kkms^{-1}. Map centre is (19h34m 35.34s, +7 27'20") (1950). Crosses mark observed positions.

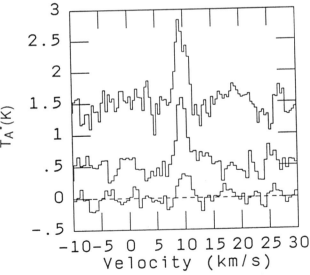

Figure 2. Profiles of the HCO$^+$(4-3)[top], (3-2) [middle], and H^{13}CO$^+$ (3-2) lines at (19h34m 35.34s, +7 27'20") (1950). The frequency resolution is 1MHz and the channel interval is 0.488MHz. The beam efficiency was 0.57 for (4-3), 0.67 for (3-2), and 0.70 for 13(3-2).

RADIO CONTINUUM ACTIVITY IN CEPHEUS A

V. A. HUGHES and G. H. MORIARTY-SCHIEVEN
Astronomy Group, Department of Physics
Queen's University
Kingston, ON K7L 3N6
Canada

Cepheus A is a condensation in the larger Cep OB molecular cloud, mapped in CO by Sargent (1979). It is known to be a source of molecular outflows (see Hughes, 1985), it shows NH_3 emission indicating a high density core, is an infrared source, and shows OH and H_2O maser emission. Of particular interest is that it consists of two strings of about 14 compact HII regions (Hughes and Wouterloot, 1982; Hughes, 1985, 1988). We have recently mapped continuum emission at 0.45 mm, chiefly from the region of the core (unpublished). It clearly appears to be a dense, dusty region where stars are forming, but so far there is no confirmed relationship between the molecular outflows, as observed with angular resolution $\sim 10''$, and the radio continuum emission with resolution $< 0''.3$, or the infrared. However, the origin of the outflows has been assumed by others to be associated with the radio Source 2, since it appears to be very compact with diameter $\sim 0''.1$ (70 au linear size), and thus could contain a star which is acting as the "power house" for the region. It has generally been assumed that the other 13 components are unlikely to contain stars, since they all appear to be about the same age. But recent observations, combined with those of up to seven years previous, have shown that a number of the more compact HII regions can be highly variable, and suggest that there may be more than one source of activity in CepA.

Observations have now been made with the VLA at wavelengths of 20cm, 6cm, and 2cm, at epochs of 1981, 1982, 1986, and 1987. The general structure is shown in the map, obtained in 1986 at C-band, resolution $0''.3$. The more compact sources are at the center of the region, while the outer regions contain the more diffuse components. What is more fascinating is that a number of the more compact components show large variability. The more diffuse components may also be variable, but this is not apparent due to their comparatively small peak flux density and large size. The highly variable sources appear to be Sources 2, 3(a), 7(a), and 8. In particular, Source 8 which was apparent in 1981, and 1986, can disappear inside a period of one year. Source 3(a) has decreased appreciably between 1982 and 1986, though it has an H_2O maser associated with it (Cohen et al., 1984); Source 3(c) shows only small variations; Source 2 appears to be optically thick, and has increased at C-Band from

2mJy in 1982, to 7mJy in 1987, while 7(a) is optically thin and has also increased by about the same amount over the same period.

A possible clue to the cause of variability comes from the fact that in order for the HII region associated with Source 8 to recombine inside a period of one year, electron densities must be at least 10^6 cm^{-3}, and we can expect there to be a change in diameter, though the source is too compact to observe this. The increase in flux density of Source 7(a) appears to be accompanied by an increase in diameter from 0".7 to 1".0, corresponding to an increase in radius at the rate of 100 km s^{-1}. No change in radius is seen in Source 2, but it is double and not resolved at 0".3 resolution. The general interpretation is that CepA is a region where stars are forming, in fact we are seeing an expansion in the HII regions, but this may be part of some form of instability occuring at this stage of stellar evolution. The rate of expansion of Source 7(a) is too high to be due to an R- or D-type ionization front, but based on the rate of expansion, and the estimated electron densities, we would expect it to contain a B0 star. It may be that this form of activity is associated with the observed molecular outflows, and early-type stars may go through a stage equivalent to a T-Tauri star.

REFERENCES

Sargent, A. I. 1979, Ap. J., 233, 163.
Hughes, V. A., and Wouterloot, J. G. A. 1982, Astr. Ap., 106, 171.
Hughes, V. A., 1985, Ap. J., 298, 830.
Hughes, V. A., 1988, Ap. J., 333, in press.
Cohen, R. J., Rowland, P. R., and Blair, M. M. 1984, MNRAS., 210, 425.

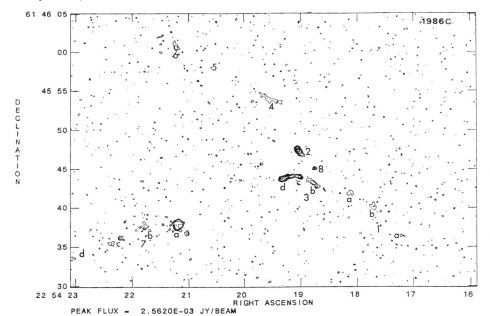

THE NEAR–STELLAR ENVIRONMENT OF COOL, EVOLVED STARS

P.G. JUDGE[1], R.E. STENCEL[2] & J.L. LINSKY[1,3]
[1] *Joint Institute for Laboratory Astrophysics,*
University of Colorado,
Boulder, CO 80309-0440, USA.
[2] *Center for Astrophysics and Space Astronomy,*
University of Colorado,
Boulder, CO 80309-0391, USA.

ABSTRACT. We discuss relationships between spectral indicators of chromospheric heating, winds and dust for "low" and "intermediate" mass stars evolving up the RGB and AGB, and suggest new observations from infrared to mm wavelengths which are needed. A full discussion of these relations, which are important for theoretical studies of heating and mass-loss processes, is in preparation.

Chromospheres and Mass Loss on the RGB/AGB

Based upon a sample of 42 nearby K, M, S and C stars, Judge (1988) discussed the near-stellar environments ($R/R_* < 10$) of stars evolving up the RGB and AGB. Here we relate that work to the observed wind properties and variability, supplementing IUE, VLA and IRAS data compiled by Judge (1988) with mm CO data (Knapp & Morris 1985; Olofsson et al. 1987, 1988; Wannier & Sahai 1986; Zuckerman & Dyck 1986), and optical and UV spectra (Sanner, 1976; Hagen et al. 1983; Drake 1985). Most of the stars are "non-variable" (spectral types K, early M), SRb (late M, some C), SRa (late M, some C) and Lb stars (C stars and 1 M giant).

We find: (i) "Warm" chromospheres persist even in "dusty" stars with relatively massive, cool circumstellar shells (as found for supergiants and late-M giants by Stencel et al., 1986). Relatively massive winds ($\dot{M} \lesssim 10^{-7}$) exist in non-dusty stars. (ii) Chromospheric heating rates drop steadily as stars evolve up the RGB/AGB, suggesting a common heating mechanism (probably acoustic, possibly magnetic). (iii) Wind velocities V_{wind} are $<< V_{escape}(R_*)$, hence most of the work in driving the wind is done against gravity (i.e. within $R \leq 2R_*$) (e.g. Holzer, 1987). (iv) V_{wind} reaches a minimum near spectral type M3-M5 III. (v) <u>Chromospheric radiative losses</u> dominate the energy balance in stars less evolved than M5 III, <u>winds</u> probably dominate for more evolved stars. (vi) Stars later than M5 III also show semi-regular and irregular pulsations of long periods ($\geq$ months).

These results suggest that fundamental changes occur near spectral type M5 III in the response of the outer atmospheres to evolution-dependent photospheric energy generation. We speculate that short-period photospheric disturbances which lead to enhanced chromo-

---
[3] Staff member, Quantum Physics Division, National Bureau of Standards

spheric heating dominate the energy balance for stars earlier than M5 III, whereas longer period disturbances dominate in cooler stars leading to the transfer of kinetic energy rather than heat and thus to enhanced mass loss (e.g. Willson & Bowen, 1986). Although mass-loss mechanisms are "known" in principle for regular pulsators (Jura, 1986), for non-Miras the problem is basically <u>unsolved</u>: point (iii) reveals the physics of the wind acceleration within the "<u>chromosphere</u>" ($R \leq 2R_*$) as the central issue of the mass-loss problem. The minimum of V_{wind} found near M5 III may be related to the decay of magnetic fields as stars evolve up the giant branch and/or a change in the nature of acoustic energy deposition, and to the additional momentum coupling of the dust observed in stars later than M5 III to the stellar radiation.

Additional observations are required between mm and infrared wavelengths: infrared CO spectra (cycle-dependent for semi-regular stars) will permit detailed studies of the velocity field in the wind acceleration zones (e.g. Hinkle et al. 1982; Wannier, 1985) and of the thermal nature of the chromospheric regions (Tsuji, 1988). mm CO line data (even upper limits) for stars <u>without substantial IR excesses</u> will reveal useful constraints on wind hydrodynamics and photochemistry, when combined with currently available data for warmer CS components. Sensitive cm and especially mm continuum observations will yield crucial constraints on ionized wind components and $sub - mm$ measurements will provide reliable angular diameters for carbon stars. Both steady-state and time-dependent theoretical models (Holzer, 1987; Willson and Bowen, 1986) should be re-examined in the light of these preliminary results.

We gratefully acknowledge grants JPL957632, NAG5-82, and the organizers for financial support.

References

Drake, S.A. (1985). In *Progress In Stellar Spectral Line Formation Theory*, eds. J.E. Beckman & L. Crivellari, pp. 351-359. Dordrecht, Holland. D. Reidel Publ. Co.
Hagen, W., Stencel, R.E. & Dickinson, D.F. (1983). Ap. J. **274**, 286.
Hinkle, K.H., Hall, D.B.N. and Ridgway, S.T., 1982. Ap. J. **252**, 697.
Holzer, T.E., 1987. In *Circumstellar Matter*, p. 289, Eds. Appenzeller I. and Jordan C. D. Reidel Publ. Co., Dordrecht, Holland.
Judge, P.G., 1988. In *The Evolution of Peculiar Giant Stars*, Eds. H.R. Johnson and B. Zuckerman, IAU Colloquium No. 106 (In press).
Jura, M., 1986. Ir. Astron. J. **17**, 322.
Knapp, G.R. & Morris, M., 1985. Ap. J. **292**, 640.
Olofsson, H., Eriksson. K. and Gustafsson, B., 1987. Astron. Astrophys. **183**, L13.
Olofsson, H., Eriksson. K. and Gustafsson, B., 1987. Astron. Astrophys. **196**, L1.
Sanner, F., 1976. Ap. J. Suppl. Ser. **32**, 115.
Stencel, R.E., Carpenter, K.G and Hagen, W., 1986. Ap. J. **308**, 859.
Tsuji, T. (1988). Astron. Astrophys. **197**, 185.
Wannier, P.G., 1985. In *Mass Loss from Red Giants*, p65. Eds. M. Morris and B. Zuckerman, Reidel: Dordrecht.
Wannier, P.G. & Sahai, R., 1986. Ap. J. **311**, 335.
Willson, L.A. & Bowen, G.H., 1986. Ir. Astron. J. **17**, 249.
Zuckerman, B. & Dyck, H.M., 1986. Ap. J. **304**, 394.

MILLIMETER AND SUBMILLIMETER OBSERVATIONS OF THE NGC 7538 MOLECULAR CLOUD

Osamu KAMEYA
Nobeyama Radio Observatory
Minamimaki, Minamisaku
Nagano 384-13, Japan

ABSTRACT. High-spatial-resolution observations of the NGC 7538 molecular cloud have been performed with the NRO 45-m telescope, JCMT, and Nobeyama Millimeter Array. We report on relationships among three high dense regions, three bipolar flows, and seven H_2O masers.

The high-density molecular cloud associated with NGC 7538 is a good object for investigating formation of OB stars. We have examined the structure of this region in detail using CO J=3-2, 1-0, ^{13}CO J=1-0, CS J=7-6, 2-1, 1-0, and H_2O $6_{16} - 5_{23}$ lines.

The distribution of peak antenna temperature of CO J=1-0 and ^{13}CO J=1-0 lines obtained by NRO 45-m telescope indicates that gas kinetic temperature and molecular column density are peaked around infrared sources (IRS1-3 complex, IRS9 and IRS11; Werner et al. 1979) and around the HII region NGC 7538. Detailed analysis of these data indicates that the high-column density region around the HII region is most likely a shocked region produced by the expansion of the ionized gas (Kameya and Takakubo 1988). Three high-column-density regions around the infrared sources are high-density cores related to each infrared sources. The apearance and the velocity structure of the cores indicate that the cores may be protostellar discs around protostars.

Figure 1 shows the maps of peak antenna temperature of CS J=7-6, 2-1, and 1-0 lines obtained with JCMT (J=7-6 line) and NRO 45-m telescope (the other lines). These maps show that CS J=1-0 emission is extended all over the high-density molecular cloud, but the higher transition emission tends to be localized around the infrared sources (IRS1-3 complex, IRS9, IRS11). Because beam size of the telescope is similar for each emission, differences between the maps are probably due to the different tracing densities of the lines: The higher transition line tends to indicate the denser part. The detailed analysis using LVG approximation suggests that molecular hydrogen density is about $10^6 cm^{-3}$ for the CS J=7-6 emitting region. These regions are probably the same high-density cores which have been found by the CO J=1-0 and ^{13}CO J=1-0 observations.

High-velocity wings of CO J=3-2 line have been observed with JCMT. Blue and Red wings in the spectra are found at IRS1-3 complex, IRS9 and IRS11, indicating three bipolar flows. These high-velocity flows are probably the same flows which have been found with NRO 45-m telescope by CS J=1-0 and CO J=1-0 lines (Kameya et al. 1986, 1989). These high-velocity flows are probably related to the high-density cores which are discussed above.

Figure 2 shows distributions of H_2O masers obtained by the Nobeyama Millimeter Array on May 1987. In addition to two previously known H_2O masers around IRS1-3 complex and IRS11, five new H_2O masers are found in the high-density molecualr cloud. A new H_2O maser is found at IRS9, but the other four H_2O masers are not associated with any infrared sources. The most interesting thing is that some of the new H_2O masers are in a ridge of the molecular cloud which seems to elongate southwest from IRS11 (see CS J=1-0 map in Figure 1). These maser clouds may be some protostars which are hidden in the dense gas and dust with high visual extinction.

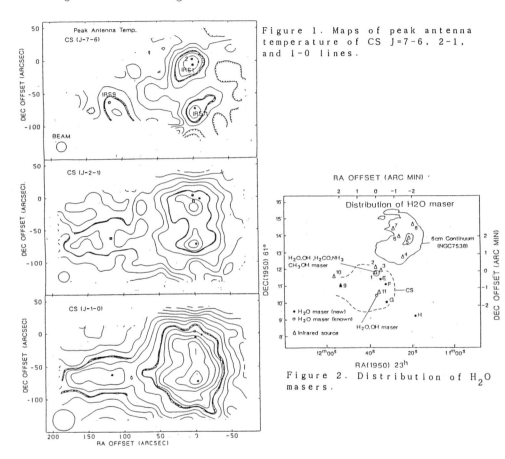

Figure 1. Maps of peak antenna temperature of CS J=7-6, 2-1, and 1-0 lines.

Figure 2. Distribution of H_2O masers.

REFERENCES

Kameya, O., Hasegawa, T.I., Hirano, N., Takakubo, K., and Seki, M. 1989, *Astrophys. J*, in press.

Kameya, O., Hasegawa, T.I., Hirano, N., Tosa, M., Taniguchi, Y., Takakubo, K., and Seki, M 1986, *Publ. Astron. Sco. Japan*, **38**, 793.

Kameya, O. and Takakubo, K. 1988, *Publ. Astron. Soc. Japan*, **40**, 413.

Werner, M., Becklin, E.E., Gatley, I., Mathews, K., Neugebauer, G., and Wynn-Williams, C.G. 1979, *Monthly Notices Roy. Astron. Soc.*, **188**, 463.

FAR INFRARED AND SUBMILLIMETER CONTINUUM OBSERVATIONS OF THE SAGITTARIUS B2 MOLECULAR CLOUD CORE

DARIUSZ C. LIS AND PAUL F. GOLDSMITH
Five College Radio Astronomy Observatory

and

RICHARD HILLS AND JOAN LASENBY
Mullard Radio Astronomy Observatory

ABSTRACT. We have observed the continuum emission from the Sagittarius B2 molecular cloud core, using the 15 m James Clerk Maxwell Telescope at Mauna Kea, Hawaii, and modeled the continuum emission from the Sgr B2(N) and (M) sources. The observed change in the peak middle-to-north flux ratio is a result of a very large opacity in the northern source and foreground extinction by the dust associated with the middle source. The results of the calculations suggest that the luminosity of the northern source may be significantly lower than that of the middle source.

1. Observations

The receiver system employed the helium cooled UKT14 bolometer together with the 1100 μ, 800 μ, 450 μ, and 350 μ filters. The data consists of four azimuth - elevation raster maps, 33 × 33 points in size with 5″ spacing, centered on the position of the Sgr B2(M) continuum source ($\alpha=17^h44^m10.5^s$, $\delta=-28°22'05''$), obtained by chopping between the source and a reference position offset by 80″ in azimuth. Uranus (less than 5° away from Sgr B2) was used for pointing and calibration. At each frequency a raster map of Uranus, with 11 × 11 points separated by 5″ was obtained before and after the Sgr B2 observation. The maximum pointing shift between two observations separated by ~30 min in time was smaller than 4″. By fitting a two-dimensional Gaussian to the Uranus maps we obtained the main beam size and the conversion from the output voltage to the unpolarized flux density. The beam was slightly elongated in the azimuthal direction (a typical semi-major to semi-minor axis ratio of ~1.2). The average beam sizes are given in Table 1.

2. Discussion

The Sgr B2(M) and (N) sources are quite well separated, and each is partially resolved. The regular beam patterns at 1100 μ and 800 μ allow for source size estimation. The observed source diameters are clearly larger than the antenna beam size, and assuming a Gaussian source shape, we derive a mean size of 18″ (0.75 pc) and 12″ (0.5 pc) for the middle and northern source, respectively. The 0.5 pc diameter of the northern source is a factor of ~2 bigger that that found by Carlstrom and Vogel (1988) from 3.4 mm interferometric measurements. This difference may be in part a result of antenna performance but may also reflect the extended low level emission which is present in the Carlstrom and Vogel data.

One of the most interesting results of previous high resolution continuum studies of Sgr B2 is a striking difference between distribution of the 53 μ continuum emission (Harvey, Campbell, and Hoffman 1977), and the 1300 μ continuum emission (Goldsmith, Snell, and Lis 1987). The 1300 μ emission peaks at the position of the northern source, where no excess 53 μ emission is observed. It has been suggested that the observed change in the middle-to-north peak flux ratio is caused by the fact that the northern source is situated behind the dust cloud associated with Sgr B2(M), and that its emission at short wavelengths is, therefore, attenuated by cold foreground dust (Thronson and Harper 1986; Goldsmith, Snell and Lis 1987).

We have modeled the dust emission from the source using the radiative transfer code of Egan, Leung, and Spagna (1988). The CO isotope observations (Lis and Goldsmith 1989) suggest that the molecular cloud associated with Sgr B2(M) has a radius of ~22.5 pc and consists of a constant density component and a power law density component with an exponent of -2 extending outward from ~1.25 pc. The continuum emission comes from two small cores embedded in the extended cloud, which due to small beam filling factor and high temperature do not show up in the $C^{18}O$ emission. Because our source size estimates may involve significant uncertainties we took source diameters as free parameters and varied them to obtain correct flux ratios in small (~19″) and large (30″) beams. Since the separation between the two sources is large compared to the core sizes, we solved the radiation transfer equation for the northern and middle sources independently, assuming no interaction between the sources. We assumed a power law grain emissivity with an exponent of 1 between 0.1 and 100 μ, and 1.5 for wavelengths longer than 100 μ. The middle source is characterized by a core diameter of 0.4 pc, 100 μ optical depth from the center to the edge of 3.7, and a central stellar luminosity of 1.5×10^7 $L_\odot$. Corresponding parameters for the northern source are 0.35 pc, 12, and 2.0×10^6 $L_\odot$, respectively. Our model successfully predicts the peak middle-to-north flux ratios (Table 1), as well as the observed spectrum in a 60″ beam at wavelengths between 30 μ and 1300 μ. The results of the calculations suggest that the observed change in the peak flux ratio is a result of the very large opacity in the northern core and attenuation by cold foreground dust associated with the middle source. Our model implies a luminosity of the northern source a factor of ~7 lower than that of the middle source. The northern source luminosity would be higher for a smaller diameter of the northern core. This would imply, however, a much higher middle-to-north peak flux ratio in a 30″ beam (the middle source would be too extended compared to the north source). A model with two very small cores is unable to reproduce the observed flux in a 60″ beam. The present model predicts a flux ratio of ~10 in a 30″ beam at 100 μ. This will be tested with high angular resolution data from the KAO to be obtained in the near future.

Table 1. JCMT Beam Sizes and Sgr B2 Source Data.

λ (μ)	FWHM(″)	M/N^a_{obs}	M/N^a_{mod}	M/N^b_{obs}	M/N^b_{mod}
1100	19	0.75	0.73	0.84	0.84
800	16	0.78	0.77	0.92	0.91
450	17	1.10	1.06	1.18	1.21
350	18	1.28	1.33	1.29[c]	1.46

[a] Referred to the telescope beam size determined from measurement of Uranus. [b] Referred to a 30″ FWHM beam. [c] The observed 350 μ flux ratio is affected by bad pixels near Sgr B2(M) and thus the ratio has to be considered as a lower limit.

References

Carlstrom, J.E., and Vogel, S.N., 1988, preprint.
Egan, M.P, Leung, C.M., and Spagna, G.F 1988, *Comput. Phys. Commun.*, **48**, 271.
Goldsmith, P.F., Snell, R.L., and Lis, D.C. 1987, *Ap. J.*, **313**, L5.
Harvey, P.M, Campbell, M.F., and Hoffmann, W.F. 1977, *Ap. J.*, **211**, 786.
Lis, D.C., and Goldsmith, P.F. 1989, *Ap. J.*, **337**, in press.
Thronson, H.A., and Harper, D.A., 1986, *Ap. J.*, **300**, 396.

SYNTHESIS IMAGING OF THE DR21(OH) PROTOCLUSTER

JEFFREY G. MANGUM[1,2], ALWYN WOOTTEN[2], AND LEE G. MUNDY[3]
[1] Department of Astronomy, University of Virginia, Charlottesville, Virginia.
[2] National Radio Astronomy Observatory, Charlottesville, Virginia.
[3] University of Maryland, College Park, Maryland.

Owens Valley Millimeter Wave Interferometer observations at $6.''9 \times 7.''7$ resolution of 2.7 mm continuum and $J = 1 \to 0$ $C^{18}O$ emission at velocity resolutions of 2.7 and 0.27 km s^{-1} have been made of the DR21(OH) molecular cloud. These observations reveal the presence of three multiple-component structures in this active star forming region. The first contains two objects which are each $\sim$ 35,000 AU in size, contain $\sim$ 100 $M_\odot$ of gas and dust, and are associated with OH and H_2O masers (see Figure 1). The binary nature of this structure was first observed at 1.4 mm by Woody et al. (1989) and Padin et al. (1989) (Woody et al. have designated the two components MM1 and MM2). These two components appear to be separated by $\sim$ 9'' on the sky and $\sim$ 3.5 km s^{-1} in velocity (see Table 1 for physical characteristics). Padin et al. point out that the two components are roughly coincident in position and velocity with the two H_2O maser emission centers (Genzel and Downes 1977). 2 cm continuum emission, though, has not been detected to a limit of 10 mJy beam^{-1} (Johnston, Henkel, and Wilson 1984), which may indicate that the newly formed massive stars which produced the H_2O maser emission have yet to emerge from the dense molecular cloud from which they were born.

The second structure, located $\sim$ 40'' south of the OH source (we have designated it DR21(OH)S), is $\sim$ 50,000 AU in size, is composed of $\sim$ 100 $M_\odot$ of gas and dust, and has no associated maser sources (see Figure 1 and Table 1). It was first detected as a source of 350 μm emission by Gear et al. (1988) and was later shown to be a thermal CH_3OH emission source by Batrla and Menten (1988). DR21(OH)S is extended in the east–west direction and has a velocity gradient of $\sim$ 3 km s^{-1}, indicating that this may be a rotating and/or multiple-component structure. To gravitationally bind this rotation requires a mass of $\sim$ 60 $M_\odot$ (assuming a source size of $\sim$ 15'' for DR21(OH)S), a value in good agreement with our line and continuum mass estimates. This source appears to be a cool ($T_K \lesssim 30$ K) pre-stellar condensation which may be evolving towards the star formation stage.

In Figure 1 there is a previously undetected continuum and line feature situated to the southwest of DR21(OH). Table 1 lists its measured physical parameters. In our $C^{18}O$ observations this source is more prominent in the high velocity resolution ($\Delta V = 0.27$ km s^{-1}) maps, indicating that it is a narrow-line condensation. This object is also extended to the north-south, indicating that it may be a multiple-component structure.

REFERENCES

Batrla, W. and Menten, K. A. 1988 *Ap. J. (Letters)*, **329**, L117.
Gear, W. K., Chandler, C. J., Moore, T. J. T., Cunningham, C. T., and Duncan, W. D. 1988, *M.N.R.A.S.*, **231**, 47p.
Genzel, R. and Downes, D. 1977, *Astr. Ap. Suppl.*, **30**, 145.
Johnston, K. J., Henkel, C., and Wilson, T. L. 1984, *Ap. J. (Letters)*, **285**, L85.
Norris, R. P., Booth, R. S., Diamond, P. J., and Porter, N. D. 1982, *M.N.R.A.S.*, **201**, 191.
Padin S., Sargent, A. I., Mundy, L. G., Scoville, N. Z., Woody, D. P., Leighton, R. B., Masson, C. R., Scott, S. L., Seling, T. V., Stapelfeldt, K. R., and Tereby, S. 1989, *Ap. J. (Letters)*, submitted.
Woody, D. P., Scott, S. L., Scoville, N. Z., Mundy, L. G., Sargent, A. I., Padin, S., Tinney, C. G., and Wilson, C. 1989, *Ap. J. (Letters)*, submitted.

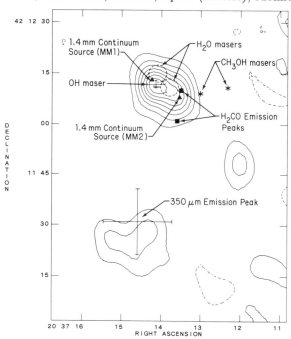

FIGURE 1

2.7 mm continuum emission. The contours are −3, 3, 6, 9, 12, 15, 18, 21, and 24 times the RMS noise of 8 mJy beam^{-1}. Positions of the CH_3OH (Batrla and Menten 1988), H_2O (Genzel and Downes 1977), and OH (Norris *et al.* 1982) masers and the 1.4 mm continuum (Woody *et al.* 1988), H_2CO (Johnston, Henkel, and Wilson 1984), and 350 µm (Gear *et al.* 1988) emission peaks are indicated.

TABLE 1

2.72 MM CONTINUUM COMPONENT CHARACTERISTICS

Component	α(1950)	δ(1950)	Angular Size (arcsec)	Integrated Flux (Jy)	Total Mass[a] ($M_\odot$)
DR21(OH)MM1	$20^h\ 37^m\ 14.^s2$	42° 12′ 12″	10.9 × 7.8	0.269	90[b]
DR21(OH)MM2	$20^h\ 37^m\ 13.^s6$	42° 12′ 9″	11.6 × 7.5	0.217	73[b]
DR21(OH)S	$20^h\ 37^m\ 14.^s7$	42° 11′ 25″	20.1 × 13.7	0.318	145[c]
DR21(OH)W	$20^h\ 37^m\ 12.^s0$	42° 11′ 48″	12.1 × 7.1	0.097	44[c]

[a] Assuming a λ^{-1} dust emmissivity law.
[b] Assuming a dust temperature of 40 K.
[c] Assuming a dust temperature of 30 K.

NEW HIGH-VELOCITY OUTFLOWS FROM THE PROTOSTAR W3 IRS5

G. F. MITCHELL[1], K. BELCOURT[1], J. P. MAILLARD[2], and M. ALLEN[3]

[1] Saint Mary's University, Halifax, N. S., B3H3C3, Canada
[2] Institut d'Astrophysique, 98 bis, boulevard Arago, F-75014 Paris, France
[3] Jet Propulsion Laboratory, Pasadena, CA 91109, USA

1. Introduction

Many young stellar objects are centres of high velocity outflows, as revealed by broad wings on CO emission lines. Emission observations at millimeter wavelengths, because of the sizeable beamwidths employed, do not reveal the properties of gas very near the central YSO. A complementary technique, the observation of absorption lines in the fundamental vibrational band of CO at 4.7 μm, has proved a powerful probe of gas flow close to embedded protostars (e.g. Geballe and Wade 1985, Mitchell et al. 1988a). We present here a high resolution infrared absorption spectrum of W3 IRS5, at $\approx 5 \times 10^5$ $L_\odot$ the most luminous known YSO, and compare with CO millimeter observations.

2. Observations

W3 IRS5 was observed on the nights of July 10 and 11, 1987, using the Fourier transform spectrometer at the Cassegrain focus of the Canada-France-Hawaii telescope on Mauna Kea, Hawaii. We employed a filter with a passband from 2080 to 2180 cm^{-1}, an aperture of 2.5 arcseconds, and a spectral resolution of 8.0 km s^{-1}. Correction for absorption lines in the Earth's atmosphere was accomplished by the observation of a standard star, α Lyr. The spectrum shows very strong absorption lines of $^{12}C^{16}O$ and weaker lines of $^{13}C^{16}O$. A composite kinematic profile employing 10 lines is shown in Figure 1. The CO absorption extends over a total velocity range of ≈ 65 km s^{-1}. Inspection of Figure 1 shows that the ^{12}CO absorption has a complex structure, consisting of two broad features, each of which is a blend of two components. The central LSR velocities of the four velocity components are -42 km s^{-1}, -53 km s^{-1}, -73 km s^{-1}, and -86 km s^{-1}. The existence of four components is confirmed by the presence of $^{13}C^{16}O$ lines at the same velocities.

3. Analysis and Discussion

Emission in the ^{13}CO J = 1-0 is centred at $v_{LSR} \approx -39$ km s^{-1} (Thronson, Lada, and Hewagama 1985) and is presumed to come from the cold gas within which W3 IRS5 is embedded. We can, therefore, identify the -42 km s^{-1} absorption component with the cloud core. The more negative velocity gas must be flowing towards us from the central source. Equivalent widths were measured from the ratioed spectrum, W3 divided by α

Lyr. We employed the ^{13}CO lines of the -42 km s^{-1} component because the ^{12}CO lines are saturated and will not yield accurate column densities. For the other three velocity components, we measured the ^{12}CO lines. Column densities were computed from the equivalent widths using a curve of growth analysis. To obtain information on the temperature of the gas, we plot $\ln[N_J/(2J+1)]$ versus the rotational energy in temperature units. If the CO energy levels are populated according to a Boltzmann distribution with a single excitation temperature, a straight line will result. The -42 km s^{-1} component is plotted in Figure 2, where we see that the high J points can be fit by one straight line and the low J points by a second straight line of steeper slope. The straight line through the low J points in Figure 2 indicates an excitation temperature of 50 ± 8 K. The higher rotational states are fit by a temperature of 470 ± 50 K and may indicate a disk or shell within ≈ 100 AU's of W3 IRS5.

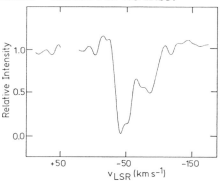

Fig. 1. A kinematic profile obtained by averaging ten lines of ^{12}CO.

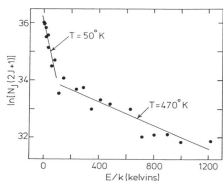

Fig. 2. A Boltzmann plot using the ^{13}CO lines at -42 km s^{-1}.

The outflowing gas (with v_{LSR} = -53, -73, and -86 km s^{-1}) is warm, with excitation temperatures from 60 K up to 400 K. If the warm gas is heated by the central source, it must be very close to it. The total ^{13}CO column density over all velocity components, on the assumption that ^{12}CO/^{13}CO = 60, is ≈ 2.2 x 10^{17} cm^{-2}. If we use the standard relation, N(^{13}CO) = 2 x 10^{15}A$_V$, the total visual extinction towards W3 IRS5 is A$_V$ ≈ 108 magnitudes, in reasonable agreement with the visual extinction implied by the large silicate optical depth of 7.64. CO 1-0 emission lines toward W3 IRS5 exhibit wings with a **total** width at the 100 mK level of 52 km s^{-1} (Bally and Lada 1983). The **outflow** velocity is, therefore, 26 km s^{-1}, only about half the maximum outflow velocity we are reporting here for warm gas near the embedded object. Another significant feature of the new outflows is their velocity structure. The existence of discrete velocity components in W3 IRS5 reinforces our previous conjecture (Mitchell et al. 1988b) that the extended outflows from YSO's, seen at millimeter wavelengths, are driven by sporadic outbursts.

References

Bally, J. and Lada, C. J. (1983), Ap. J., **265**, 824.
Geballe, T. R. and Wade, R. (1985), Ap. J. (Letters), **291**, L55.
Mitchell, G. F., Allen, M., Beer, R., Dekany, R., Huntress, W., and Maillard, J.-P. (1988a), Ap. J. (Letters), **327**, L17.
_____ (1988b), Astr. Ap., **201**, L16.
Thronson, H. A., Lada, C. J., and Hewagama, T. (1985), Ap. J., **297**, 662.

SOLAR MASS CLUMPS IN THE B5 CORE

Y. Pendleton, J. Davidson (NASA Ames), S. Casey, A. Harper, R. Pernic (Yerkes Observatory), and P. Myers (Harvard-Smithsonian)

Results from IRAS have shown that four compact sources exist within the central 0.°5 of the dark cloud Barnard 5 (B5) (Beichman et al.,1984, *Ap. J. (Letters)*,**278**, L45). The source denoted IRS1 by those authors is the only IRAS source located within the core ($\sim$ 5' x 10'). IRS1 appears pointlike in all bands except the 100 μm band. The NH_3 maps of B5 (Benson, 1984, Ph.D thesis; Benson and Myers 1988, private communication) show two peaks of emission in the core of B5. The southernmost of the two peaks in NH_3 emission corresponds to the location of the IRAS source while the northern peak, located approximately 2' north of IRS1, has no IRAS counterpart. Boss (1985, **ApJLett,288**, L25) suggested that a second protostellar source could be present at the position of the northern NH_3 peak based on his models of binary protostellar formation. We report observations of the B5 core at 160 and 360 μm which were made in order to test Boss' hypothesis and to further study the environment of a known star forming region. The 360 μm observations were made at the NASA Infrared Telescope Facility and the 160 μm observations were made from the Kuiper Airborne Observatory using the University of Chicago submillimeter 32 detector array camera and Far-IR 32 detector array camera. Figures 1 and 2 are photometric maps of the B5 core at 360 and 160 μm, respectively, made using a 45" beam. In figure 1, contour levels represent 1Jy/beam with the peak contour being 9 Jy/beam. In figure 2, contour levels represent 1.2 Jy/beam with the highest contour level at 10.8 Jy/beam. The (0,0) position corresponds to $\alpha(1950) = 03^h 44^m 28.^s7$ and $\delta(1950) = 32°44'30''$. Both maps display a crescent shaped feature similar to that seen in the outer regions of B5 in the $C^{18}O$ observations (Goldsmith, Langer, and Wilson, 1986,*Ap. J. (Letters)* ,**303** , L11) (GLW). The NH_3 map (Benson and Myers, 1988) correlates much better with the 160 μm map than with the 360 μm map. Four regions of relatively high density are observed along the crescent ridge, which we have labelled A, B, C, and D. Region A corresponds to the position of the IRAS (IRS1) point. Regions B and C have not been previously identified, while D corresponds to the northernmost NH_3 peak. The background flux from the ridge is $\sim$ 4 JY/beam at 360 μm and $\sim$ 3 JY/beam at 160 μm. After background subtraction and beam deconvolution, both A and B appeared approximately gaussian in shape with FWHM values of 44" and 32", respectively, at 360 μm. The total flux densities for A and B at 360 μm are 12 Jy and 5 Jy, respectively, and total flux densities at 160 μm for A and B are 16 Jy and $\leq$1.5 Jy.

To summarize:

1) The intensity of the ridge area (figures 1 and 2) is consistent with heating by the interstellar radiation field (ISRF), which has a total intensity of $\sim 2\times10^{-6}$ Wm^2sr^{-1}, as calculated by Mathis, Metzger, and Panagia (1983,*Astr. Ap.*, **128**, 212). The 360 μm map also indicates a total gas mass of 50$M_\circ$ which agrees very well with the result from the $C^{18}O$ observations by GLW. This mass is far greater than the Jeans Mass of the core, which, in the absence of significant magnetic and rotational support, will lead to gravitational instablility and fragmentation.

2) The broad spectrum of clump A (not shown) implies the presence of more than one component. For wavelengths $\leq$ 100 μm, a hot component ($\geq$ 50K) dominates the spectrum. IRAS measurements have determined a total luminosity ($\lambda\leq$ 100 μm) of $\sim$ 8$L_\circ$ for this

component Beichman et al., 1984). A colder component (~ 15K) dominates the spectrum at wavelengths longer than 100 μm.

3) The ISRF is insufficient to explain the heating of the cold component, however, our results are consistent with additional heating from the nearby source IRS1. Perhaps more intriguing is the possibility that the cold component is associated with the small disk-like structure around IRS1 recently seen in HCN by Fuller (reported elsewhere in these proceedings). The number density (2×10^5cm^{-3}) and visual extinction ($A_v=26$) values calculated from the continuum data given here agree very well with the number density implied by the HCN measurements of Fuller and the A_v value derived from near infrared measurements of IRS1 by Myers (private communication).

4) Clump B appears at 360 μm but not at 160 μm, implying a low dust temperature (≤ 11K). Taken together, the temperature estimate from the 360 and 160 μm maps and the intensity measured from the 360 μm map, imply a mass of ~ 3M$_\circ$, and are consistent with heating by the ISRF. Clump B appears to be gravitationally unstable based on our measurements which imply that it has a Jeans Mass of only 0.6M$_\circ$. Clump B is, therefore, indicative of the earliest stages of star formation, where fragmentation and isothermal contraction are occuring.

5) Although NH$_3$ is generally a good tracer of cold ($\sim$10K) dust, in regions of protostellar formation (such as B5) where temperatures can be $\leq$10K, submillimeter observations can reveal information about very cold dust that NH$_3$ observations would miss. This is evidenced by the fact that the NH$_3$ map of B5 correlates much better with the 160 μm map than with the 360 μm map.

6) Although our observations suggest a density clump in region D, these results cannot definitively answer the original hypothesis by Boss which predicted the presence of a second protostellar source in the B5 core at the location of the northernmost NH$_3$ peak. However, these observations have revealed additional clumps which are possible protostellar objects (most notably clump B), a result which is entirely consistent with Boss' theoretical models of fragmentation (Boss, private communication).

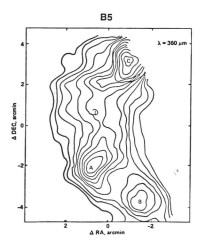

Fig. 1

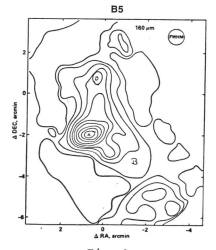

Fig. 2

INFRARED AND SUBMM OBSERVATIONS OF THE RHO OPHIUCHI DARK CLOUD

Derek Ward-Thompson & Ian Robson,
Lancashire Polytechnic, Preston, UK
Dolores Walther and Bill Duncan, JACH, Hilo HI 96720
Mark Gordon, NRAO, Tucson AZ 85721-0655

We present IRAS and submm data of the S1/IRS1 (Grasdalen et.al 1973, Harvey et.al 1979) region of the Rho Oph dark cloud. We have identified IRAS source 16235-2416 with IRS1. We announce the discovery of a new submm source, which is a cooler condensation of dust, and which we call SM1. It's mass and density are such that it is a candidate protostar. We announce the first positive detection of grain mantles at such a low visual extinction in this cloud.

IRAS Raw Data

To obtain better resolution than SKYFLUX, and improved accuracy for an extended source than the Point Source Catalogue, IRAS Raw Data was used, and a new method of background subtraction was employed. Each IRAS scan (2 degrees long) had the sky subtracted individually. This was done by obtaining a mean pixel value and standard deviation for the whole scan. Any pixel whose value lay more than 2 sigma above the mean, or more than 5 sigma below the mean, was set invalid. This initial cut was to exclude point sources and real extended emission. The remaining pixels were then input to an iterative process, which, at each iteration, calculated a 2-D polynomial surface, of specified order (in this case a constant), as a fit to the remaining valid pixels, and calculated a new standard deviation from the surface. Any pixel whose value lay more than 2 sigma from the surface was then rejected. 7 iterations were sufficient for the routine to 'settle' on the sky. The order of the fit necessary was taken to be the lowest which can remove the characteristic striping of IRAS images.

A map of the region around 16235-2416 at 60 microns is shown in Fig.1a. (the source is marked). There is more emission to the west of this than to the east, as suggested by Harvey et.al (1979). S1, IRS1 & 16235-2416 are indistinguishable at IRAS resolution. The source was found to have a grey-body spectrum with T=36+/-5K and B=1+/-0.1.

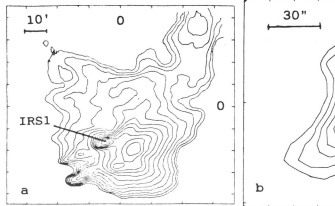
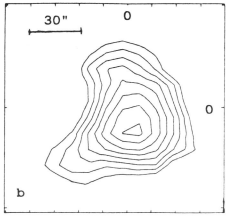

Fig.1a Rho Oph dark cloud at 60microns. Origin is 16h23m0s -24 0'0" **b** SM1 at 350microns. Origin is 16h23m26s -24 17'0"

Submm Data

A new peak of emission at submillimetre wavelengths has been discovered at R.A.(1950) = 16h 23m 25.4s, Dec.(1950) = -24 17' 16". This source is here named SM1. Fig.1b. shows a map of SM1 at 350 microns. The object can be fitted with a grey-body spectrum of T = 15 +/- 5K and B = 2.2 +/- 0.3. A high value of B indicates grain mantles (Aanestad 1975). We are publishing (Ward-Thompson et al. 1988) a spectrum confirming the presence of ice mantles in this region, as well as maps at 800 & 1100 microns. This is the first positive detection of ice mantles at such a low visual extinction (12mag) in the Rho Oph dark cloud.

Discussion

The mass of dust within an emitting region can be obtained from the submm flux (Hildebrand 1983), on the assumption that the cloud is optically thin at these wavelengths. Employing a typical gas-to-dust ratio, the total mass can then be calculated. The mass within a 0.14pc diameter of SM1 was found to be 0.01-0.1Mo. The Jeans mass at this density was calculated to be 0.7Mo, making SM1 a candidate protostellar object.

References

Aanestad P.A.(1975) Ap.J **200**,30
Grasdalen G.L.,Strom K.M. & Strom S.E.(1973) Ap.J **184**,L53
Harvey P.M.Campbell M.F.& Hoffmann W.F.(1979)Ap.J **228**,445
Hildebrand R.H. (1983) Q.J.R.A.S. **24**, 267
Ward-Thompson D., Robson E.I., Whittet D.C.B., Walther D.M, Duncan W.D. & Gordon M.A. (1988) MNRAS In press

Outflows from Massive, Pre-Main Sequence Objects:
Results from the OVRO Millimeter Wave Array

M. Barsony
Department of Astronomy
University of California
Berkeley, CA 94720 USA

NO DISK IN S106

In Figure 1 below the OVRO CS integrated emission is superposed on the 1.3 cm VLA radio map of S106. The tiny white speck in the very center is S106 IR, the source of the ionized lobes. Its wind is characterized by v_w=220 km s^{-1} and $\dot{M}$=1.7 $\times$ 10^{-6} M$_\odot$ yr^{-1}. The equatorial gap in this source has previously been attributed to a disk structure (Bieging, 1984). The higher resolution OVRO map resolves the ''disk'' into two shells of swept-up gas. The upper mass limit for any undetected molecular disk set by the OVRO observations is 0.01 M$_\odot$ (Barsony *et al.* 1989)

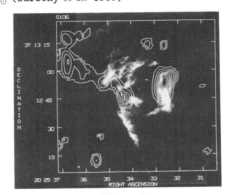

Figure 1

S87 — A MASSIVE MOLECULAR OUTFLOW

In Figure 2 the OVRO CS emission (white contours) is overlaid on the 6 cm VLA radio emission from S87 (the diffuse white structure). The molecular emission is red-shifted to the north and blue-shifted to the south of the VLA peak by 4 km s^{-1}. Lower limits to the lobe masses are 19 M$_\odot$ in the red wing and 23 M$_\odot$ in the blue wing. The bipolarity of this source is determined by the surrounding cloud density structure (Königl 1982) and not by an intrinsically bipolar ionized wind (Barsony 1989).

The Hα line profile of S87/IRS1 is presented in Figure 3. The velocity resolution is 25 km s^{-1}, with the 0.4" $\times$ 30" slit oriented along a north-south line passing through the position of IRS1. Essentially all the Hα emission is blue-shifted with respect to the cloud's rest velocity, at V_{LSR} = 22 km s^{-1}. The line profile is asymmetric because the red-shifted ionized wind component (v_w = 160 km s^{-1}, $\dot{M}$ = 1.8 $\times$ 10^{-5} M$_\odot$ yr^{-1}) is obscured by the intervening molecular cloud.

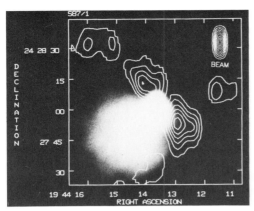
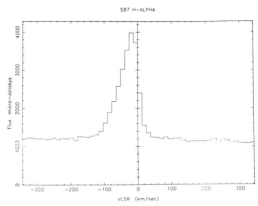

Figure 2 Figure 3

LkHα101 : A Relic Molecular Outflow?

LkHα101 is an emission line star at the apex of a reflection nebula. In Figure 4, the 6 cm VLA contours (Becker and White 1988) are overlaid on a Gunn r image of the $2.5' \times 4.0'$ field centered on LkHα101. Although LkHα101 is the source of a powerful, ionized, stellar wind ($v_w = 350$ km s^{-1} and $\dot{M} = 1.1 \times 10^{-5}$ M$_\odot$ yr^{-1}), it is <u>not</u> currently driving a molecular outflow. The millimeter emission from this source is shown overlaid on a Gunn i image in Figure 5. The central contours delineate the 110 GHz continuum emission, originating from the unresolved, ionized wind, whilst the remaining contours represent the integrated ^{13}CO emission amounting to 23 M$_\odot$ of molecular gas (Barsony *et al.* 1990).

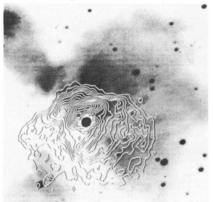
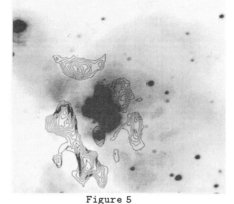

Figure 4 Figure 5

References

Barsony, M. 1989, *Ap.J.*, 345, 268.
Barsony, M., Bally, J., Scoville, N. Z., and Claussen, M. 1989, *Ap.J.*, 345, 268.
Barsony, M., Scoville, N. Z., Schombert, J. M., and Claussen, M. J. 1990 *Ap. J.* in press.
Becker, R. H. and White, R. L. 1988, *Ap.J.*, 324, 893.
Bieging, J. H. 1984, *Ap.J.*, 286, 591.

Section V

Galaxies

THE EVOLUTION OF STARBURST GALAXIES TO ACTIVE GALACTIC NUCLEI

Nick Scoville
Owens Valley Radio Observatory
California Institute of Technology

Recent observations of luminous IRAS galaxies ($L_{IR} > 10^{11}$ $L_\odot$) have revealed that virtually all are extremely rich in molecular gas with H_2 abundances 5-20 times that of the Milky Way, and deep CCD imaging indicates that most are recent galactic mergers. Interferometric observations with the Owens Valley Millimeter-Wave Interferometer at 2.6 and 1.3 mm for five of these galaxies demonstrate that approximately half of the interstellar matter is contained in the central kpc. Interferometry on the most luminous galaxies ($L_{IR} \geq 10^{11} L_\odot$) reveals that approximately half of the total interstellar matter is contained in the central kpc with mean densities of several hundred H_2 cm^{-3}. Such gas concentrations should result in the very rapid formation of stars, i.e. a central star burst yielding a massive central star cluster. The deep potential of the central star cluster and the high density of interstellar gas will ensure that virtually all of the gas lost during late stellar evolution sinks to the center of the cluster, building up a central, massive black hole. For a coeval star cluster of 4x10^9 $M_\odot$, a mass of approximately 1.5x10^9 $M_\odot$ will accumulate within approximately 10^8 years and accretion at an average rate of 7 $M_\odot$ yr^{-1} over this time will result in a mean accretion luminosity of 10^{13} $L_\odot$. This luminosity, radiated at X-ray and uv wavelengths from the inner accretion disk ionizes the mass loss envelopes of the surrounding red giant stars providing an origin for the broad emission line regions of QSO's.

1. Luminous Infrared Galaxies

In the nearby bright IR galaxies, high resolution millimeter-wave interferometry and single dish observations have revealed a variety of morpohologies in the neutral gas. Three of the galaxies first mapped with the Owens Valley millimeter-wave inteferometer showed elongated bar-like distributions for the molecular gas in the central kpc. The results for IC 342 and NGC 6946 have been published by Lo et al. (1984) and Ball et al. (1985). The more recent CO interferometry for NGC 253 consisting of a mosaic of seven 1′ fields (Canzian, Mundy, and Scoville 1988) shows a massive bar of molecular gas aligned with the stellar bar seen in optical and near infrared maps (Scoville et al. 1985).

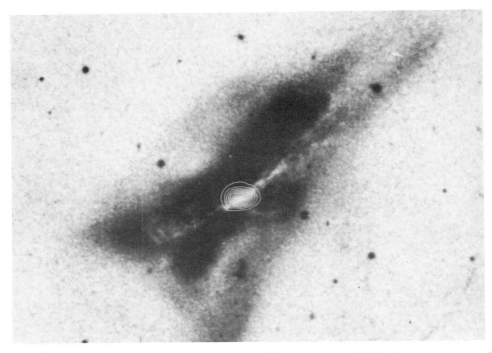

Figure 1: Integrated intensity map of CO emission in NGC 520 superimposed on the optical photograph from the Arp atlas. Contour levels are 10% of peak. The hatched beam symbol indicates both the position angle and size of the synthesized beam (Sanders *et al.* 1988).

A rather different morphology is found in the nearby Seyfert II galaxy NGC 1068. In this case, approximately $4 \times 10^9 M_\odot$ of molecular gas resides in a ring at the outer edge of the bright optical disk (Myers and Scoville 1986). This ring of neutral gas situated just outside the stellar bar recently discovered in the near infrared (Scoville *et al.* 1988) is somewhat surprising in view of the abundant evidence for a high rate of star formation in the interior optical disk. On the other hand, the kinematics of the molecular gas indicate a substantial component of radial motion suggesting that at times in the past, there has been an abundance of star forming material within the central disk.

Perhaps most dramatic in terms of star burst activity are the high luminosity and ultraluminous galaxies discovered as a result of the IRAS survey. At the higher luminosities, one sees a high preponderance of double nuclei and/or extended tidal tails indicative of strong galactic interactions or the merging of two galaxies. It is also evident that the optical spectra of the ultraluminous galaxies are dominated by non-thermal emission characteristic of a narrow line AGN or Seyfert nucleus rather than thermal HII region-type spectra seen in the lower luminosity galaxies. This qualitative assessment of the optical data strongly suggests that *the highest luminosities are initiated by galactic collisions, and the dominant energy source may in fact be a non-thermal AGN*. Virtually all the luminous IRAS galaxies have also been shown to be extremely rich in interstellar gas, predominantly molecular

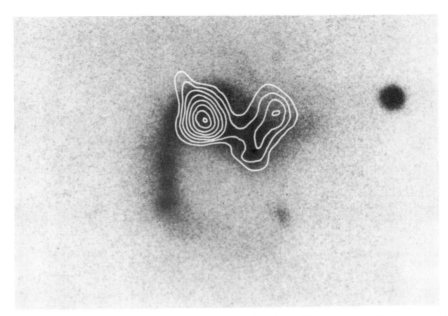

Figure 2: Integrated CO insensity map for Arp 55 superposed on an optical photograph; contours are 10% of peak intensity (Sanders et al. 1988).

hydrogen (e.g. Sanders, Scoville, and Soifer 1987).

The molecular gas is also highly concentrated in their nuclei. Over the last two years, the Millimeter Wave Interferometer at Owens Valley Radio Observatory has been used for aperture synthesis mapping of the CO emission in eight of the luminous galaxies (Scoville et al. 1986, Sargent et al. 1987, Sanders et al. 1988, Scoville et al. 1989). In Figures 1 and 2, the OVRO maps are shown for NGC520 and Arp 55. In each case, 30-70% of the total CO emission is confined to a region $\leq 10''$ in size, centered on one of the galactic nuclei. Most spectacular is Arp 220 ($L_{IR}=1.3 \times 10^{12}$ $L_\odot$) where 10^{10} $M_\odot$ of H_2 is contained within the central $R \leq 750$ pc. The OVRO measurements of the dust continuum at 1.3 mm indicate that approximately 40% (140 mJy) of the total continuum is contained within a 4" region (Woody et al. 1988). The mean molecular gas density *averaged* over a spherical volume of this size is approximately 200 H_2 cm^{-3}. The ratio of far infrared luminosity to H_2 mass is 100 $L_\odot/M_\odot$. For comparison, the mean value of this luminosity-to-mass ratio in Galactic GMCs is 3 $L_\odot/M_\odot$ and the maximum value, obtained in localized areas immediately adjacent to galactic HII regions, is 40 $L_\odot/M_\odot$.

2. Stellar Evolution in Nuclear Star Bursts

We have recently undertaken a theoretical investigation of the evolution of interacting gas-rich galaxies (Norman and Scoville 1988, Scoville and Norman 1988). Our model starts with the formation of a single coeval stellar cluster of total mass 4×10^9 $M_\odot$. This mass is distributed among stars with a Salpeter initial mass func-

Molecular Gas in the Nuclei of High Luminosity IRAS Galaxies

	$<CZ>_{CO}$ (km s^{-1})	Distance[a] (Mpc)	OVRO beam (″)	R_{beam} (kpc)	M_{gas}^{b} ($10^9 M_\odot$)	M_{dyn}^{c} ($10^9 M_\odot$)	M_{gas}/M_{dyn}
NGC 520	2261	29	6	0.42	1.9	8.0	0.24
IC 694	3030	42	5	0.51	1.4	5.6	0.25
NGC 7469	4963	66	6	0.96	4.5	12.5	0.36
Arp 220	5452	77	4	0.75	9.0	25.1	0.36
Arp 55	11957	163	7	2.8	14.4	23.0	0.63

[a] Assuming a Virgocentric flow model with H_o = 75 km s^{-1} Mpc^{-1}.

[b] H_2 + He mass assuming a Galactic CO to H_2 conversion ratio of 3.6×10^{20} H_2 cm^{-2} (K km s^{-1})$^{-1}$.

[c] Dynamical mass inside the interferometer beam calculated from $M_{dyn} = 2.1 \times 10^5 \Delta V^2 R_{kpc}$ where ΔV is the CO half-power line width.

[d] OVRO interferometer data from Scoville et al. (1986), Sargent et al. (1987), and Sanders et al. (1987).

tion ($\alpha=2.35$) over the range 1-50 $M_\odot$. The total number of stars in the cluster is 1.4×10^9 and evolution of the stellar population is followed. The small radius of the cluster (10-50 pc) implies an extremely high escape velocity ($>10^3$ km s^{-1}) ensuring that all mass-loss occurring during the late stellar evolution phases (red giant mass-loss and supernovae) will be trapped in the cluster and eventually sink, dissipatively, to the center. Thus, based on standard stellar evolution and an assumed initial mass function, it is possible to predict the growth rate of the central black hole and its accretion luminosity.

In Figure 3, the black hole luminosity (0.1 $\dot{M}$ c^2), the total stellar luminosity and the main sequence stellar luminosity are shown as a function of time. Over most of the time span 2×10^7 - 5×10^9 years, all three luminosity components decay as power laws in time. The dominant luminosity contribution, provided by accretion onto the black hole, decays as t$^{-1.1}$ (assuming a Salpeter IMF).

The bulk of the accretion luminosity will be radiated at uv and x-ray wavelengths. This high flux of ionizing radiation will have immediate, observable effects on the mass-loss envelopes of red giant stars in the surrounding stellar cluster. For a star with $\dot{M} = 10^{-5} M_\odot$ yr^{-1} at distance 10^{18} cm, the uv radiation will penetrate to a depth of 10^{14} cm and cause complete ionization in the zone outside this point. The structure of the mass-loss envelope will be that of an inverted Stromgren sphere with maximum electron density 10^9 cm^{-3} and size 10^{14} cm. These parameters are nearly identical to those derived for the broad emission line regions in QSOs, and Scoville and Norman (1988) suggest that the ionized mass-loss envelopes are the source of the broad line emission in AGNs. Astronomical research with the OVRO Interferometer is supported in part by NSF grant AST 87-14405.

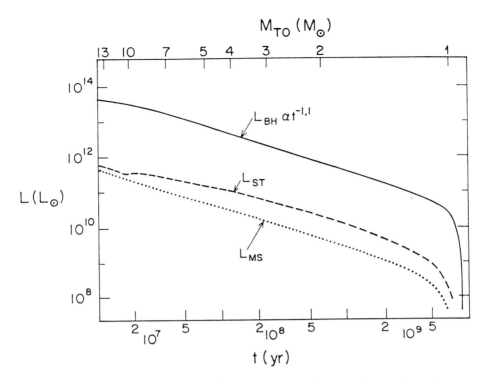

Fig. 3. The accretion luminosity, total stellar luminosity, and main sequence stellar luminosity are shown as a function of time for a coeval population of total mass 4×10^9 $M_\odot$ (Norman and Scoville 1988). At all times $<10^9$ years, the black hole accretion luminosity exceeds the total stellar luminosity by more than a factor of 25 and the accretion luminosity decays as $t^{-1.1}$.

3. REFERENCES

Ball, R., Sargent, A.I., Scoville, N.Z., Lo, K.Y., and Scott, S.L. 1985, *Ap.J. (Letters)*, **298**, L21

Canzian, B.J., Mundy, L.G., and Scoville, N.Z. 1988, *Ap.J.*, **333**, 157

Lo, K.Y. et al. 1984, *Ap.J. (Letters)*, **282**, L59

Myers, S.T. and Scoville, N.Z. 1986, *Ap.J. (Letters)*, **312**, L39

Norman, C.A. and Scoville, N..Z. 1988, *Ap.J.* **332**, 163

Sanders, D.B., Scoville, N.Z., Sargent, A.I., and Soifer, B.T. 1988, *Ap.J. (Letters)*, **324**, L55

Sargent, A.I., Sanders, D.B., Scoville, N.Z., and Soifer, B.T. 1987, *Ap.J. (Letters)*, **312**, 235

Scoville, N.Z., Matthews, K., Carico, D., and Sanders, D.B. 1988, *Ap.J. (Letters)* **327**, L61

Scoville, N.Z. and Norman, C.A. 1988, *Ap.J.*, **332**, 163

Scoville, N.Z., Sanders, D.B., Sargent, A.I., Soifer, B.T., Scott, S.L., and Lo, K.Y. 1986, *Ap. J. (Letters)*, **311**, L47

Scoville, N.Z., Soifer, T., Neugebauer, G., Young, J.S., Matthews, K., and Yerka, J. 1985, *Ap.J.*, **289**, 129

Scoville, N.Z., Tinney, C., Sanders, D.B., Sargent, A.I., and Soifer, B.T. 1989, *Ap.J.* (submitted)

Soifer, B.T., *et al.*. 1984, *Ap. J.*, **283**, L1

Woody, D.P., Scott, S.L., Scoville, N.Z., Mundy, L.G., Sargent, A.I., Padin, S., Tinney, C.G., and Wilson, C.D. 1988, *Ap.J. (Letters)*, (in press)

SUBMILLIMETRE OBSERVATIONS OF ACTIVE GALAXIES

Walter K. Gear
Royal Observatory
Edinburgh
Scotland
EH9 3HJ

1:INTRODUCTION

It is a little over a year since I gave a review on this topic at the Stirling Summer School (Gear 1987) and as I predicted then, the field has since mushroomed in terms of the number of people making mm and submm observations of Active Galaxies, as can be seen from the number of papers in this area at this symposium.

That review and other papers in the proceedings of the Summer School contain a great deal of essential technical and theoretical background for any relative newcomer to the field, so in this review I will concentrate on results, and specifically the most recent results to emerge from the new generation of large mm and submm telescopes.

2: GAS MASSES AND STARBURST GALAXIES

Stretching the definition of activity slightly to include galaxies undergoing very active star-formation in their nuclei, this field is one where submillimetre continuum observations can be very powerful, and one where there have been several talks and posters at his conference.

For sources where one is convinced that the submm emission is thermal we can use the submm measurements to determine the mass of interstellar material in the galaxy make deductions about the rates and efficiencies of star formation compared to other sources. This is because the submillimetre continuum emission, unlike the vast majority of molecular line tracers, is always optically thin, so the emission from the whole galaxy is being sampled equally. The deduction of column densities and masses from submillimetre continuum data has been summarized very elegantly by Hildebrand (1983), I have also presented a slightly modified analysis in Gear (1987).

The Bonn group have made 1.3 mm measurements of a large sample of Markarian galaxies (Krugel et al 1988a,b) and from the derived masses found a relationship between total luminosity and gas mass $L/L_\odot = 3.4 \times 10^4 (M_g/M_\odot)^{0.74}$. They find that this relationship holds very well for all 'active' galaxies including star bursts, but that in a sample of 'inactive' galaxies the ration L/M_g is a factor 20 lower. This result, if true, suggests that a galaxy can be in one of 2 states, with some trigger mechanism providing a switch.

When looking at these results it is important to bear in mind, however, that the uncertaintites involved are really very large and often get swept under the carpet. The relative calibrations of gas masses derived from continuum data and from molecular line data, particularly CO, has been known to be a problem for some time. It is worth remembering that all the calibration procedure is based on observations of a very small number of *galactic* regions which are thick enough to measure continuum optical depth but thin enough to get accurate optical extinction measurements. The relative calibration of optical extinction to molecular hydrogen mass is then based on measurements of the line of sight to 9 stars by the Uhuru satellite in the early 1970's. These calibrated relationships are then taken to hold not just in other regions in our own galaxy but in other galaxies as well. Then there are the assumptions involved in the fitting procedure; inaccuracies of up to an order of magnitude are possible if one is not careful about fitting temperatures and dust emissivities properly.

Clearly some accurate cross calibrations of the different methods for determining gas masses needs to be done for different types of galaxy. An interesting example for which data is available is the classical starburst galaxy M82 (Smith et al 1988 in prep). The CO (J=1-0)/(J=2-1) line ratio in this source is very unusual and similar to that expected if the CO emission is in fact optically thin, whereas the $^{12}CO/^{13}CO$ ratio suggests that it may in fact be due to a temperature gradient effect. Recent 450 μm data shows that the continuum morphology is different from the CO morphology (Figure 1), showing that the two things may *not* in fact be tracing the same thing, i.e. the H_2 column density which they are both supposed to.

Figure 1: (a) CO Map of M82

(b) 450 micron continuum Map of M82

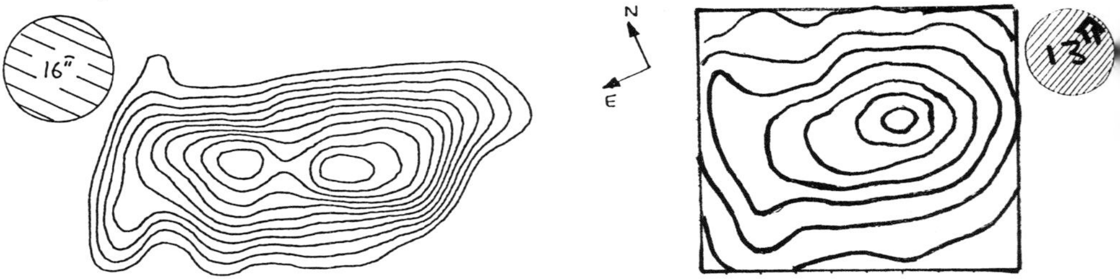

3: RADIO-LOUD QUASARS AND BLLACS

This class of AGN has the longest history of mm observations, dating back to the first heroic 1mm measurements of 3C273 and 3C279 by Low (1965). The main reason why these have been the best-studied AGN up until now is that they are the brightest ! with at least 2 dozen objects being brighter than 1 Jy at 1mm. They are also rewarding for long-term study because of their often dramatic varibility behaviour. The paper by Ian Robson in this volume also discusses some of the latest results on these objects.

It is well-known that compact, core-dominated radio-loud quasars, and the Bllac objects, have flat radio spectra. This flatness is caused by the superposition of a number of self-absorbed synchrotron components, which in many cases can be seen on VLBI maps. The more compact these regions are, the higher the frequency at which they become self-absorbed, and the higher the frequency at which they emit the bulk of their energy. Observations in the submm range sample the emission from the *most compact* cores of these sources, regions which are unresolved on even the highest frequency VLBI maps.

Detailed monitoring of samples of these sources (Robson et al 1983; Gear et al 1986; Brown et al 1988) has shown very clearly that the submm emission arises in a distinct, highly variable component, whose emission is superimposed on the more slowly varying "quiescent" longer wavelength components. Flares originate in this region and propagate to radio wavelengths. The reader is referred to Figures in the paper by Robson in this volume, space does not allow their reproduction here.

Detailed monitoring of individual flares (Robson et al 1983, Gear et al 1986) has also shown that their evolution is not consistent with earlier, relatively simple source models and has led to the development of more successful models in which the flaring emission is interpreted in terms of shock waves in adiabatically expanding relativistic jets (Marscher and Gear 1985; Gear 1987).

Observing sources through 'anti-flares' is also important, as has been demonstrated for NGC1275 (3C84). This source has all the radio properties typical of a compact radio source, and was first observed in the mm/submm by Longmore et al (1984). At that time the flux was relatively high and was clearly a pure non-thermal spectrum. Over the early 1980's however it gradually faded and in 1983 when IRAS observed it its combined submm-far-IR-IR spectrum looked rather different, as shown in Figure 2 taken from Gear et al (1985). By subtracting off the power law spectrum obtained by extrapolating from 1mm to 1 μm we get the spectral shape of the excess component, Figure 3. The dashed line in the Figure is not a fit to the spectrum but is in fact the shape of the spectrum of NGC1068, the classical thermal IR-mm source, normalized to the 100 μm flux of NGC1275. This shows that both thermal and non-thermal emission can exist in the same source. Fortunately NGC1275 is at the lower end of the luminosity range for compact radio sources, if a similar thermal excess existed in 3C273 for instance it would never be detectable because the non-thermal luminosity is two orders of magnitude greater.

Figure 2: IR-mm Spectrum of NGC1275(3C84) in 1983. Note the FIR excess component.

Figure 3: The spectrum of the excess component, from Gear et al (1985).

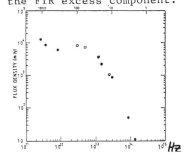
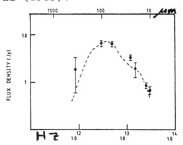

4: RADIO-QUIET QUASARS AND SEYFERT GALAXIES

Radio-loud sources are of course very much a minority amongst AGN. With the recent increases in sensitivity available with new detector systems and telescopes such as IRAM, JCMT and CSO, there has been an enormous amount of interest over the past 2 years in making observations of radio-quiet AGN in order to determine the nature of their IR-mm emission and to try to understand why they have no radio emission when their UVOIR emission appears so similar to the radio-loud objects.

IRAS unfortunately did not provide an answer to this question, as it showed that the rising IR spectra continue out to at least 60 μm in most sources and even 100 μm in some cases. It was already known that the spectra of radio quiet AGN must turnover before 1mm (Ennis et al 1982; Robson et al 1985), the best method of determining the nature of the emission therefore is to measure the spectral slope of the submm emission as it rises from the radio to meet the IRAS measurements. A self-absorbed synchrotron source can have a slope no steeper than 2.5, while optically thin dust emission can have a slope of 3-4, depending on the emissivity of the dust grains. In principle it is possible to have an even steeper cut-off caused by external free-free absorption of a synchrotron source, but in order for this to explain all radio-quiet AGN, the covering factor of the broad-line cloud region would have to be almost unity, which is not consistent with optical and UV data.

The deepest measurements have been made by the Bonn group, and they in fact detected 1.3 mm emission from the IRAS-discovered quasar 13349+2438 at the level of 4 mJy. However as can be seen from Figure 4 taken from their paper, the gap between 1300 μm and 100 μm is sufficiently large that it was possible to fit both a thermal and non-thermal model to the spectrum. In order to make a more unambiguous case for one process over the other it is neccessary to make higher frequency sub-millimetre measurements and preferably to detect a source at more than one frequency beyond the turnover.

A great deal of telescope time on all the new dishes has been dedicated to this task, and there are several papers at this conference discussing the results. Figure 5 shows the results of a an observation of NGC4151 at 450 μm by Edelson, Malkan, Robson and myself (Edelson et al 1988). We used UKT14 on JCMT over a period of 3 nights to achieve an upper limit of 200 mJy, which is by far the deepest limit that has been obtained at this frequency. This point is shown along with FIR data including a 155 μm detection by Engargiola et al 1988. This limit is very significant as it shows categorically that in this object at least the submm emission is dominated by thermal emission from heated dust, as the 155 μm to 450 μm spectral index is limited to being steeper than 3.0. If there is a significant nonthermal contribution to the infrared emission from NGC4151 then it must turnover shortward of 100 μm.

The Bonn group have come to a similar conclusion on the basis of very deep 1.3 mm detections and upper limits. And Dr Chini is giving a paper at this symposium on the results. Clearly however this point is not completely answered yet though as we really need some submillimetre *detections* and spectral measurements to be absolutely sure. Very recently Robson, Hughes and myself have detected at 800 μm a Markarian Seyfert Galaxy detected at 1.3 mm by Chini et al, the slope of this source is $\nu^{4.0}$ so it is clearly a thermal source. However this source is again at the lower luminosity end of the range, the more extreme luminosity sources need to be investigated in more detail.

Figure 4: Spectrum of the radio quiet QSO 13349+2438 from Chini et al (1987). Note that the 0.1 to 1.3 mm spectrum can still be fit by either a thermal (solid) or non-thermal (dashed) model.

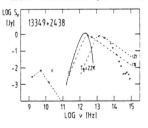

Figure 5: IR-mm spectrum of NGC4151 from Edelson et al (1988). A non-thermal source model is ruled out by the 450 micron upper limit.

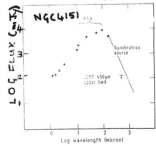

5: FUTURE OBSERVATIONS

5.1 Galaxies : Clearly a high priority will be to make detailed high resolution maps of all galaxies close enough to resolve, and to make detailed comparisons between continuum and molecular line observations.

Photometric surveys of larger samples of galaxies will also be important. Combining submm and IRAS data allow the temperature, luminosity, dust properties and total masses to be determined as a function of galaxy type and location (e.g. the effect of being in a cluster such as Virgo).

5.2 Blazars and Radio-Loud Quasars : The increased sensitivity of the new large dishes will allow investigation of the spectral shape of larger samples than has previously been possible. It will be important to make coordinated monitoring programmes, so that flares can be monitored in detail from their earliest stages through to the point where they can be seen on cm VLBI maps. Anti-flares also need to be followed, as was described above for 3C84.

The very exciting developments in mm-VLBI described elswhere in these proceedings are of tremendous importance for studying very compact sources. mm-VLBI needs to be pushed to even higher frequencies however as even at 90 GHz many core sources are still optically thick.

If at all possible it would be highly desirable to simultaneously monitor the millimetre and X-ray emission of a highly variable source continuously for a lengthy period, in order to test synchro-compton models. If these models are valid there should be a very high degree of correlation between the two wavebands as the photons that are compton-scattered to the x-ray region originate at the synchrotron self-absorption peak.

5.3: Radio-Quiet AGN :   The question of thermal or non-thermal emission needs to be settled, not just for one or two objects, but for a wide range of luminosity.

REFERENCES
Brown, L.M.J. et al 1988, Ap.J. in press.
Chini, R., Kreysa, E. and Salter, C.J. 1987, A.A. 182,L63.
Edelson, R., Malkan, M., Robson, E.I. and Gear, W.K. 1988, Nature in press.
Engargiola, G., Harper, D.A., Elvis, M. and Willner, S.P. 1988, Ap.J. 332,L19.
Ennis, D., Werner, M. and Neugebauer, G. 1982, Ap.J. 262,460.
Gear, W.K. et al 1985, MNRAS 217,281.
Gear, W.K. et al 1986, Ap.J. 291,511.
Gear, W.K. 1987 in Proceedings os Stirling Summer School on Millimetre and Submillimetre Astronomy, eds R.D. Wolstencroft and W.B. Burton, Kluwer Academic Publishers.
Hildebrand, R.H, 1983, Q.J.R.A.S 24,267.
Krugel, E., Chini, R., Kreysa, E. and Sherwood, W.A. 1988a, A.A. 190,47.
Krugel, E., Chini, R., Kreysa, E. and Sherwood, W.A. 1988b, A.A. 193,L16
Longmore, A.J. et al 1984, MNRAS 209,373.
Low, F.J. 1965, Ap.J.
Marscher, A.P. and Gear, W.K. 1985, Ap.J. 298,114.
Robson, E.I. et al 1983, Nature 305,194.
Smith., P., Puxley, P.J., Gear, W.K. and Mountain, W.K. 1988 in prep.

MILLIMETER OBSERVATIONS OF LUMINOUS IRAS GALAXIES

J. Keene, D. P. Carico, G. Neugebauer, and B. T. Soifer
Division of Physics, Mathematics and Astronomy
California Institute of Technology
Downs Laboratory of Physics, 320-47
Pasadena CA, 91125
USA

We have observed a group of luminous IRAS galaxies in the continuum at 1.25mm with the Caltech Submillimeter Observatory in 1988 May and September. Table I lists the galaxies and their measured flux densities in a 30" beam. The observing conditions for these observations were excellent and stable. Uranus was the fundamental flux density calibrator; it was assumed to have a brightness temperature of 100 K at 1.25 mm (Hildebrand et al. 1985; Ulich 1974). Integration times were typically a half hour per night, repeated for a total of two or three nights. The bolometer NEFD is 0.5 Jy Hz$^{-1/2}$ giving statistical errors of 8 to 12 mJy. Systematic uncertainties have not been included in the error estimates, but we believe that they are small, at most 20%.

TABLE I

Galaxy	F_ν(1.25mm) (mJy)	Galaxy	F_ν(1.25mm) (mJy)
Arp 220	225 ± 11	Zw 049.057	47 ± 12
NGC 3690	95 ± 12	UGC 2982	42 ± 12
NGC 1055	70 ± 10	NGC 1143/4	37 ± 8
NGC 7771	67 ± 8	NGC 958	33 ± 8
NGC 7541	65 ± 8	Mrk 231	29 ± 8

For an optically thin source, the observed infrared continuum dust emission, $F(\nu)$, is proportional to the integral over temperature of the emission function for an individual dust grain, $Q(\nu)B(\nu,T)$, times the number of dust grains at each temperature. We have calculated the amount of dust within discrete temperature intervals by deconvolving this integral. The amount of dust was used to obtain the luminosity contributed by dust at each temperature, $L(T)$. The dust emissivity, $Q(\nu)$, was assumed to be proportional to ν shortward of 100 μm and ν^2 longward. Shown in Figure 1 are the spectra of Arp 220 and NGC 3690 from 3 μm to 1.25 mm, along with the derived dust luminosity as a function of temperature. The results for the two galaxies are strikingly different. Arp 220 is well fit by nearly isothermal dust, perhaps because of the extremely compact and optically thick nature of its source, whereas NGC 3690 has a much broader distribution of luminosity with temperature. The effects of finite optical depths have not been included in the analysis. These are known to be important in compact sources, and, when included, will increase the derived temperatures and narrow the distributions somewhat.

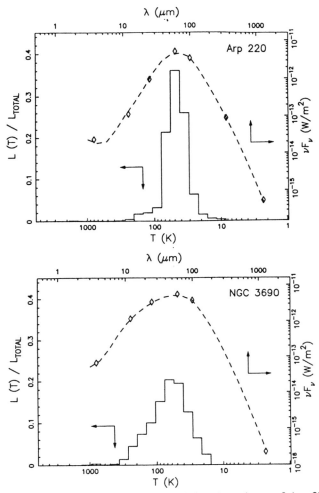

Figure 1. The open diamonds mark the observed infrared continuua of Arp 220 and NGC 3690. The histogram at the bottom of each panel gives the luminosity as a function of temperature derived as described in the text. Dashed lines through the data show the model continuua generated by the luminosity distributions. Arrows indicate the appropriate axes for each curve. Data at wavelengths shortward of 1mm are from Carico et al. (1988) except for Arp 220 at 350 μm, which is from Eales et al. (1988).

The CSO is supported by NSF grant # AST 83-11849. D.P.C., G.N. and B.T.S. are supported in part by the NSF and by the IRAS Extended Mission Program.

References

Carico, D. P., Sanders, D. B., Elias, J. H., Matthews, K., and Neugebauer, G. 1988, *Astron. J.*, **95**, 356.
Eales, S. A., Wynn-Williams, C. G., and Duncan, W. D., 1988, preprint.
Hildebrand, R. H., Loewenstein, R. F., Harper, D. A., Orton, G. S., Keene, J., and Whitcomb, S. E. 1985, *Icarus*, **64**, 64.
Ulich, B. L. 1974, *Icarus*, **21**, 254.

JCMT OBSERVATIONS OF CO(3-2) FROM M82

K. Y. Lo,[1] S. Stephens,[1]
E. Rosenthal,[2] S. Eales,[2] C. G. Wynn-Williams[2]
[1]Astronomy department, University of Illinois
[2]Institute for Astronomy, University of Hawaii

INTRODUCTION AND OBSERVATIONS

The inner 500 pc of M82 is a nearby proto-typical starburst region.[1] A starburst requires at least three conditions: (i) an adequate supply of gas, (ii) gathering of the gas within the star-burst region, typically ≤ 1 kpc in extent in the nuclear region, and (iii) a triggering mechanism for the onset of star formation throughout the region. It is clearly important to understand the physical conditions of molecular gas in such regions.

By necessity sofar, observations of extragalactic molecular gas have relied principally on the CO(1-0) transition alone. Inferences on the H_2 column density depend <u>entirely</u> on the <u>assumption</u> that molecular gas properties are identical to those found in the disk of our Galaxy. In starburst regions, the physical conditions are clearly different. For example, throughout the starburst region in M82, the far-UV flux is ~10^5 $L_\odot/pc^2$, comparable to that at 0.25 pc from the Trapezium stars in Orion.[2] This may be quite typical in all starburst regions.

Thus, direct determination of the gas properties in external galaxies is necessary. Ideally, this requires observations of more than one transition from a variety of molecules at resolutions sufficient to resolve individual molecular clouds. J=2-1 and J=1-0 transitions of CO from M82 have been observed at high resolution.[2,3,4]

We have used the 15-m James Clerk Maxwell Telescope to observe CO(3-2) transitions from the inner region of M82, in order to obtain further constraints on the gas properties in the starburst region. The telescope beam at $\lambda 0.87$mm is ~14", with η_{mb}~0.6 and η_{fss}~0.7. The receiver used is a Schottky diode receiver with a carcinotron as local oscillator, and had a typical DSB $T_r \leq 1000$ K. During the observations (Feb., 1988), the zenith opacity at 346 GHz was < 0.3.

RESULTS AND DISCUSSION

We obtained spectra at 18 locations sampling incompletely the central 1' x 0.5' region. The general profile shapes of the CO(3-2) emission and the integrated intensity distribution are similar to those of the 1-0 and 2-1 transitions of CO obtained at similar resolution.[3,4] The T_R^* of the CO(3-2) emission within the region mapped is ≤ 5 K, where $T_R^* = T_A^*/\eta_{fss}$.[6]

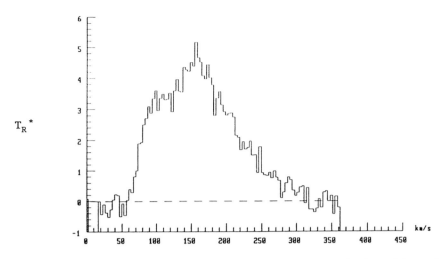

Figure 1: CO(3-2) emission at ~12" to the west of the M82 nucleus along the major axis.

In determining line ratios, there are a few practical difficulties, such as calibration errors and differences in angular resolutions. Also, different J transitions may be sampling gas at different excitations at varying locations within the beam, and have different amount of self-absorption.[5] To obtain constraints on the mean properties of the gas, we have chosen to compare the temperatures of the different J transitions at a point 12" to the west of the nucleus along the major axis. We use the 1-0 measurements from the Nobeyama 45-m (17" resolution) and the 2-1 measurements from IRAM 30-m (13" resolution).[3,4] The 2-1/1-0 ratio of T_R^* is ~3,[4] whereas the 3-2/1-0 ratio is ~1. If we assume that all the J-transitions have the same distribution within the beam, then the 3-2/1-0 ratio indicates that the CO(3-2) emission cannot be optically thin, since the expected ratio of T_R then is $9 \exp(-16.5/T_{ex})$.[7] This would imply that the mean T_{ex} of the gas within the beam of is not very high and that the filling factor of the CO(3-2) emission may be smaller than that of the lower J transitions. This research is partially supported by the U.S. National Science Foundation.

REFERENCES

1. Rieke, G., et al, 1980, Ap. J., **238**, 24.
2. Lo, K. Y., et al, 1986, Ap. J., **312**, 574.
3. Nakai, N., et al, 1987, Publ. Astron. Soc. Japan, **39**, 685.
4. Loseau, N., et al, 1988, in Molecular Clouds in the Milky Way and External Galaxies, eds. R. Dickman, J. Young, (Springer-Verlag), in the press.
5. White, G., et al, 1986, Ast. & Ap., **159**, 309.
6. Kutner, M, Ulich, B., 1981, Ap. J., **250**, 341.
7. Ho, P., Turner, J., Martin, R., 1987, Ap. J. (Lett.), **322**, L67.

CO(2→1) / CO(1→0) OBSERVATIONS OF LUMINOUS INFRARED GALAXIES

D. B. SANDERS, A. I. SARGENT, N. Z. SCOVILLE, T. G. PHILLIPS
Division of Physics, Mathematics and Astronomy
California Institute of Technology
Downs Lab of Physics, 320-47
Pasadena, CA 91125

CO(2→1) spectra of several luminous infrared galaxies have been obtained with the Caltech Submillimeter Observatory (CSO). For those galaxies distant enough to be unresolved with the CSO beam, we have compared the CO(2→1) profile with the corresponding, unresolved CO(1→0) profile in order to investigate the contribution of optically thin CO to the total CO emission.

CO(2→1) observations of NGC 2623 (= Arp 243), Markarian 231, and NGC 6090 were made during 1987, November with the 10.4m telescope and 500 MHz AOS under excellent weather conditions. Total integration times were between 20 and 40 min. These three galaxies are each very luminous in the infrared, with L_{fir} values ranging from 10 - 100 times the far-infrared luminosity of the Milky Way. All three are advanced mergers as evidenced by the appearance of long tidal tails extending from nearly overlaping disks. CO(1→0) observations have previously indicated that these galaxies are rich in molecular gas (Sanders et al. 1986). Total H_2 masses from $1 - 2 \times 10^{10}$ $M_\odot$ have been computed using the relation $M(H_2)[M_\odot] = 5.8\ L_{CO}$ [K km s^{-1}pc^2] derived for molecular clouds in the Milky Way (c.f. Scoville and Sanders 1987). The $L_{fir}/M(H_2)$ ratios for these galaxies range from ~ 4 – 40 times the average value of 4 $L_\odot M_\odot^{-1}$ for molecular clouds in late-type spirals (e.g. M51, IC 342, NGC 6946, Milky Way).

Galaxy	RA (1950) (h m s)	Dec (1950) (° ′ ″)	log L_{fir}[a] ($L_\odot$)	$L_{fir}/M(H_2)$ ($L_\odot M_\odot^{-1}$)
NGC 2623	08 35 25.3	25 55 50	11.48	50
Mrk 231	12 54 04.8	57 08 38	12.33	140
NGC 6090	16 10 24.0	52 35 11	11.36	16

[a] $L_{fir} = L(40 - 400\mu m)$, assuming $H_o = 75$ km s^{-1}Mpc^{-1}

H_2 mass estimates based on CO(1→0) luminosities alone can be subject to large errors if a significant fraction of the CO emission is optically thin. Therefore, it is important to measure the (2→1) / (1→0) ratio to obtain a first order estimate of the CO optical depth. Figure 1 compares the CO(2→1) and CO(1→0) profiles in T_R^* units. Since the molecular disk in spiral galaxies is nearly always smaller than half the optical diameter, D_{25}, the CO emission regions are unresolved in both transitions. Both profiles have similar shapes, and the line ratios are fairly constant across the profile. Large variations in the ratio that occur over small velocity ranges near the edges of the profiles may be partially due to differences in smoothing, and small errors in previously computing the correct heliocentric velocity.

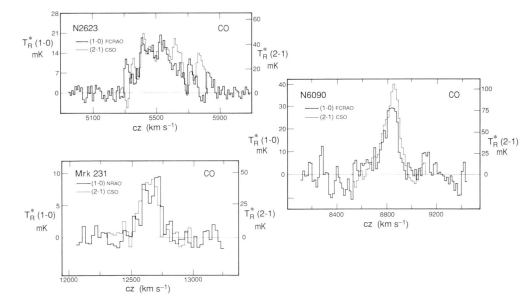

Figure 1. Comparison of CO(2→1) spectra of luminous infrared galaxies obtained at the CSO with CO(1→0) spectra obtained with either the 12 m NRAO or 14 m FCRAO telescope.

Galaxy	D_{25} ('')	$L_{CO(2-1)}^a$	$L_{CO(1-0)}$	R
		(K_{mb} km s^{-1}pc^2)		
NGC 2623	40	4.2×10^9	3.6×10^9	1.2
Mrk 231	55	3.9×10^9	4.6×10^9	0.9
NGC 6090	30	1.2×10^9	1.5×10^9	0.8

[a] $L_{CO} \equiv \pi/4 \, d_b^2 \int T_{mb}(CO) \, dv$, where d_b is the diameter of the beam on the source in pc.

The observed (2→1) / (1→0) luminosity ratios are in the range R = 0.8 - 1.2, consistent with optically thick CO emission (R = 4 in the optically thin limit). Our ratio of 1.2 for NGC 2623 is not as low as the 0.5 - 0.7 range obtained from IRAM 30m maps (Casoli *et al.* 1988), but the difference can be explained if the IRAM data were not properly converted to main beam brightness temperatures. The (2→1) / (1→0) ratios for the luminous infrared galaxies are similar to what is found for the bulk of the molecular material in the Milky Way. This suggests that previous H_2 msss estimates based on (1→0) alone are not seriously in error. The CSO is supported by NSF grant AST 83-11849.

REFERENCES

Casoli, F., Combes, F., Dupraz, C., Gerin, M., Encrenaz, P., and Salez, M. 1988, *Astron. Ap.*, **192**, L17.

Sanders, D. B., Scoville, N. Z., Young, J. S., Soifer, B. T., Schloerb, F. P., Rice, W. L., and Danielson, G. E. 1986, *Ap. J. (Letters)*, **305**, L45.

Scoville, N. Z., and Sanders, D. B. 1987, in *Interstellar Processes*, eds. D.J. Hollenbach, and H.A. Thronson (Dordrecht: Reidel), P. 21.

Millimetre and submillimetre observations of Blazars

Ian Robson,
School of Physics and Astronomy, Lancashire Polytechnic
Preston, PR1 2TQ, U.K.

Like many extragalactic continuum observers at this conference, I have spent most of my mm/submm telescope time during the past year integrating at great length on blank field radio quiet AGN's in the quest for the holy grail; to determine whether the far IR-short radio emission is due to thermal reradiation from dust or electron synchrotron radiation. In spite of teething troubles expected from a new telescope, we have detected a Markarian Seyfert at 800 microns (Hughes et al in preparation) which is clearly a dusty source, and probably typical of most IRAS selected AGN's as evidenced from upper limits at submm wavelengths and IRAS data (eg Edelson et al. 1988). I am instead going to discuss something different and convince you all that there is great merit in looking at some very bright objects! I refer to radio loud Blazars. Instead of taking many hours of integration to get a 4 or 5σ detection on the radio quiet's, here one can obtain oodles of signal-to-noise in no time whatever. The limiting uncertainty is dominated by the difficulty of atmospheric extinction correction and absolute calibrations.

What can these high S/N mm/submm measurements shed on the study of Blazars? The answer is twofold. Firstly, results of our recently completed multifrequency observational programme (Brown et al. 1988a,b) showed that the mm-UV spectrum is from a single synchrotron emitting component, being optically thin from about 1 mm to the UV with a self-absorption turnover at about 1mm. Previous observations (eg Landau et al 1986), lacked the mm/submm coverage and clearly missed this mm peak, instead inferring that the entire cm-UV was from a single component. Fig 1 shows that this is clearly not the case for the majority of the objects in our sample of Blazars; there are two distinct components, one peaking at about 1 mm, the other at cm wavelengths. We attributed the cm component to synchrotron emission from an underlying relativistic jet which varies on a timescale of years; the mm-UV component is from the innermost part of the jet and is more highly variable on timescales of weeks and months. This is due to re-injections or re-accelerations of relativistic electrons. More work needs to be done on a much larger sample of radio loud quasars and BL Lac objects to determine whether this is a general property of all such sources, or whether it reflects a fundamental difference, maybe related to the beaming angle. Obviously this is an exciting regime for future research.

The second result concerns monitoring the variability of the mm/submm spectrum. Flare emission becomes optically thick in the mm/submm region and the precise manner in which the flare spectrum develops with time is a very critical test of emission models. Indeed, it was by monitoring such a flare development in 3C273 (Robson et al. 1983) that all existing emission models were shown to be unacceptable and a new model based on a relativistic shock was derived by Marscher and Gear (1985). Monitoring programmes are therefore of critical importance. The data of Fig 1 were from UKIRT, the JCMT allows these to be improved dramatically with a great increase in sensitivity and the ability to obtain more detailed spectral coverage using the 2.0, 1.3, 1.1, 0.8, 0.6 and 0.45 mm bandpass filters. These improvements will enable physical parameters such as source size and magnetic fields to be derived with much greater accuracy

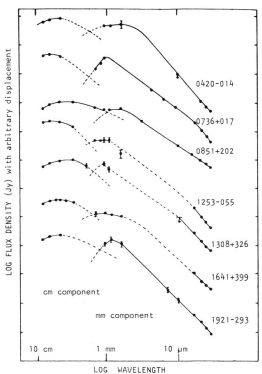

Fig 1. Examples of the mm and cm synchrotron components of Blazars adapted from Brown et al. 1988a.

References
Brown et al. 1988a,b. Ap.J. in press, (May 1989 issue).
Edelson, Gear, Malkan & Robson. 1988. Nature, in press.
Landau et al. 1986. Ap.J. 308, 78.
Marscher and Gear. 1985. Ap.J. 298,114.
Robson et al. 1983. Nature, 305,194.

COLD DUST IN GALAXIES

S. A. EALES and C. G. WYNN-WILLIAMS
Institute for Astronomy, University of Hawaii

W. D. DUNCAN
United Kingdom Infrared Telescope

Models for dust heated by the interstellar radiation field of a normal galaxy show that most of the thermal emission from the dust should be occurring in the submillimeter waveband (100 μm < λ < 1000 μm) (Spitzer 1978; Draine and Lee 1984; Draine and Anderson 1985). The *IRAS* survey confirmed that because most galaxies are emitting more strongly at 100 μm than at 60 μm, much of a galaxy's emission should be in this wavelength region. We have used the UKT14 bolometer on the United Kingdom Infrared Telescope to observe the submillimeter emission from a sample of 11 spiral galaxies that are bright at 100 μm. The instrumental beam of the bolometer system has an approximately Gaussian shape with size (FWHM) of ≈80 arcsec. We used a beam separation of 136 arcsec E-W. The flux calibration and the extinction coefficients were obtained from observations of Saturn, Uranus, OMC-1, and IRC+10216. More details are given in Eales, Wynn-Williams, and Duncan (1989).

We detected 5 out of 11 galaxies at both 350 and 450 μm. Figure 1 shows the mean continuum energy distribution for the 5 galaxies. The error bars at each wavelength show the variance about the mean at that wavelength and hence the range of continuum energy distributions exhibited by this subsample of 5 galaxies.

The 12–1100 μm flux densities cannot all be fitted by thermal emission from dust at a single temperature. We decided not to fit a multicomponent model to the *IRAS* and submillimeter data, but instead to see whether the 60–1100 μm flux densities alone could be fitted by thermal emission from dust at a single temperature, because (1) the emission at 12 and 25 μm may be dominated by nonequilibrium emission (Boulanger et al. 1989) and (2) this procedure should show whether the submillimeter data provide any evidence for a dust component in addition to those radiating in the *IRAS* bands. To do this, we fitted thermal spectra to the *IRAS* 60 and 100 μm flux densities and to our submillimeter flux density (excluding upper limits) using a range of dust emissivity functions. We found that the 60–1100 μm flux densities could always be fitted by emission from dust at a single temperature but that the data were too poor to distinguish between the different emissivity laws. The temperatures we obtained are typically between 30 and 50 K, considerably higher than the temperature (≈10–20 K) predicted for dust in the interstellar medium of a normal galaxy (Spitzer 1978; Draine and Lee 1984; Draine and Anderson 1985). A possible explanation is that although the submillimeter and far-infrared fluxes for a galaxy

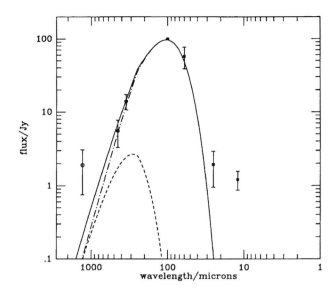

Fig. 1.—The mean fluxes of the five galaxies that were detected at both 350 and 450 μm. Before calculating the mean, the energy distributions of the individual galaxies were normalized to 100 Jy at 100 μm. The IRAS measurements are shown by filled squares, our submillimeter measurements by filled circles, and the 1300 μm measurements of Chini et al. (1986) by an open circle. Beam corrections have been made so that all measurements are for the same region of each galaxy. The solid line shows the result of fitting a two-component dust model to the 60–450 μm measurements: dust at 35 K and at 10 K, in the proportions by mass of 1:5. The dashed line shows the emission from the 10 K dust, and the dotted dashed line shows the emission from the 35 K dust.

are best fitted by a warm thermal spectrum, this does not rule out the existence of colder dust in substantially larger quantities than the warm dust we have detected. The flux densities in Figure 1 are best fitted by a thermal spectrum with a temperature of 35 K, but they are also adequately fitted by a two-component model: dust at 35 K and a second component consisting of 5 times as much dust at 10 K. The gas masses that we estimated from our submillimeter flux densities (Hildebrand 1983) are as much as 8 times lower than the gas masses estimated from the brightness of the CO 1-0 line. In view of the problems with both methods, in particular the ease with which one can underestimate the amount of cold dust when using the submillimeter method, the discrepancy is not too surprising.

This work was partly supported by NSF grant AST 86-15684.

Boulanger, F., Beichman, C., Desert, F. X., Helou, G., Perault, M. and Ryter, C. 1989, *Ap. J.*, in press.
Chini, R., Kreysa, E., Krigel, E. and Mezger, P. G. 1986, *Astr. Ap.*, **166**, L8.
Draine, B. T., and Anderson, N. 1985, *Ap. J.*, **292**, 494.
Draine, B. T, and Lee, H. M. 1984, *Ap. J.*, **285**, 89.
Eales, S. A., Wynn-Williams, C. G. and Duncan, W. D. 1989, *Ap. J.*, **339**, in press.
Hildebrand, R. H. 1983, *Quart. J.R.A.S.*, **24**, 267.
Spitzer, L. 1978, *Physical Processes in the Interstellar Medium.* New York: John Wiley.

CO (3-2) OBSERVATIONS OF SPIRAL GALAXIES

J. L. TURNER
UCLA, Los Angeles, CA 90024 U.S.A.
R. N. MARTIN
Steward Observatory, Tucson, AZ 85721 U.S.A.
P. T. P. HO
Center for Astrophysics, Cambridge, MA 02138 U.S.A.

ABSTRACT. We report the detection of the CO J=3-2 line in several nearby galaxies, including M82, M83, and NGC 253. The line intensities indicate that the nuclear gas in these galaxies is very warm (T $\gtrsim$ 30–50K), and often optically thin, over scales of hundreds of parsecs.

1. Introduction

The J=3-2 transition of CO at 346 GHz offers several advantages over the 1-0 line in the study of large star-forming complexes in galaxies. It is more sensitive to warm (T>20K) gas, attaining enhancements of up to a factor of 9 in integrated intensity over the 1-0 line for low optical depths. It is a tracer of relatively high density gas; the critical density for 3-2 is a factor of 30 higher than for 1-0. Finally, the higher frequency offers spatial resolutions of ~20″ for 10-15m dishes, comparable to the 30m and 45m telescopes at 100 GHz.

Evidence is accumulating for the presence of warm gas on large scales in the cores of spiral galaxies. Ammonia measurements indicate T_k ~70K in IC 342 (Martin and Ho 1987); interferometer maps of CO (1-0) in nearby galaxies reveal T_b ~10-15K over ~200 pc sizescales (Lo et al. 1984, 1987; Canzian, Mundy, and Scoville 1988). In M82, the presence of warm, optically thin CO emission is well-established (Knapp et al. 1980; Sutton, Masson, and Phillips 1983; Olofsson and Rydbeck 1984; Stark and Carlson 1984.) The advantage of studying warm gas is that it traces the spatial distribution and kinematics of gas that is directly associated with star formation.

2. Observations

The observations were made with the NRAO 12m telescope during February 1986-88. System temperatures, including the atmosphere, were ~4500-10000 K during the runs. We used 256 2-MHz filters to obtain 1.7 km s^{-1} velocity resolution. The FWHM beamsize of the 12m is ~22″ at 346 GHz; the fraction of forward power in the main beam η_{mb}, is 0.23. The contribution of the error beam (FWHM ~5′) is negligible for source sizes $\lesssim$ 2′. Pointing errors at the 12m have an rms scatter of ~10″, most of which is day-to-day variation. Our pointing checks indicated an accuracy of 2-3″ over a several-hour mapping session.

3. Results

We detected NGC 1068, Maffei 2, and M51; we have mapped IC 342 (Ho, Turner, and Martin 1987), M82, M83, and NGC 253. Nondetections include NGC 2903, Arp 220, NGC 3690, NGC 6240, and NGC 6946 (cf. Rosenthal, this volume) at an rms noise level of ~ 0.05–0.1K. The detected galaxies have peak T_a^* of 0.4–1.5 K; the mapped emission covers regions of ~ 40–$60''$, or ~ 0.5–1 kpc in extent. For comparison with 1–0 data, we calculate main beam brightness temperatures assuming unity filling factor, $T_{mb} = T_a^*/\eta_{mb}$. (T_{mb} represents the antenna temperature in the main diffraction beam only, which is appropriate since the sources are small compared to the error beam.) The observed T_{mb} are high, ~ 10–13 K. We can estimate the optical depth of the CO emission by comparing $T_{mb}(3-2)$ with the $T_{mb}(1-0)$ measured with the Nobeyama 45m telescope at $14''$ resolution. From the 1–0 measurements of Nakai et al. (1987) and Martin et al. (1988), we derive 3–2/1–0 intensity ratios of at least 2–3 for M82 and IC 342, ratios which may in reality be higher since the 1–0 beam is smaller than our 3–2 beam. In these galaxies, the CO emission is likely to be at least partially optically thin. However, since the 3–2/1–0 intensity ratios are less than the limiting value of 7–7.5 for low optical depths and $T_k \lesssim 100$–150K, it is possible that a significant fraction, 30–50%, of the 1–0 emission may arise in a cool, optically thick component.

The main beam brightness temperatures are a lower limit to the true brightness temperature of the emission. We can derive an estimate of the true $T_b(3-2)$ by estimating the size of the emitting region. To do so we assume that the 3–2 spatial distribution is the same as that seen in the 1–0 interferometer maps (Lo et al. 1984, 1987; Canzian, Mundy, and Scoville 1988). We reason as follows: since the 3–2 emission has a higher critical density, it seems unlikely that the 3–2 emitting region is more extended than that producing the 1–0 emission, most of which is detected by the interferometer maps (A. Eckart, private communication). Peak T_b derived in this way range from ~ 10K in NGC 253 to ~ 35K in M82 and ~ 40K in IC 342. These temperatures are consistent with the high 3–2/1–0 intensity ratios. It appears that warm, moderately optically thin gas may be common on hundred-parsec scales in the centers of galaxies, unlike in Galactic disk regions, where warm gas is confined to parsec-scale molecular cores.

4. References

Canzian, B., Mundy, L. G. and Scoville, N. Z. 1988, *Astrophys. J. (Lett.)*, **333**, 157.
Martin, R. N. and Ho, P. T. P. 1986, *Astrophys. J. (Lett.)*, **308**, L7.
Ho, P. T. P., Turner, J. L., and Martin, R. N. 1987, *Astrophys. J. (Lett.)*, **322**, L67.
Knapp, G. R., Phillips, T. G., Huggins, P. J., Leighton, R. B., and Wannier, P. G. 1980, *Astrophys. J.*, **240**, 60.
Lo, K. Y. et al. 1984, *Astrophys. J. (Lett.)*, **282**, L59.
Lo, K. Y. et al. 1987, *Astrophys. J.*, **312**, 574.
Martin, R. N., Turner, J. L., and Ho, P. T. P. 1988, in preparation.
Nakai, N., Hayashi, M., Handa, T., Sofue, Y., and Hasegawa, T. 1987, *Publ. Astron. Soc. Japan*, **39**, 685.
Olofsson, H., and Rydbeck, G. 1984, *Astron. Astrophys.*, **136**, 17.
Stark, A. A. and Carlson, E. R. 1984, *Astrophys. J.*, **279**, 122.
Sutton, E. C., Masson, C. R., and Phillips, T. G. 1983, *Astrophys. J. (Lett.)*, **275**, L49.

CO(2→1) EMISSION FROM NGC 3256: AN INTERACTING PAIR OF GALAXIES

A. I. SARGENT, D. B. SANDERS, T. G. PHILLIPS
Division of Physics, Mathematics and Astronomy 320-47
California Institute of Technology
Pasadena, CA 91125

At optical wavelengths, characteristic "tails" and chaotic nuclear appearance clearly identify NGC 3256 as an interacting system (Toomre 1977; Schweizer 1986). The newly-constructed, 10.4 m diameter, Leighton telescope of the Caltech Submillimeter Observatory (CSO) has been used to acquire observations of this merger in the CO (2→1) transition at 230 GHz. The turbulent nature of NGC 3256 is immediately obvious in Figure 1, a high contrast photograph provided by F. Schweizer, where circles representing the beam (30″ FWHM) indicate the positions observed.

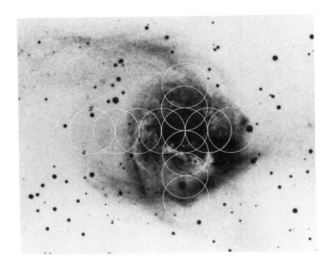

Figure 1. A high-contrast photograph of the core of NGC 3256. Positions observed are indicated by circles of diameter $\sim 30''$.

Individual CO (2→1) spectra are presented in Figure 2. The peak antenna temperature occurs at the nucleus but enhanced integrated intensities are also seen immediately east and south of center, reflecting high column densities and very disturbed gas motions in these regions. CO emission extends over an area 7.9 × 6.5 kpc in size, comparable to the 10 μm dust distribution (Graham et al. 1987). The enhanced luminosity detected by IRAS, 3.2×10^{11} $L_\odot$, is therefore plausibly accounted for by a burst of star formation. Although the total mass of gas, 2×10^{10} $M_\odot$, is a factor of two higher than in Arp 220, the luminosity is an order of magnitude lower, and L_{FIR}/M_{H_2} ~ 15 $L_\odot/M_\odot$, also typical of a starburst galaxy.

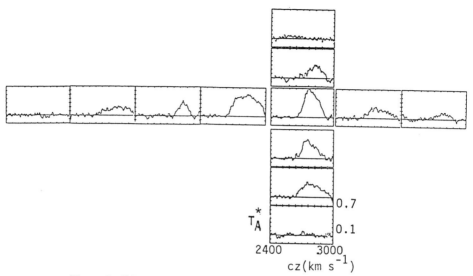

Figure 2. CO (2→1) spectra of NGC 3256.

For Arp 220, L_{FIR}/M_{H_2} is much higher, ~ 130 $L_\odot/M_\odot$, suggesting that in NGC 3256 the merger may be less advanced (c.f. Joseph and Wright 1985). This hypothesis is supported by the presence of *two* nuclear condensations, separated by only 4″, in photographic plates provided by J. Bergeron. On the other hand, Graham et al. (1987) assert that within 5 kpc of the galaxy center the distribution of old red stars is described by the $r^{1/4}$ profile characteristic of elliptical galaxies. If this is indeed the case, it implies that relaxation has already taken place and that the merger is far advanced. Both high resolution imaging and kinematic studies are needed to resolve this question. The CSO is supported by NSF grant AST 83-11849.

References

Graham, J. R., Wright, G. S., Joseph, R. D., Frogel, J. A., Phillips, M. M., and Meikle, W. P. S. 1987, in *Star Formation in Galaxies*, ed. C. J. Lonsdale (U. S. Government Printing Office), p. 517.
Joseph, R. D., and Wright, G. S. 1985, *M. N. R. A. S.*, **214**, 87.
Joy, M., Lester, D. F., Harvey, P. M., and Ellis, H. B. 1988, *Ap. J.*, **326**, 662.
Schweizer, F. 1986, *Science*, **231**, 227.
Toomre, A. 1977, in *The Evolution of Galaxies and Stellar Populations*, (Yale University Observatory, New Haven), p. 401.

CO (2-1) STUDIES OF CENTAURUS A

T. G. Phillips, D. B. Sanders and A. I. Sargent
*Division of Physics, Mathematics and Astronomy
California Institute of Technology
Downs Lab of Physics, 320-47
Pasadena, CA 91125*

Using the Caltech Submillimeter Observatory (CSO), Phillips et al. (1987) detected CO emission from the nearby radio galaxy Centaurus A. They reported emission from the dust lane region and showed that the material rotated very much like a disk galaxy imbedded within the larger elliptical system. They also showed that the heliocentric systemic velocity revealed by the CO observations at 547 ($\pm$ 5) km/s was close to the value of 538 ($\pm$ 10) km/s found for the elliptical component by Wilkinson et al. (1986).

Further observations have been carried out at the CSO, and some of the conclusions have been improved and revised. A more extensive map of the dust lane has been made and compared with the HI map made at the VLA by van Gorkom (1987). Figure 1 shows the velocity - dust lane major axis plot, for both CO and HI. The CO appears in the inner part of the disk where the velocity gradient is high, but the HI emission continues to outer parts of the dust lane where the velocity flattens out.

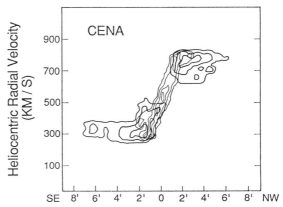

Figure 1. CO and HI rotation curve for the dust lane region. HI - heavy contours; CO - light contours.

Also, we have mapped the CO emission across the dust lane (dust lane minor axis) and have found a CO extent considerably greater than that for the visible dust. This effect can also be seen in the data of Joy et al. (1988) where the 100 μm dust emission was mapped. Figure 2 displays the 100 μm dust and the CO emission for both the major and minor axes of the dust lane. They agree well and demonstrate that the gas and dust extend far out into the elliptical component, and also that the disk is well centered on the radio nucleus and must be very nearly edge-on.

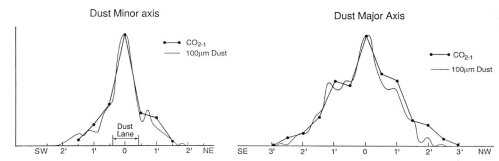

Figure 2. 100 μm dust and CO (2–1) emission profiles for the major and minor axes of the dust lane.

We have also inspected the nuclear region more carefully, with better pointing accuracy and better signal-to-noise ratio than previously, and now are able to see an absorption feature, which may be compared with that seen in HI by Gardner and Whiteoak (1976) and van der Hulst et al. (1983), or in C_3H_2 by Bell and Seaquist (1988). A CO (1–0) absorption feature has been detected by Israel et al. (1988) (see also van Dishoeck et al., this publication, for a detailed discussion). Figure 3 shows the CO (2–1) nuclear spectrum, together with the original HI absorption profile of Gardner and Whiteoak.

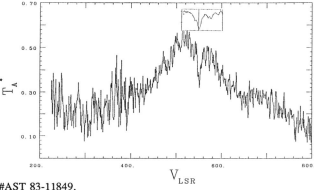

Figure 3. CO (2–1) emission profile at the nuclear position. HI profile (inset).

The CSO is supported by NSF grant #AST 83-11849.

References

Bell, M. B. and Seaquist, E. R. 1988, *Ap. J.*, **329**, L17.
Gardner, F. F. and Whiteoak, J. B. 1976, *Proc. Astr. Soc. Australia*, **3**, 63.
Israel, F. P. et al. 1988, preprint (see also van Dishoeck et al., these proceedings).
Joy, M., Lester, D. F., Harvey, P. M. and Ellis, H. B. 1988, *Ap. J.*, **326**, 662.
Phillips, T. G. et al. 1987, *Ap. J.(Letters)*, **322**, L73.
van der Hulst, J. M., Golisch, W. F. and Haschick, A. D. 1983, *Ap. J.(Letters)*, **264**, L37.
van Gorkom, J. 1987, *IAU Symp. #127*, p. 421.
Wilkinson, A., Sharples, R. M., Fosbury, R. A. E. and Wallace, P. T. 1986, *M.N.R.A.S.*, **218**, 297.

H_2 EMISSION AND CO ABSORPTION TOWARD THE NUCLEUS OF CENTAURUS A: A CIRCUMNUCLEAR DISK*?

E.F. van Dishoeck[1], F.P. Israel[2], J. Koornneef[3]
F. Baas[4], J.H. Black[5], and Th. de Graauw[6]

[1] Princeton University Observatory, Princeton NJ 08544
[2] Sterrewacht, P.O. Box 9513, 2300 RA Leiden (NL)
[3] STScI, Johns Hopkins Univ., Baltimore MD 21218
[4] Lab. Astrophys., P.O. Box 9504, 2300 RA Leiden (NL)
[5] Steward Observatory, Univ. of Arizona, Tucson, AZ 85721
[6] Lab. Space Research, P.O. Box 800, 9700 AV Groningen (NL)

The nucleus of Centaurus A has been studied extensively at infrared and radio wavelengths, yet the physical properties of the material surrounding it are not well known. We report here observations of the H_2 and CO molecules, which provide significant constraints on the nature of the molecular gas close to the nucleus. The results are discussed in more detail in Israel et al. (1988).

H_2 emission at 2.1 μm has been detected toward the nuclear radio position (see Figure 1). The emission is unresolved, indicating a size of the H_2 emitting region <95 pc if a distance of 5 Mpc is adopted. The observed H_2 line ratios suggest that the molecule is mostly collisionally excited, with an excitation temperature between 1000 and 2000 K. The surface brightness of the (1,0) S(1) line indicates high densities, $n_H \geq 10^4$ cm^{-3}. The absence of detectable Brackett–γ emission significantly limits the number of early–type stars in this region.

^{12}CO and ^{13}CO $J=1\rightarrow 0$ observations were made at the 15m SEST telescope (see Figure 2). The spectra of both species show extended line wings over 350 km s^{-1} in velocity, deep narrow absorption, and a significant continuum at $T_A^* \approx 0.2$ K. The enhanced wing emission implies the presence of rapidly rotating molecular material with a mass of order 10^7 M$_\odot$ close to the nucleus. Based on VLBI measurements at lower frequencies (Meier et al. 1988), we conclude that the millimeter continuum can only be due to the very compact (0.5 milliarcsec) core, and not to the more extended jets. The CO absorption most likely occurs against this continuum. The ^{13}CO optical depth, together with an estimated extinction $A_V \approx 27$ mag toward the nucleus, indicate a rather high (but uncertain) excitation temperature, $T_{ex}(^{13}$CO$)\approx 30$ K. Thus the absorption may occur close to the nucleus, in contrast with the CO emission, which arises mostly in the dust band far away from the nucleus.

The ^{12}CO profile shows a main absorption feature close to the systemic velocity, and a shoulder at a velocity blue–shifted by about 10 km s^{-1}. These absorptions coincide with those found for C_3H_2 (Bell and Seaquist 1988) and H_2CO (Gardner and Whiteoak 1976). Similar absorption features are seen in the CO 2–1 profile of Phillips et al. (this volume). In contrast, H I absorption against the nucleus (van der Hulst et al. 1983) is strong only at the systemic velocity and at red–shifted velocities. Thus the blue–shifted gas appears rich in molecular material, and the red–shifted gas, poor.

Based on our detected H_2 point source emission, our observed CO absorption and wing emission, as well as unresolved far–infrared emission found by Joy et al. (1988), we suggest the presence of material in a disk or annulus around the nucleus. The observations are consistent with a disk that has an outer edge at $r \approx 100$ pc, an inner edge at $r \approx 25$ pc, a thickness of about 50 pc, and a density distribution $n \propto r^{-1}$.

Based on observations collected at the European Southern Observatory (ESO) and the Swedish–ESO Submillimetre Telescope at La Silla, Chile

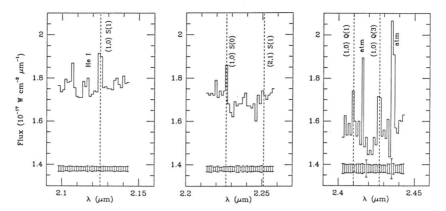

Figure 1. Infrared spectrum at $2\mu m$ of the nucleus of Cen A obtained with the ESO infrared cooled grating spectrometer (IRSPEC) on the 3.6 m telescope with a 6″ aperture and $\lambda/\Delta\lambda \approx 900$. The various H_2 lines are identified. Note the strong continuum.

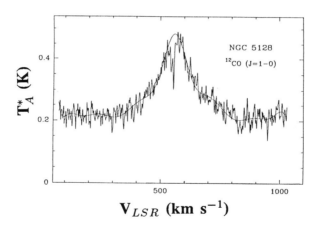

Figure 2. SEST ^{12}CO $J=1\rightarrow 0$ spectrum toward the nucleus of Cen A.

REFERENCES

Bell, M.B., and Seaquist, E.R. 1988, *Ap. J. (Letters)*, **329**, L17.
Gardner, F.F., and Whiteoak, J.B. 1976, *M. N. R. A. S.*, **175**, 9p.
Israel, F.P., van Dishoeck, E.F., Koornneef, J., Baas, F., Black, J.H. and de Graauw, Th. 1988, *Astr. Ap.*, submitted.
Joy, M., Lester, D.F., Harvey, P.M., and Ellis, H.B. 1988, *Ap. J.*, **326**, 662.
Meier, D.L. *et al.* 1988, preprint.
Van der Hulst, J.M., Golisch, W.F., and Haschick, A.D. 1983, *Ap. J. (Letters)*, **264**, L37.

CO 3-2 Observations of NGC 253

F.N. Bash[1], J.H. Davis[1], D.T. Jaffe[1], W.F. Wall[1], E.C. Sutton[2]

[1] University of Texas, Astronomy Dept., Austin, Tx. 78712
[2] University of California, Space Sciences Laboratory, Berkeley, Ca. 94720

ABSTRACT. We have observed the J=3-2 spectral line in ^{12}CO and the J=2-1 line in ^{12}CO and ^{13}CO at a few positions in NGC 253 using the Caltech Submillimeter Observatory and James Clerk Maxwell Telescopes. As a result we have observed three transitions with virtually identical beam sizes and can estimate the gas density and kinetic temperature, assuming a single component gas. In addition, ^{13}CO data suggest a ^{12}CO/^{13}CO abundance ratio that is 2 to 5 times lower than the solar value.

Introduction

NGC 253 is a barred spiral galaxy with a bright nucleus in the far infrared ($3 \times 10^{10} L_\odot$, Telesco and Harper 1980). Determining molecular gas excitation conditions in the central region of NGC 253 can help in understanding this high level of nuclear activity. We have observed the ^{12}CO 3-2, ^{12}CO 2-1, ^{13}CO 2-1 lines, each with 20" (or 330 pc for a distance of 3.4 Mpc) beams, toward a few positions in NGC 253. The J=3-2 observations were carried out at the Caltech Submillimeter Observatory (CSO). The ^{12}CO 3-2 line was previously detected in NGC 253 by Turner et al. (1987). The J=2-1 data were obtained with James Clerk Maxwell Telescope (JCMT) observations. We report here on temperature and density estimates of the molecular gas in the central region of NGC 253.

Physical Interpretation

The data from all three lines with identical spatial resolution have allowed us to place temperature and density limits on the molecular gas in the central 330 pc of NGC 253. It is possible that a pointing offset of 4 to 8" exists between the two data sets. This amounts to less than a 20% uncertainty in the ^{12}CO 3-2 to ^{12}CO 2-1 line intensity ratio for our 20" beams. The physical argument presented below is insensitive to this 20% uncertainty. Only an unlikely combination of pointing and calibration uncertainties would invalidate our non-LTE models.

The ^{12}CO 3-2 and ^{12}CO 2-1 spectra are shown in Figure 1. The main beam brightness temperatures of the lines are 3.5 K for ^{12}CO 3-2, 6.7 K for ^{12}CO 2-1, 0.4 K for ^{13}CO 2-1. A single component, LTE gas model is not consistent with these intensities. The ^{12}CO 2-1 brightness requires $T_k > 11$ K, where T_k is the kinetic temperature of the gas. But the ^{12}CO 3-2 to ^{12}CO 2-1 line intensity ratio requires $T_k = 6$ K. Hence, it is possible that the J=3-2 transition is subthermally excited, requiring $n(H_2) \leq 10^4 cm^{-3}$. Indeed, single component non-LTE modeling yields $n(H_2) \approx 100 - 300 cm^{-3}$ and $T_k \geq 30$ K. The non-LTE models were constructed using the Large Velocity Gradient algorithm (de Jong, Chu, Dalgarno 1975). The models were reasonably successful in describing the intensity ratio of the two

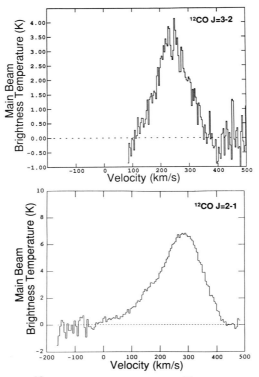

Figure 1: The ^{12}CO 3-2 spectrum (top) and ^{12}CO 2-1 spectrum (bottom) towards the center of NGC 253 are shown above.

^{12}CO lines as well as their individual intensities.

The ratio of the ^{13}CO and ^{12}CO intensities can be used to set a lower limit on the optical depth in the J=2-1 lines. This follows from the assumption that the excitation temperature of a ^{13}CO transition is lower than that in the corresponding ^{12}CO transition due to the lower radiative trapping in ^{13}CO lines. Our data suggest $\tau \geq 0.06$ in the ^{13}CO 2-1 line, which implies a lower limit to τ in the ^{12}CO 2-1 line for some ^{12}CO/^{13}CO abundance ratio. The single component non-LTE models require a ^{12}CO/^{13}CO abundance ratio of ~20 to 40 to permit the model ^{12}CO 2-1 opacities to exceed this lower limit and to account for the observed ^{13}CO 2-1 intensity.

References

De Jong, T., Chu, S.I., Dalgarno, A. 1975, *Ap. J.* **199**, 69
Telesco, C.M., Harper, D.A. 1980, *Ap. J.* **235**, 392
Turner, J.L., Ho, P.T.P., Martin, R.N. 1987, in: *Star Formation in Galaxies*, p383, NASA Conference proceedings pub. 2466, Ed. C.J.L. Persson

DUST EMISSION FROM RADIO–QUIET QUASARS

Rolf Chini
Max-Planck-Institut für Radioastronomie, Auf dem Hügel 69
5300 Bonn 1, F.R.G.

ABSTRACT. Continuum emission at 1300μm has been investigated for all radio-quiet quasars with good IRAS observations. From the spectral index between 100 and 1300μm, which is close to or even larger than 2.5, we suggest that dust emission on kpc scales, powered by the active nucleus, explains the FIR spectrum.

1. Introduction

Since IRAS data have become available the crucial question has been to determine the origin of the FIR emission in radio-quiet quasars. From their spectra which generally keep rising from the NIR to 100μm and from the absence of significant radio emission in the cm regime it becomes clear that the spectral turnover must occur at mm/submm wavelengths.

2. Observations and Results

The observations were at the IRAM 30m MRT, using the MPIfR bolometer system. To obtain meaningful upper limits for the 1300μm flux density of the entire sample of 25 radio-quiet quasars we integrated down to a limit of 3mJy (3σ); 15 objects were detected above the 3σ level.
The most important quantity that can be derived from our observations is the spectral index α between 100 and 1300μm defined by $S \sim \nu^{\alpha}$. For synchrotron radiation the maximum value for α is 2.5 in the case of self-absorption. The mean α of our sample is 2.18$\pm$0.32; four of the quasars exceed the limit of 2.5 and another eight come so close to it that synchrotron self-absorption can be clearly ruled out for the interpretation of the FIR spectra.
There are several ways of explaining a positive spectral index steeper than +2.5 like the Razin-Tsytovich effect, free-free absorption in a screen or synchrotron screen absorption. They all require, however, physical parameters which seem to be unlikely. Dust emission, on the other hand, may explain the observed energy distribution very well, conditions as they are known from Sey 1,2 type galaxies: A fit to the spectra from 60 to 1300μm gives an average dust temperature of 36.3$\pm$5.6K, similar to the 33.0$\pm$3.7K of the active galaxies studied by Krügel et al (1988a,b). Converting the optically thin submm emission into a dust and further into a gas mass one obtains values between 7 10^7 2

$10^{10} M_\odot$. Plotting the luminosity L from 0.3 to 1300μm as a function of gas mass M (see Fig.1), most objects form a relation which is similar to that of active galaxies (Krügel et al., 1988b). This raises the question whether the luminosity and thus the heating of dust comes from a starburst or from an active nucleus.

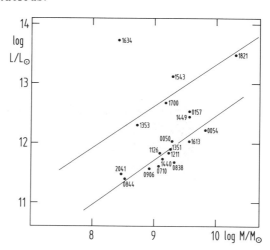

Figure 1: Far infrared luminosity versus gas mass as inferred from the spectrum assuming a single-temperature dust component to be dominant, and the "normal" dust-to-gas ratio.

3. Discussion

The luminosities emerging in the FIR and X-ray range are remarkably similar for the quasars in the present sample. It follows immediately that the X-ray emission has to be isotropic and cannot be relativistically boosted. The temperature and luminosity in the FIR suggest a scale of the order of several kpc, comparable to the extent of the narrow-line region, while X-ray variability suggests a scale of a fraction of a parsec. We conclude that radio-quiet quasars are akin to strong Sey 1 galaxies with the active nucleus powering a large, kpc-scale, dusty region and giving rise to weak radio emission akin to that of a radio galaxy in addition to the normal disk-radio-emission. Radio-quiet quasars appear to be the active nuclei in large gas-rich galaxies. The FIR emission is thermal emission by dust. If the NIR to X-ray emission is nonthermal, then it has to be nearly isotropic in sharp contrast to standard models, otherwise the good FIR-X correlation cannot be understood.

References

Chini, R., Kreysa, E., Salter, C.J.: 1987, Astron.Astrophys. **182**, L63
Krügel, E., Chini, R., Kreysa, E., Sherwood, : 1988a,
    Astron.Astrophys. **190**, 47
Krügel, E., Chini, R., Kreysa, E., Sherwood, : 1988b,
    Astron.Astrophys. **193**, L16
Neugebauer, G., Miley, G.K., Soifer, B.T., Clegg, P.E.: 1986, Ap.J.**308**, 815

PHOTOMETRY OF THE HOTSPOTS OF CYGNUS A

S.A. Eales[1], W.D. Duncan[2], P. Alexander[3]
[1]*Institute for Astronomy, University of Hawaii, Hawaii*
[2]*Joint Astronomy Center, Hilo, Hawaii*
[3]*MRAO, Cavendish Laboratory, Cambridge, England*

ABSTRACT. We have observed the hotspots of Cygnus A with the JCMT[1] at 1100 and 800µm. We find that the spectra of both hotspots are power laws from ~1 GHz to ~370 GHz.

1. Observations

We observed the Np and Sf hotspots of Cygnus A on the 7th June 1988 (UT) using the JCMT on Mauna Kea, Hawaii with UKT14 (Duncan *et al.*, *in preparation*). We centred the telescope's beam on each hotspot by moving the telescope to the position given by Wright and Birkinshaw (1984) and then making a pointing correction. We used a beam throw of 40" during chopping and typical integration times of 500 seconds. We calibrated our data by making observations of Mars and Uranus over a range of airmass. The brightness temperatures used for these planets are given in Orton *et al.* (1986) and Griffin *et al.* (1986). The transparency of the atmosphere was approximately 80% at 800µm. The main uncertainties in the final flux densities come from calibration uncertainties which we estimate give errors of 15% and 10% for the 800µm and 1100µm flux densities respectively.

2. Discussion

Despite the 19" beam of the JCMT it is safe to assume that there is essentially no contribution to our fluxes from the radio lobe as the synchrotron break frequencies are known to be of order a few Gigahertz for regions just outside the hotspot (Alexander *et al.* 1984). Spectra for the A+B and the D hotspots are shown in Fig. 1 using the values from Alexander *et al.* (1984) and Wright & Birkinshaw (1984): the spectra are consistent with a power-law slope of $\alpha \sim 1$ in each hotspot.

Standard shock acceleration models for the radiating electrons (e.g. Bell 1978) predict a power-law spectrum ($\log I(v) \propto -\alpha_0 \ln v$) where α_0 is termed the injection index. Although we have observed a power-law spectrum with $\alpha = 1$ from ~1 GHz to ~400 GHz in both hotspots this is unlikely to be the injection index since:

(1) It is considerably steeper than the 0.5→0.6 predicted from standard theory and the low-

---

[1] The James Clerk Maxwell Telescope is operated by the Royal Observatory Edinburgh on behalf of the Science and Engineering Research Council of the United Kingdom, the Netherlands Organisation for Pure Research and the National Research Council of Canada.

frequency spectral index for the complete sample of Laing, Riley & Longair (1983).

(2) An injection index of 0.6 was inferred by Winter *et al.* (1980) based on the need to fit synchrotron-loss spectra over the range 151 MHz to 5 GHz. The low-frequency spectral index of the source is nearer to 0.7 (Baars *et al.* 1977).

The spectrum is, however, consistent with that expected from an equilibrium distribution of electrons with continuous acceleration combined with ageing. This gives a spectrum with a low-frequency spectral index α_0 at electron energies where expansion losses dominate, which steepens at higher frequencies to $\alpha_0 + 0.5$ (Pacholczyk, 1970). Any spectral break must be above ~370 GHz in *both* hotspots; this implies that either the hotspots currently have a supply of zero-age electrons or that injection ended less than 5000 yrs ago (taking B = 10 nT). Therefore if the jet is intrinsically one-sided we would need or 'flip-flop' time-scale of less than 5000 yrs.

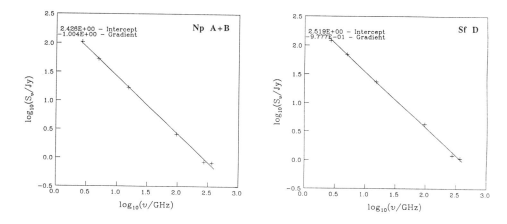

Figure 1. *Spectra of the A+B and D hotspots at 4" resolution.*

References

Alexander, P. Brown, M.T. and Scott, P.F., 1984. MNRAS, **209**, 851.
Baars, J.W.M., Genzel, R., Pauliny-Toth, I.I.K. & Witzel, A., 1977. AA, **61**, 99.
Bell, A.R., 1978. MNRAS, **182**, 1471.
Griffin, M.J., *et al.*, 1986. Icarus, **65**, 244.
Laing, R.A., Riley, J.M. & Longair, M.S., 1983. MNRAS, **204**, 151.
Orton, G.S., *et al.*, 1986. Icarus, **67**, 289.
Pacholczyk, A.G., 1970. *Radio Astrophysics,* W.H. Freeman & Co., San Francisco, pp 139.
Winter, A.J.B., *et al.*, 1980. MNRAS, **192**, 931.
Wright, M.C.H. & Birkinshaw, M., 1984. AA, **281**, 135.

MOLECULAR GAS IN IC 10

M. Hauschildt, J.H. Fairclough, G.S. Wright, T.M. Walker
Joint Astronomy Centre
665 Komohana Street
Hilo, HI 96720
USA

ABSTRACT. CO J=2-1 observations of the irregular galaxy IC10 with the JCMT show an outflow of molecular gas near the stronger of the two IRAS point sources, the gas in this area is hot and optically thin. The gas near the other IRAS point source is cold and optically thick, implying that this is an evolved star formation region.

IC10 is a bright nearby irregular type galaxy with two centres of activity, visible in radio continuum maps and as IRAS sources. The south eastern of these sources is the strongest water vapour maser source observed in a compact irregular galaxy (Henkel et al. 1986). IC10 is one of the very few magellanic irregular galaxies detected in CO J=1-0 (Henkel et al., 1986, Ohta et al., 1988). To our knowledge it is the only magellanic irregular galaxy (apart from the Magellanic Clouds) detected in CO J=2-1 up to now. Following up the JCMT discovery of the CO J=2-1 line in this galaxy we mapped part of the molecular cloud associated with the stronger one of the two IRAS point sources (IC10A) and observed the weaker IRAS point source (IC10B) as well. Our results on the CO J=2-1 emission are used to address two questions: Firstly is the gas optically thin or not? Secondly, our 20" resolution map allows us to study the distribution of molecular gas around the 10" diameter water vapour maser coinciding with IC10A.

Observations of the CO J=2-1 line in the two IRAS sources in IC10 have been carried out in June 1988 with the James Clerk Maxwell Telescope in very good weather conditions.

Figure 1 shows the grid map of the CO J=2-1 emission from IC10A as observed here. CO J=1-0 observations have been carried out with a beam width similar to ours by Ohta et al. (1988), so we can directly compare overlapping points to find out whether the molecular gas around IC10A is optically thin. We detected the 2-1 line everywhere they detected CO J=1-0 and also detected it in some places where they didn't. The peak of our map is a 1-0 non-detection to a level of ~0.2K. We can safely assume that the emission in this part of IC10 is from optically thin CO. As the gas becomes optically thin to the east of the dust lane (see their figure 2), away from the optically visible galaxy, the important question arises: What heats the gas? We see a red wing on the spectrum at (0,-20") which is close to the H2O maser. The nearest position observed by Ohta et al. shows a very similar spectrum. In CO 2-1 a second velocity component, redshifted by ~100km/s, becomes apparent to the east of this position. These spectra suggest an outflow of molecular gas in this region, associated with the activity in the maser source and heating the gas in this region.

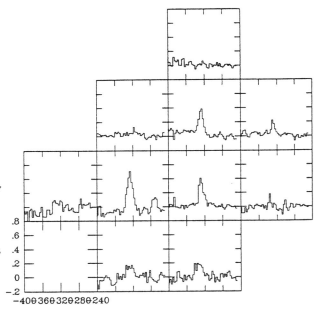

Like Ohta et al. (1988), we observed only one point in IC10B, shown in figure 2 which is the result of 5.5 hours of integration giving a noise limit of 14mK. The peak temperature reported by Ohta et al. is 0.23K as compared to our 0.09K. This implies that the molecular gas in IC10B is cold and optically thick. IC10B is a radio as well as far infra red source, though with a completely different far infra red continuum distribution to IC10A. The cold molecular gas, IRAS and radio emission are consistent with the suggestion by Ohta et al. that IC10B is an evolved star formation region.

REFERENCES

Henkel, C., Waterlout, J.G.A., Bally, J., 1986,
   Astron.Astrophys. 155, 194
Klein, U., Graeve, R., 1986, Astron.Astrophys. 161, 155
Ohta, K., Sasaki, M., Saito, M., 1988, Nobeyama Radio Observatory
   Report No. 195 (submitted to Publ.Astron.Soc.Japan)

MODELS OF COLLIDING GALAXIES: KINETIC ENERGY AND DENSITY ENHANCEMENTS

S. A. LAMB
University of Illinois
1011 W. Springfield Avenue
Urbana, IL 61801, U.S.A.

R. A. GERBER
University of Illinois
1110 W. Green Street
Urbana, IL 61801.

R. H. MILLER
University of Chicago
5640 Ellis Avenue
Chicago, IL 60637.

B. F. SMITH
NASA/AMES Research Center
245-3
Moffett Field, CA, 94035.

1. INTRODUCTION

Collisions and close encounters between galaxies have a profound effect upon the galaxies involved, for example, morphological changes can be pronounced. A very obvious effect of interaction on galaxies is a mean increase in their far infrared luminosities as detected by IRAS (Kennicutt et al. 1987; Bushouse, Lamb, and Werner 1988). This observed excess in infrared luminosity has often been invoked as evidence of increased star formation in these systems, but may in some cases be attributable to radiation from shocks in the gas occurring as a direct result of the collision (Harwitt et al. 1987). Of particular interest is the possibility that colliding galaxies are sites of unusual star formation.

2. N-BODY MODELS

Here we report on progress we have made in using N-body models to obtain information which will be useful in probing some aspects of the relationship between galaxy collisions, star formation, and radiation from shocks. The models we use are those reported in Miller and Smith (1980) and in Gerber, Lamb, Miller, and Smith (1988). A number of different encounters are simulated with varying values of initial orbital angular momentum and two different values of initial orbital energy. There is a parabolic sequence with zero initial total orbital energy and finite initial orbital kinetic energy and a hyperbolic sequence with both finite orbital energy and orbital kinetic energy. Within each of these two sequences, there are experiments with different values of initial orbital angular momentum. The collisions range from almost head-on, low angular momentum encounters to those in which the impact parameter is approximately half a galactic radius.

3. KINETIC ENERGY AND DENSITY ENHANCEMENTS

We use the self-consistent calculations to answer the questions, what is the increase in kinetic energy of the material due to the collisions, over what timescale is this energy enhanced, and what is the behavior in time of the central density in a colliding galaxy? Our calculations model point masses representative of the stellar component of a galaxy; they do not include a representation of the gaseous component. However, we suggest that basic quantities such as kinetic energy and density increases per unit mass of the system, calculated for the first passage of a pair, before the gas has had time to dissipate much of its acquired energy, give one a basic input into future calculations of both the star formation rates and the emission due to shock heating in such encounters.

We calculate the kinetic energy over all particles in a galaxy and include that due to orbital motion. The maximum increase in this kinetic energy occurs shortly after closest approach of the galaxies. It is almost independent of the speed of impact of the two galaxies (i.e. the values are almost identical for the parabolic and hyperbolic series). The increase in kinetic energy is sensitive to the impact parameter, that is, to the initial orbital angular momentum. By far the largest increase in kinetic energy is experienced by the less massive galaxy in an almost head on collision with a more massive galaxy. In the cases in which a disk is embedded in a spherical galaxy, the relative orientation of the disks at collision also has some effect upon the kinetic energy increase within that disk, with "face on" collisions showing the largest increase. The width of the energy peak ranges between 3.5×10^7 years and 7.0×10^7 years for galaxies with mass around 5×10^{10} $M_\odot$. This gives the basic timescale for transference of orbital energy of motion into internal energy of motion. It provides an overall timescale for the dissipation of energy in shocks and for star formation. Both the kinetic energy and central density can double in an individual galaxy during a head-on collision. However in an encounter in which the galaxies overlap for a time (as in all of our experiments), the maximum density achieved may not be at the center of either galaxy. This suggests the possibility of significant amounts of star formation occurring in localized regions away from the galactic nuclei, which has been noted to occur in several systems observed optically (see Bushouse, 1986).

REFERENCES

Bushouse, H. A. (1986) Thesis, University of Illinois.
Bushouse, H. A., Lamb, S. A., and Werner, M. W., (1988) Ap. J., Dec 1st issue.
Gerber, R. A., Lamb, S. A., Miller, R. H., and Smith, B. F. (1988) in preparation.
Harwitt, M., Houck, J. R., Soifer, B. T., and Palumbo, G. G. C. (1987) Ap. J., 315, 28.
Kennicutt, R. C., Keel, W. C., van der Hulst, J. M., Hummell, E., and Roettiger, K. (1987) Astron. J., 93, 1011.

Millimeter Wave Molecular Line Observations of Galaxies

N. Nakai
Nobeyama Radio Observatory
Minamimaki, Minamisaku
Nagano 384-13
Japan

ABSTRACT. More than eighty galaxies and QSOs have been observed with the Nobeyama 45-m telescope since 1983. Recent results of CO mapping of nearby galaxies are presented.

1. NGC 1097

NGC 1097 is a barred spiral galaxy at the distance of 16 Mpc. The CO distribution (Gerin et al. 1988) shows a ring structure at a radius of 800 pc (Fig.1). The detected H_2 mass (1.3 x 10^9 $M_\odot$) is abnormally high for such a nuclear region. The presence of such a molecular ring can be due to the secular action of the bar enhanced by the interaction with a companion galaxy NGC 1097A.

2. NGC 1068

CO observations of the nuclear region of a Seyfert galaxy NGC 1068 (Kaneko et al. 1989) have revealed two peaks of molecular gas on the molecular disk (Fig.2). The two peaks coincide with regions of strong 10 m emission. The CO line width is broad at the positions of two radio lobes, which suggests interaction between the nuclear ejecta and the ambient molecular gas.

3. M83

The region of 3'5 x 1' along the bar of M83 has been mapped in CO (Handa et al. 1989). Spectra in the nuclear region show very broad and weak wings. The broad wings of this face-on galaxy cannot be explained only by galactic rotation and turbulent motion but strongly suggest outflow (or inflow) of molecular gas from the disk (V_{flow} = 100 km s^{-1}).

4. Maffei 2

The central region of 1'5 square has been mapped (Nakai et al. 1988). CO(1-0) spectra around the nucleus show high and low velocity components (broad and weak wings) whose velocity is different from the galactic rotation by 100 km s^{-1}. Comparison with CO(2-1) spectrum shows that the peculiar components are optically thin (Fig.4).

References

Gerin,M., Nakai,N., and Combes,F. (1988) Astron.Astrophys., in press.
Handa,T., Nakai,N., Sofue,Y., Hayashi,M., and Fujimoto,M. (1988) submitted to Publ.Astron.Soc.Japan.
Kaneko,N., Morita,K., Fukui,Y., Sugitani,K., Iwata,T., Nakai,N., Kaifu,N., and Liszt,H.C. (1989) Astrophys.J., in press.
Nakai,N., Handa,T., and Sofue,Y. (1988) in preparation.
Sargent,A.I., Sutton,E.C., Masson,C.R., Lo,K.Y., and Phillips,T.G. (1985) Astrophys.J., $\underline{289}$, 150.

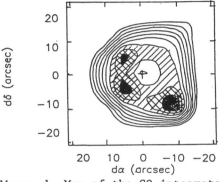

Figure 1. Map of the CO integrated intensity of NGC 1097.

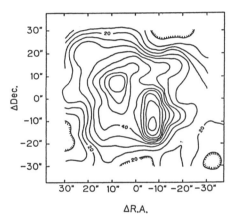

Figure 2. Map of the CO integrated intensity of NGC 1068.

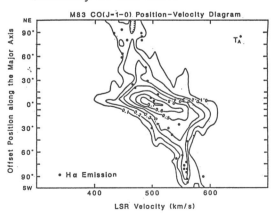

Figure 3. Position-velocity diagram along the major axis of M83.

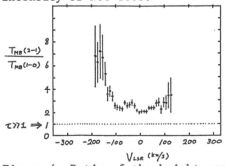

Figure 4. Ratio of the brightness temperature between CO(1-0) and CO(2-1) (Sargent et al. 1985) at the center of Maffei 2.

MOLECULAR SPIRAL STRUCTURE IN M51

R. J. Rand and S. R. Kulkarni
Dept. of Astronomy, 105-24
California Institute of Technology
Pasadena, CA 91125

ABSTRACT. We present a mosaic map of M51 in the ^{12}CO $1 \to 0$ line made with the Owens Valley mm-interferometer. The molecular arms we see are narrow and show an offset relative to the Hα emission. They also show coincidence with the dust lanes. Several interarm emission regions have also been detected, but with very little accompanying Hα emission. We briefly discuss the evidence for density-wave triggering of star formation.

1. The Observations and the Mosaic Map

Our map is a mosaic of 22 fields observed at Owens Valley, each with resolution $10" \times 7"$, and primary beam 65" (FWHP). Eight fields were mapped by S. N. Vogel (Vogel et al. 1988, hereafter VKS), and the remainder by R. J. Rand. The mosaic technique takes into account the primary beam response in each field, and no emission beyond the half-power points in any field has been included. The map is shown in Fig. 1a, and in Fig. 1b the contours of CO are overlaid on a continuum-subtracted Hα image. The contour interval is 15% of the peak of 45 Jy km s^{-1}. The optical image were taken with the 60" telescope at Palomar Observatory.

Notable features of Fig. 1 are: *a)* The inner arm seen in the map of VKS is now seen to extend further in toward the nucleus (the nucleus itself has yet to be mapped by us); *b)* the outer arm in the VKS map has also been extended in both directions; *c)* the other arm is now seen as a strong feature extending out from the nuclear region; *d)* the CO ridges are narrow and are offset from the Hα ridges; *e)* interarm Giant Molecular Associations (GMAs) have been detected for the first time. An off-Hα image (not shown) shows that the CO ridges coincide very well with the visible dust lanes. There seem to be only a few faint H II regions associated with the interarm complexes. This could be due to extinction, but the 6-cm continuum map of van der Hulst et al. (1988) confirms that there is generally not much H II in these regions.

2. Discussion

The main issue to be addressed using our CO and optical data is the relationship between the density wave and star formation. The narrow appearance of the CO arms, the well-defined offset between the CO and Hα ridges, and the coincidence of the dust lanes with the CO emission are all features of both hydrodynamical (it e.g. Lin and Shu 1964) and cloudy-ISM (*e.g.* Roberts and Hausman 1984) models of density waves. These models also predict that star formation is triggered by the passage of the density wave, leading to a higher star formation *efficiency* (SFE) in the arms than in the interarm regions.

Other interpretations of the star formation process are possible, however. Elmegreen (1987) suggests that different components of the molecular gas react differently to density waves. The lower velocity dispersion diffuse gas may shock in the density wave and produce dust lanes, while the denser, star-forming clouds may move in ballistic orbits and not shock. Thus, star formation may not be triggered by density waves. In this scenario, the substantial part of the molecular gas we have detected must belong to the shocked, diffuse component, since it is narrow and coincident with the dust lanes.

In Elmegreen's scenario, the ballistic cloudy component should show the same arm-to-interarm contrast as the Hα tracer, which we measure to be 7.5. In order to be consistent with the observed arm-interarm contrast of the CO, which we determine to be about 2.4, the cloudy component could therefore be no more that about 18% of the total molecular gas. This would result in a very high SFE, but little arm-interarm SFE contrast. Judging from the lack of Hα emission in the vicinity of the interarm GMAs we have detected, the SFE there must be substantially lower than on the arms. Lord (1987) and VKS have already presented evidence that the SFE is in fact higher in the arms. In any case, we should expect the arm-interarm contrast of the cloudy component to be less than that of the diffuse component, given the expected reaction of the two components to the density wave. Thus, there are major problems with Elmegreen's (1987) scenario, in which density waves do not trigger star formation. We therefore conclude that *it is extremely difficult to avoid the necessity for density wave triggering of star formation.*

4. References

Elmegreen, B. G. in *I.A.U. Symposium No. 115, Star Forming Regions* (ed. Peimbert M. and Jugaku, J.) 457-481 (Reidel, Dordrecht, 1987).
Hausman, M. A. and Roberts, W. W. 1984, *Ap. J.* **282**, 106.
Lin, C. C. and Shu, F. H. 1964, *Ap. J.* **140**, 646.
Lord, S. Ph.D. thesis, Univ. Massachusetts (1987).
van der Hulst, J. M., Kennicutt, R. C., Crane, P. C. and Rots, A. H. 1988, preprint.
Vogel, S. N., Kulkarni, S. R. and Scoville, N. Z. 1988, *Nature* **334**, 402-406.

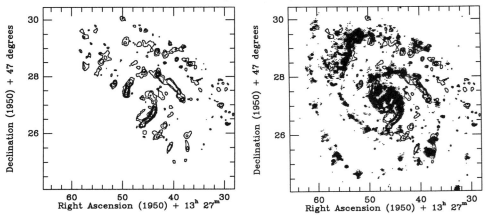

Figure 1. *a)* Contour map of CO emission in M51 detected with the interferometer; *b)* contours of CO overlaid on a continuum-subtracted Hα image of M51.

CO (3-2) Emission from the Nucleus of NGC 6946

E. Rosenthal, S. Eales
Institute for Astronomy, University of Hawaii

S. Stephens, K.Y. Lo
Astronomy Department, University of Illinois

ABSTRACT. We have observed the ^{12}CO (J= 3-2) emission from the nucleus of the nearby spiral galaxy NGC 6946 using the James Clerk Maxwell Telescope (JCMT) at a resolution of 15" (0.7 kpc). CO emission is clearly detected at the nucleus, 12" south along the molecular bar (Ball et al. 1985), and 12" west perpendicular to the molecular bar. The peak beam-averaged brightness temperature (T_{mb}) is 1.4 ± 0.5 K at the nucleus and 1.0 ± 0.4 K on average at the other two positions. Comparison with 1-0 mapping from Nobeyama *using the same beamsize* (Doi et al. 1988) shows that the (3-2)/(1-0) intensity ratio is 1.0 ± 0.5 on average at the three positions.

NGC 6946 has been observed extensively in the CO (1-0) line (most recently by Tacconi and Young 1986; Young and Sanders 1986; see Verter 1985 for a summary) and in the 2-1 line (e.g. Lo et al. 1980; Turner et al. 1986). In order to place further constraints on the physical conditions of the interstellar gas in this galaxy, we have mapped the CO (3-2) emission from its nucleus.

The observations were made from 14-20 February 1988 using Receiver B and the AOS at the JCMT. Five positions along the molecular bar (Ball et al. 1985) and two positions perpendicular to it were observed, with the map centered on the nucleus. The grid cell size used was 12" by 12", and typical integration times were 15 min/scan (30 cycles of 15 sec on source/15 sec on sky). Pointing offsets were repeatable from night to night, but there exists a systematic error uncertainty ≤ 10", primarily due to inaccuracies in the telescope pointing model and also to daytime seeing problems. A linear fit was done to the baselines in order to correct for instrumentally induced baseline tilt. The data were then calibrated and binned to ~ 7 km s^{-1} resolution. The calibration assumes a main beam efficiency η_{mb} = 0.6 ± 0.15, such that T_{mb} = T_A^*/η_{mb}.

Comparison with 1-0 mapping by Doi et al. (1988) from Nobeyama *using the same beamsize* shows that the (3-2)/(1-0) intensity ratio is 1.0 ± 0.5 on average at our three detected positions (see Figure 1). A possible detection at position (-12",0") shows a (3-2)/(1-0) ratio of 0.4 ± 0.3; it is also possible that there is a detection at (0",12"), but the baseline is extremely uncertain. The ratio from the three clear detections is consistent with mapping at 30" resolution (Lo et al. 1980), which showed the nuclear (2-1)/(1-0) intensity ratio to be ~ 1. Our three lines all show a line-center V_{LSR} ~ 90 km s^{-1}, and the CO (3-2) FWHM is ~ 100 km s^{-1} at the three positions.

Despite the incomplete coverage of our JCMT map off the molecular bar, the T_{mb} (3-2) appears to drop off more rapidly from the nucleus than the T_{mb} (1-0) in the Nobeyama map, although the CO (3-2) emission is apparently extended on a scale of at least ~ 500 pc. The

CO (3-2) emitting gas must have an excitation temperature of at least ~ 10 K. If the CO is optically thick in the 3-2 line, then it also must have a beam filling factor ≤ 10% in order to match our observed 3-2 brightness temperature of ~ 1 K. The line ratios are consistent within the errors with any T_{ex} ≥ 10 K in the optically thick case. However, models involving at least some portion of the CO emitting gas being optically thin cannot be ruled out with the current data. The CO (3-2) appears to be associated with the molecular bar in the nucleus of NGC 6946, which is probably associated with nuclear star formation activity (Ball et al. 1985). This situation is similar to that seen in IC 342 (Ho et al. 1987), although in that case the brightness temperatures of the CO (3-2) lines are much higher.

References

Ball, R., Sargent, A.I., Scoville, N.Z., Lo, K.Y., and Scott, S.L. 1985, *Ap. J. (Letters)* **298**, L21.
Doi, M., Ishizuki, S., Sofue, Y., Nakai, N., and Handa, T. 1988, *Pub. Astr. Soc. Japan*, in press.
Ho, P.T.P., Turner, J.L., and Martin, R.N. 1987, *Ap. J. (Letters)* **322**, L67.
Lo, K.Y., Phillips, T.G., Knapp, G.R., Wootten, H.A., and Huggins, P. 1980, *Bull. Am. Astr. Soc.* **12**, 859.
Tacconi, L.J., and Young, J.S. 1986, *Ap. J.* **308**, 600.
Turner, J.L., Ho, P.T.P., and Martin, R.N. 1986, in *Star Formation in Galaxies*, ed. C.J. Lonsdale Persson (NASA Conference Publication 2466), p. 383.
Verter, F. 1985, *Ap. J. Suppl.* **57**, 261.
Young, J.S., and Sanders, D.B. 1986, *Ap. J.* **302**, 680.

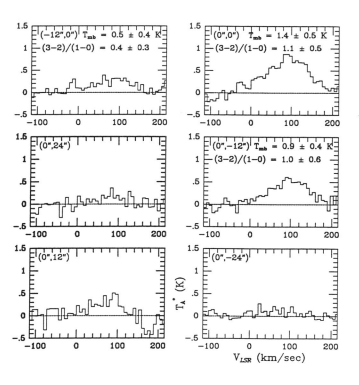

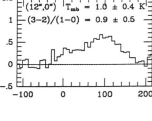

Figure 1. NGC 6946 CO (3-2) map. The binned spectra plotted on the T_A^* scale. Each spectrum is typically the result of 2 coadded scans. The spectra are individually labelled with their corresponding T_{mb} and (3-2)/(1-0) intensity ratio. Note that positions (0",12") and (0",24") outlined in bold have true locations above the nuclear (0",0") position.

The Azimuthal Distribution of the ISM in NGC 6946

Linda J. Tacconi
Netherlands Foundation for Research in Astronomy
7990 AA Dwingeloo, The Netherlands

We have completed a study of the atomic and molecular gas distribution of the ISM in NGC 6946. The main goal of the program is to gain insight into how the presence of a spiral potential affects the gas surface densities and the massive star formation efficiencies in this galaxy. In our galaxy, it is believed that most of the massive star formation occurs in the spiral arms (Georgelin and Georgelin 1976). Recently, it has been suggested that the primary effect of a spiral density wave is to organize the ISM into a global spiral pattern (*e.g.* Scoville, Sanders, and Clemens (1986)). The observed spiral variations in the OB star formation efficiency on the spiral arms are thus due to cloud orbit crowding in the spiral potential, with star formation in the arms resulting from cloud-cloud collisions (*e.g.* Scoville and Hersh (1979); Kwan and Valdes (1983)).

As part of the program we have made ^{12}CO observations with the 14 meter antenna of the Five College Radio Astronomy Observatory (HPBW = 45" at 115 GHz) at 110 positions in the galaxy. The majority of these observations were made in a series of concentric rings about the center of this galaxy with the furthest beams lying 3.'00 from the center. H_2 suface densities have been derived from the CO intensities by applying a standard galactic conversion factor. We present here a comparison of the azimuthal variations in the CO data in the northeast quadrant of the galaxy with those in a 21-cm HI map (Tacconi and Young 1986), in B- and I-band images (generously given to us by D. Elmegreen and B. Elmegreen), and also with Hα fluxes from the HII region catalogue of Bonnarel, Boulesteix, and Marcelin (1986), all smoothed to a resolution of 45". The full data sets and analysis are presented elsewhere (Tacconi and Young 1989a, 1989b).

Arm and interarm regions for NGC 6946 were determined from the 45" smoothed I-band image. We have computed arm-interarm contrasts for the NE region by taking the peak values over each I-band arm azimuth range as the "arm" values and have adopted the averages over both I-band interarm regions as the "interarm" values. The resulting constrasts are thus the maximum arm-interarm enhancements which are derived. The spiral arms in the northeast quadrant have enhancements in the CO integrated intensities, Hα surface brightnesses, and the $\mu(H\alpha)/\sigma(H_2)$ ratio. We interpret the Hα/H_2 ratio as a measure of the massive star formation efficiency, and infer that the massive star formation efficiency is greater on the spiral arms than in the interarm regions of NGC 6946. Figure 1 shows the derived arm-interarm contrast ratios as a function of radius for the NE arm complex. Radial increases in the B- and I-band spiral arm amplitudes are evident. Large radial increases in the arm-interarm contrasts of the Hα surface brightness and massive star formation efficiency are also prominent, with smaller increases observed in the H$_2$ surface density contrasts. Assuming that the I-band arms are tracing the true mass surface density enhancements, we suggest that the massive star formation efficiency depends on the strength of the spiral arms in NGC 6946.

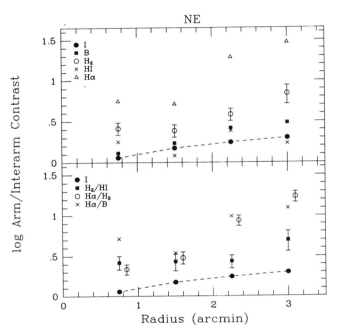

Figure 1: The arm-interarm contrast ratios as a function of radius for the northeast arm complex. The points are coded as indicated to distinguish between different quantities. Error bars are shown where the uncertainties are larger than the points.

Acknowledgements: We thank D. Elmegreen and B. Elmgreen for the B- and I-band images. The FCRAO is operated with support from the National Science Foundation under grant AST-82-12252 and with permission of the Metropolitan District Commission, Commonwealth of Massachusetts.

References

Bonnarel, F., Boulesteix, J., and Marcelin, M. 1986, *Astr. Ap. Suppl.*, **66**, 149.
Georgelin, Y.M., and Georgelin, Y.P. 1976, *Astr. Ap.*, **49**, 57.
Kwan, J. and Valdes, F. 1983, *Ap.J.*, **271**, 604.
Scoville, N.Z. and Hersh, K. 1979, *Ap.J.*, **229**, 578.
Scoville, N.Z., Sanders, D.B., and Clemens, D.P. 1986, *Ap.J. (Lett.)*, **310**, L77.
Tacconi, L.J. and Young, J.S. 1986, *Ap.J.*, **308**, 600.
———— 1989a, in press, *Ap.J. Suppl.*.
———— 1989b, submitted, *Ap.J.*.

DETECTION OF CO J = 1 → 0 EMISSION FROM AN *IRAS* SELECTED SAMPLE OF S0 GALAXIES

Leslie J. Sage and J.M. Wrobel
Astrophysics Research Center, Department of Physics
New Mexico Institute of Mining and Technology
Socorro, NM 87801
U.S.A.

ABSTRACT. We have obtained CO fluxes or upper limits for eleven S0 galaxies with *IRAS* 100 μm flux density $\geq$ 3 Jy. The observations were made with the NRAO 12m antenna at Kitt Peak. Five of eleven galaxies were detected at levels of 5σ or greater in integrated line intensity. These observations contradict the traditional view of S0 galaxies as being devoid of molecular gas. The ratios of FIR to CO luminosity among the sample are similar to those among normal spirals. The major difference between S0 and spiral galaxies is the smaller spatial extent of the S0's CO emitting region. The molecular gas traced by the CO emission presumably fuels the star formation inferred from *IRAS* observations. The origin of the molecular gas is uncertain at present.

1. INTRODUCTION. The discovery of far infrared (FIR) emission from elliptical (E) and lenticular (S0) galaxies was quite surprising (see *e.g.* Knapp *et al.* 1988, Ap. J. Suppl. submitted). Thronson and Bally (1987, Ap. J. (Letters), **319**, L63) found that some E and S0 galaxies have infrared colors consistent with emission from dusty regions surrounding young stars. Star formation requires the presence of interstellar gas, usually in molecular form. It has been assumed in the past that E and S0 galaxies have little or no interstellar gas, yet recent surveys have detected HI emission from some (Knapp, Turner, and Cunniffe 1985, A. J., **90**, 454; Wardle and Knapp 1986, A. J., **91**, 23). However, as HI emission is poorly correlated with global FIR emission in spiral galaxies (Young *et al.* 1986, Ap. J., **304**, 443), it probably does not trace star forming regions as well as the H_2 mass indicator CO.

2. OBSERVATIONS AND RESULTS. The sample selection criteria and data acquisition/reduction procedures are outlined by Sage and Wrobel (1988, Ap. J.(Letters), submitted). For most galaxies, only the nucleus was observed (the 'central pointing'). Four galaxies (NGC 205, NGC 404, NGC 4710 and NGC 5195) were mapped along the major axis. NGC 5195 was also observed at several additional positions. Of the 4 S0's searched for off-center emission, only 1 (NGC 5195) was detected. The integrated CO intensity for the central pointings is $\sim 5\sigma$ or better for NGC 404, NGC 3665, NGC 4526, NGC 4710, and NGC 5195 (see Figure 1). Formally, no CO was detected from NGC 205, NGC 3245, NGC 4429, NGC 4435, NGC 4459 and NGC 5363, although the spectra of NGC 205 and NGC 4459 and favorable comparisons with stellar and HI velocities suggest weak central detections.

3. DISCUSSION. Table 3 in Sage and Wrobel (1988) compares various properties of the S0 sample with those of spiral galaxies. This table shows that L_{FIR}/L_{CO} and S_{100}/S_{60} of the S0 samples are quite similar to those of the spiral sample, which suggests that in S0 galaxies the dust and gas are coextensive, as in spirals.

A major difference between S0 and spiral galaxies is the extent of the molecular gas. Among spirals CO is typically observable out to $D_{25}/4$ (Solomon and Sage 1988, *Ap. J.*, in press), while among the 4 S0's that were mapped along the major optical axis, only 1 exhibited off-center CO. The differences in size of the CO/FIR emitting regions are further supported by comparing L_{CO} and L_{FIR} to the blue luminosity L_B, calculated using using total blue magnitudes from RC2 and $M_B(\text{Sun}) = 5.41$. Table 3 from Sage and Wrobel shows that the ratios of L_{CO} and L_{FIR} to L_B are an order of magnitude lower among S0's than spirals. This can be understood as a consequence of molecular gas and dust occupying a smaller fraction of an S0 than a spiral.

For most of the 11 S0 galaxies, the ratio S_{100}/S_{60} can be understood as arising from current star formation (Helou 1986, *Ap. J. (Letters)*, **311**, L33). If this process is indeed responsible for the FIR emission, then the star formation rate (SFR) can be estimated from SFR $\simeq 2 \times 10^{-10} \times L_{FIR}$, where SFR will have units of $M_\odot \text{yr}^{-1}$ if L_{FIR} is in solar luminosities and the 'cirrus' contribution is neglected (Gallagher and Hunter 1986, in *Star Formation in Galaxies*, ed. C. Persson (NASA: CP-2466), p. 167). Except for NGC 205 and NGC 404, the SFRs vary from ~ 0.1 to 1 $M_\odot \text{ yr}^{-1}$.

The origin of the molecular gas in S0 galaxies is not certain. The gas is probably not left over from galaxy formation, as the inferred amounts of H_2 will last for only a small fraction of a Hubble time given the calculated SFRs. This suggests that quasi-continuous or sporadic resupply of molecular gas is required. For example, molecular gas could be acquired in a quasi-steady fashion via stellar mass loss with subsequent cooling (Faber and Gallagher 1976, *Ap. J.*, **204**, 365) or via cooling flows (Thomas 1986, *M.N.R.A.S.*, **220**, 949). It would be useful to have reliable predictions of the CO velocity fields for these two models. Predictions for the cooling flow model alone vary greatly (Thomas 1986; Bregman *et al.* 1988, *Ap. J. (Letters)*, **330**, L93). Alternatively, the molecular gas in an S0 could be sporadically augmented by capture from a gas-rich companion, as suggested for the HI (Wardle and Knapp 1986).

This work was partially supported by the National Science Foundation (NSF) through grant AST-8611247 to JMW. NRAO is operated by Associated Universities, Inc., under contract with the NSF.

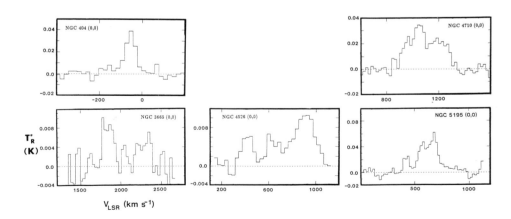

Figure 1. Reduced CO spectra (central pointing) for the 5 detected galaxies.

RADIAL DISTRIBUTION OF ATOMIC AND MOLECULAR GAS IN DISK GALAXIES

Zhong Wang
Institute for Astronomy, University of Hawaii; and
Astronomy Department 105-24, California Institute of Technology

Lennox L. Cowie
Institute for Astronomy, University of Hawaii

ABSTRACT. The observed radial surface mass density distributions of both atomic and molecular gas in 15 nearby spiral galaxies are analyzed. They are found to be consistent with a disk evolution model in which molecular cloud formation and star formation follow prescriptions similar to that discussed by Young (1987). Based on these and the assumption of "local condition determination", we propose a scaling law to describe large-scale gas distribution in disk galaxies of differing mass and size.

I. Introduction

In order to explore the basic characteristics of atomic and molecular gas distributions in disk galaxies, we analyzed HI and CO data of 15 nearby spirals. The sample is chosen based on the completeness of radial mappings and availability of corresponding rotation curves. Among the questions we try to address in this study are: (a) What are the main factors that determine the large-scale gas distribution in disk galaxies? (b) Can an one-zone disk evolution model fit the observed radial profiles of both atomic and molecular gas? and (c) Is there a general pattern of gas distribution in disk galaxies? If yes, how should it relate to the total mass, rotation curve and Hubble type of a galaxy?

2. A Brief Description of the Model

Following the suggestion of Young (1987), we assume that the disk gas fuels star formation in a two-step process: (i) atomic gas converts to molecular gas forming giant molecular clouds, this can be expressed as $CFR \propto \sigma_{HI}^m$; (ii) GMCs collapse to form stars, described as $SFR \propto \sigma_{H_2}^n$. Furthermore, the actual rates of these processes at any given radius of a disk also depend on the local gravitational stability criteria (Toomre, 1964; Balbus and Cowie, 1985), total (stellar and gaseous) mass density, and the rate at which the spiral arms pass by (Wyse, 1986). In this model, the disk's total mass distribution has an exponential dependence on galactic radius, and most of the mass is in atomic state at the beginning of disk evolution. A relatively small amount of mass returned to the ISM by massive stars is evaluated based on a conventional IMF, and is assumed to be atomic. Models with and without radial gas flows are both tried. The adjustable parameters are chosen by comparing with observations of our own Galaxy and other late-type galaxies. In particular, we choose

$m = 2$ and $n = 1$ which give a satisfactory fit to the available data.

It is found as a typical result of the model: the radial distribution of HI is flat to a large radius, beyond which it drops in the same way as the disk's total surface mass density σ_{tot}; H_2, on the other hand, can be described as a fast decreasing exponential function of radius, with its effective scale length smaller than that of σ_{tot} by a factor of about 1.5. However, this model can not fit a centrally depleted molecular gas profile (such as in our own Galaxy) without allowing a moderate amount of radial flow. This flow can be the result of, for example, accretions due to shearing viscosity of gas in the gravitational potential of the central bulge (see Icke, 1978; Sanders, 1979).

3. Analysis and Comparison

For each of the sample galaxy we compare the radial distributions of σ_{HI}, σ_{H_2} (derived from CO observations assuming a constant $N(HI)/I_{CO}$ ratio), and σ_{tot} (derived from the observed rotation curve out to about half of the disk optical radius). The main gas distribution characteristics of the theoretical model are seen in all of the sample galaxies, even though their mass and size scales are vastly different. This seems to indicate that the perception of "local condition determination" of the disk gas contents may be valid for these galaxies, and that the two-step star formation process as advocated by Young and others is probably a reasonable approximation. Large variation in terms of gas concentrations is seen in the central regions of these galaxies, but this can be explained if the effect of the central bulge mass component is considered.

We propose a scaling law for the galactic-scale gas distribution based on the theoretical and observational evidence. Since many other disk properties such as the shape of spiral arms and density contrast of arm and interarm regions seem to make only minor difference to the azimuthally averaged results, it is possible to build a standard model (or templet) of galactic disk evolution. Other disks with different mass and length scales can be scaled to the templet by these two parameters, thus the radial profiles of their gas contents as a result of the evolution processes are predictable. This idea is tested on the sample galaxies by plotting the observed M_{gas}/M_{tot} versus M_{tot} for HI and CO, and comparing those with the predicted values (here M_{tot} is the total mass of the disk out to half of its optical radius, obtained by integrating σ_{tot}). The results are in satisfactory agreement with the theory. We further speculate that this idea may also apply to dwarf irregular galaxies, so that it can explain the lack of CO gas seen in these low mass disks.

4. References

Balbus, S. A. and Cowie, L. L. 1985, *Ap. J.*, **297**, 61.
Icke, V. 1978, *Astr. Ap.*, **78**, 21.
Sanders, R. H. 1979, in *IAU Symp. No. 84*, ed. W. B. Burton, p383.
Toomre, A. 1964 *Ap. J.*, **139**, 1247.
Wyse, R. 1986, *Ap. J. (Letters)*, **311**, L41.
Young, J. S. 1987, in *Star Forming Regions*, ed. M. Peimbert and J. Jugaku, p557.

CO(1-0) IN A NEWLY-BORN ELLIPTICAL GALAXY : NGC 7252

C. DUPRAZ, F. CASOLI, F. COMBES, M. GERIN
Ecole normale supérieure
24, rue Lhomond
F-75231 Paris Cedex 05
France

ABSTRACT. We present CO(1-0) observations of the galaxy NGC 7252, made with the SEST 15 m. The molecular mass, $2.9 \ 10^9 \ M_o$, is unexpectedly high for an old post-starburst merger remnant. We conclude that the starburst triggered by the merging of two galaxies does not necessarily exhaust their gas content.

1. NGC 7252 : a Post-Starburst Merger Remnant

NGC 7252 (Fig. 1: Schweizer, 1982 = S82) is the best candidate for an object in the late stages of a merger. Its two tails and loops ensure that it is built up from two disk galaxies, while the presence of a single nucleus and $r^{1/4}$ light distribution show that it is old enough ($\approx 10^9$ years) to have already dynamically relaxed. Therefore, *NGC 7252 appears as a newly-born elliptical galaxy*. The luminosity of the object is $L_B = 4.7 \ 10^{10} \ L_o$.

However, the stellar population of NGC 7252 is comparable to that of a late-type spiral. The color indexes, U-B $\approx$ 0.17, B-V $\approx$ 0.66, are bluer than those of an elliptical, and strong Balmer absorption lines reveal the presence of many A-type stars. As these young stars, a bright disk of ionized gas of radius r $\approx$ 2.5 kpc (S82) is a probable relic of the burst of star formation that occured at the time of the merging. The infrared luminosity of NGC 7252 is $L_{IR} = 3.8 \ 10^{10} \ L_o$. It is larger than that of normal ellipticals, much lower than that of ongoing mergers, and actually similar to that of normal spirals (Soifer *et al.*, 1987).

2. $^{12}CO(1-0)$ Observations

We have obtained a $^{12}CO(1-0)$ spectrum towards the nucleus of NGC 7252 with the Swedish-ESO 15 m Submillimeter Telescope (Fig. 2):
- $\alpha_{1950} = 22^h 17^m 57.9^s$, $\delta_{1950} = -24°55'50"$
- beamsize at 115 GHz = 45",
- Schottky receiver, T_{rec} (SSB) = 300 K,
- mean atmospheric opacity = 0.2,
- system temperature $T_{sys} \approx 550$ K,
- AOS backend, 700 x 690 kHz channels,
- integration time of two hours on source using a beam-switching procedure,
- r.m.s. noise of 4 mK in the spectrum smoothed to 5.5 MHz,
- linear baseline.

The characteristics of the CO line are the following:
- heliocentric velocity $V_{hel} = 4710$ km/s,
- full linewidth at half maximum $\Delta V_{50} = 250$ km/s,
- integrated CO emissivity $\int T_A^* \ dV = 3.7$ K.km/s.

3. The Massive Molecular Component of NGC 7252

The close agreement between CO and optical (4749 km/s: S82) velocities proves that the molecular gas that we detect is physically associated with NGC 7252. Assuming a distance to the object of 62.8 Mpc (H_0 = 75 km/s/Mpc), and using a CO-to-H_2 conversion ratio of 2.8 10^{20}, we derive a molecular mass of 2.9 10^9 M_o within a radius of 7.7 kpc around the center.

Due to the limited resolution of our observations, we cannot ascertain the molecular-gas distribution of NGC 7252. However, the cold gas is bound to have settled down into a disk-like configuration within a few 10^8 years after the merging event. Since the width of the CO line (250 km/s) is in good agreement with the range of velocities measured in Hα ($\approx$ 200 km/s), it is then likely that the molecular gas is associated with the disk of ionized gas (Sect. 1).

The molecular mass we measure is rather unexpectedly high for an "old" merger remnant like NGC 7252. Indeed, it is often argued that the starburst triggered by the merging should starve out in less than a few 10^8 years by depletion of the molecular "fuel". Our result shows that, at least in some cases, the starburst is not so efficient: the molecular gas of NGC 7252 is far from being exhausted to date (there is more left there than in the Milky Way).

The star-formation activity of NGC 7252 has nowadays (10^9 years after the merging event) settled down. Indeed, its $L_{IR}/M(H_2)$ ratio of 13 is much lower than that of typical starburst galaxies, and is rather similar to that of normal spiral galaxies. For instance, the difference of a factor two between NGC 7252 and the Milky Way is easily accounted for by the strong interstellar radiation field that bathes the central regions of the former, elliptically-shaped object.

The molecular mass of NGC 7252 is also huge compared to that of "normal" ellipticals: $M(H_2) \leq$ a few 10^8 M_o (e.g., Bregman and Hogg, 1988). To become a bona fide elliptical galaxy, NGC 7252 should then ultimately rid itself of its high molecular content.

References:

Bregman, J.N., Hogg, D.E. (1988) *A. J.* **96**, 455-457
Schweizer, F. (1987) *Ap. J.* **252**, 455-460 (S82)
Soifer, B.T., Houck, J.R., Neugebauer, G. (1987) *Ann. Rev. Astr. Ap.* **25**, 187-203

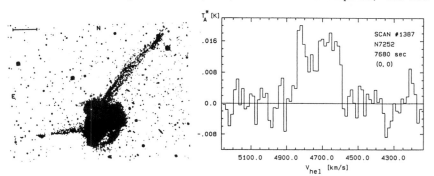

Figure 1 (left): Blue photograph of NGC 7252, taken from Schweizer (1982)
Figure 2 (right): CO(1-0) spectrum of the nucleus of NGC 7252, taken with SEST

VIRGO GALAXIES WITH EXTREME CO/HI RATIOS

JEFFREY D.P. KENNEY
Astronomy Department 105-24
California Institute of Technology
Pasadena, CA 91125

ABSTRACT. We discuss the nature of three S0-Sa galaxies in the Virgo cluster which are extremely HI-poor, yet have significant CO and FIR emission. The high CO/HI flux ratios of these galaxies indicate that they have 10-30 times more molecular gas than atomic gas. One of these galaxies, NGC 4419, has a significantly asymmetric CO distribution. We suggest that these galaxies are undergoing or have undergone severe interactions with the intracluster medium, perhaps accelerating their evolution from spirals to lenticulars.

1. Introduction

Although many spiral galaxies in the Virgo cluster are known to be extremely HI-poor and have asymmetric HI distributions (Haynes, Giovanelli and Chincarini 1984; Warmels 1986) nearly all of the large spirals have normal CO properties (Kenney and Young 1986, 1989a; Stark et al. 1986). This is commonly understood to be the result of interactions between the gas in the Virgo intracluster medium (ICM) and the interstellar medium (ISM) of the galaxies which race through it. Low density HI gas is stripped from the outer disks of Virgo spirals, while the H_2 gas in the inner disks survives. Although the survival of H_2 lessens the impact of HI stripping on the subsequent evolution of the galaxy, it may still be possible for a spiral galaxy to become an S0 galaxy as a result of stripping. Here we discuss the properties of 3 early type Virgo disk galaxies which appear to have undergone severe ISM-ICM interactions, and which are the best candidates for galaxies whose morphological evolution from spiral to lenticular has been accelerated by stripping.

2. CO/HI Flux Ratios

We have detected moderately strong CO ($J=1\rightarrow0$) emission from three Virgo galaxies with extremely weak HI emission. NGC 4293, NGC 4419, and NGC 4710 all have unusually high CO/HI flux ratios of 280-620, with inferred $M(H_2)/M(HI)$ ratios of 13-29 (Kenney and Young 1989b). Their HI masses are only $\sim.0005-.003 M_{total}$, thus it is small HI fluxes rather than large CO fluxes which is responsible for these high CO/HI ratios. In the Milky Way, there are roughly equal amounts of HI and H_2, while other HI-deficient Virgo spirals have $M(H_2)/M(HI)\sim 0.5-5$ (Kenney and Young 1989a). Although such large ratios are unusual for an entire galaxy, they are typical of the central regions of large spiral galaxies (Young and Scoville 1982). The CO/HI ratios therefore are easily explained if all the gas resides in the central few kiloparsecs, as would result from a severe stripping event.

Of the three galaxies, the strongest orbital constraints can be placed on the Sa galaxy NGC 4419. Its radial velocity of 1250 km s^{-1} with respect to the rest of the cluster puts it at the extreme end of the Virgo radial velocity distribution function. Since NGC 4419 is located only 2.3° from M87, its orbit carries it directly through the dense part of the ICM at a high velocity, ensuring a strong ICM-ISM interaction. The Sa NGC 4293 and the S0 NGC 4710 are each located 6° from M87, outside the active 'stripping zone'. If they indeed passed through the cluster core, they would take $\sim 2 \times 10^9$ years to reach their observed positions, assuming that they are moving with the mean cluster velocity.

3. The Asymmetric CO Distribution in NGC 4419

We have obtained single dish CO (J=1→0) maps of NGC 4419 with 17" and 45" resolution, and an interferometer map of the central 1' with 7" resolution (Kenney et al. 1989). Outside of the central kiloparsec, there is 3-4 times more CO emission on the northwest side than on the southeast side. Optical images show no evidence to suggest that the asymmetry is the result of an interaction with another galaxy. The excess CO flux in the northwest region of the galaxy represents $\sim 1.4 \times 10^9$ $M_\odot$, a mass which is ~ 20 times larger than the largest individual gas complexes known in either the Milky Way (Elmegreen and Elmegreen 1987) or the LMC (Cohen et al. 1988), and thus is likely consist of many smaller entities. The lifetime of the CO asymmetry in NGC 4419 is limited to $\sim 10^8$ years by differential rotation, indicating that the asymmetry must be short-lived or constantly regenerated.

An asymmetric gas distribution is an expected outcome of an strong ISM-ICM interaction in which the galaxy is moving nearly edge-on through the ICM (Kritsuk 1983). Because NGC 4419 is probably moving through the ICM at a large inclination angle, clouds on the side of the galaxy rotating into the ICM will be subject to a ram pressure force $\sim (1450$ km s$^{-1}/1050$ km s$^{-1})^2 \sim 2$ times stronger than on the opposite side. However, giant molecular clouds with column densities of $\sim 10^{22}$ H_2 cm^{-2} are themselves too dense to be significantly perturbed by ram pressure, even on the approaching side of NGC 4419. Thus if an ICM-ISM interaction is responsible for the asymmetry, then the molecular gas which is perturbed has column densities of $\sim 10^{21}$ H_2 cm^{-2}.

4. References

Cohen, R. S., et al. 1988, Ap.J.(Letters), **331**, L95.
Elmegreen, B. G., and Elmegreen, D. M. 1987, Ap.J., **320**, 182.
Haynes, M. P., Giovanelli, R., and Chincarini, G. 1984, Ann.Rev.Astr.Ap., **22**, 445.
Kenney, J. D., and Young, J. S. 1986, Ap.J.(Letters), **301**, L13.
Kenney, J. D. P., and Young, J. S. 1989a, Ap.J., in press.
Kenney, J. D. P., and Young, J. S. 1989b, in preparation.
Kenney, J. D. P., Young, J. S., Hasegawa, T., and Nakai, N. 1989, in preparation.
Kritsuk, A. G. 1983, Astrofizika, **19**, 263.
Stark, A. A., et al. 1986, Ap.J., **310**, 660.
Warmels, R. H. 1986, Ph.D. thesis, Groningen.
Young, J. S., and Scoville, N. Z. 1982, Ap.J., **260**, L11.

Section VI

Molecular Studies

Galactic Cloud Spectroscopy

Glenn J. White
Department of Physics, Queen Mary College,
Mile End Road, London E1 4NS, England

Introduction

Towards a number of high-mass star formation regions, there is a close association between the neutral gas in the star forming molecular cloud, and ionised gas resulting from the photo-dissociation caused by the UV emitted from massive early type stars. At the interface between such an HII region and a molecular cloud, is a neutral gas layer which is subject to both an intense radiation field, and to shocks arising from the expansion of the ionisation front of the HII region. The gas in these regions is highly excited, hot, and may be fairly dense, and as such we expect that the submm- λ transitions will be moderately strong. In order to study these interface regions, we have made studies of several edge-on ionisation fronts lying adjacent to dense molecular clouds.

Recent theoretical modelling which calculates the energy balance and chemical equilibrium of dense molecular clouds illuminated by high uv-fields and shocks, has contributed considerably to our understanding of the physical environment in the interface regions between ionised and neutral gas lying at the edge of star formation regions. Similarly new observational studies have provided additional impetus to such modelling with the rather surprising detection of significant amounts of hot (T ~ 200-400 K) gas at the edges of clouds, and the anomalous excitation of several molecular species in these interface regions. In addition to the chemical models previously mentioned, the interaction of the ionisation shocks with the molecular cloud material can result in a strong enhancement of the molecular abundances of some specific species, and lead to dynamical effects which will affect the velocity field of the material close to the interface regions, which should be detectable with submm- λ spectroscopic techniques. These models have been moderately successful in their general predictions, but not all of the predictions are supported by recent observational material.

Using new high angular resolution JCMT 15m observations, covering large areas with both good sampling and high dynamic range, it is possible to directly image star formation regions in a relatively unbiased way, to study the large scale effects of star formation on a molecular cloud.

Observations

The data shown in this paper were obtained during the period 1987 and 1988, with the James Clerk Maxwell Telescope at Mauna Kea, Hawaii. For the CO J=2-1 observations of Orion, which were carried out as part of the scientific comissioning of the telescope, the rms accuracy of the surface was estimated to be ~ 75 µm. The beamsize was measured to be 22 arc seconds, and the value of η_{fss} estimated as 0.75. The later observations of M17 were carried out after improvements to the surface, for which the surface error had been reduced to ~ 38 μm rms.The tracking and pointing of the telescope, based on observations of planets, was believed to be better than 5 arc seconds. The observations were obtained using a 230 and 345 GHz cryogenic Schottky system (the so-called 'System A $_{lower}$' and 'System B' respectively), with either an autocorrelator or AOS backend. A more complete description of this, and other JCMT instrumentation has recently been given by White (1988).

CO J=3-2 Observations of M17

The observations of integrated J=3-2 ^{12}CO emission (White and Greaves *in preparation*) close to the interface with the HII region are shown in Figure 1.

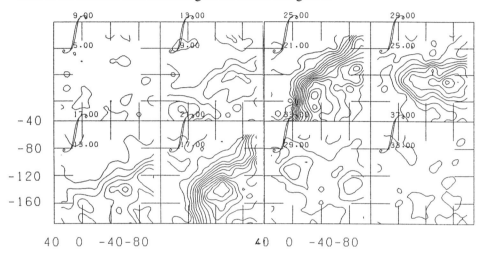

Figure 1 CO J=3-2 map of part of the M17 ridge, where the 0,0 coordinate is RA (1950) = $18^h 17^m 32.5^s$, Dec (1950) = $-16°10'40"$. The maps show the velocity ranges 5 to 37 km s^{-1}.

The ridge of CO emission seen in a more extensive J=1-0 map (White et al in preparation) extends approximately 6 arc minutes along an SE-NW direction. The CO J=3-2 map of integrated emission contains evidence for several clumps which lie close to a CO peak seen in the lower resolution data of Rainey et al (1987). The eastern edge of the complex is sharply bounded, with the CO emission falling from T $_R^*$= 30 - 40 K to less than 10 K over distances of 20-30 arc seconds (0.2 - 0.3 parsecs). To the east of this sharp boundary of the cloud lies the ionisation bar which is seen most clearly from radio wavelength studies (Felli et al 1984). The proximity of an ionisation bar to the neutral cloud material, and the rapid decline in CO emission going across the front, suggests that the neutral gas may be dissociated or depleted close to the interface region, and raises the possibility of external heating by the OB cluster which ionizes most of the M17 region.

^{13}CO J=3-2 of M17

The ^{13}CO observations show a rather different spatial distribution to that of the CO maps, in particular the emission is more concentrated into a smaller number of resolved clumps, and the eastern edge of the cloud is considerably more sharply bounded, lying about 40 arcseconds (~ 0.5 pc) further into the main molecular cloud than is indicated by the CO observations. The map of integrated ^{13}CO emission is shown in Figure 2 for various velocity ranges.

CO J=2-1 observations of the Orion Molecular Cloud

The Orion Nebula is one of the most intensively studied areas of high mass star-formation in the Galaxy. The large scale structure of the molecular material in the region has been shown to consist of a highly clumped and filamentary medium (Kutner et al 1977, Bally et al 1987, Batrla et al 1983, Gillespie and White 1980, Loren 1979, White and Phillips 1988). The central region covering about 10 square arc minutes around the core of the optical nebula contains many interesting objects, exhibiting a wide range of morphological structures. This central core may contain a luminous (~10^5 L_0) infrared object, IRc2, which is believed to be an ~ 50 M_0 protostar.

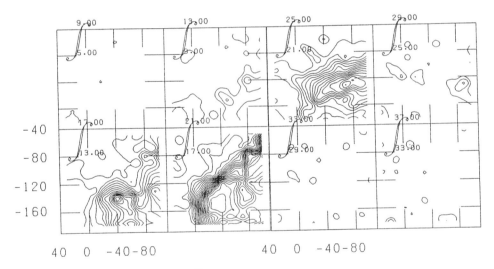

Figure 2 ^{13}CO J=3-2 map of part of the M17 ridge, where the 0,0 coordinate is RA (1950) = $18^h17^m32.5^s$, Dec (1950) = -16°10'40". The maps show the velocity ranges 5 to 37 km s^{-1}.

The JCMT map of the Orion Nebula, shown in Figure 3, contains about 400 individual spectra, with good spatial and velocity coverage. In this paper I will concentrate on the more interesting regions observed in this mapped area, concentrating on a) the high velocity molecular outflow region, b) the Bright Bar, and c) the molecular streamer(s);

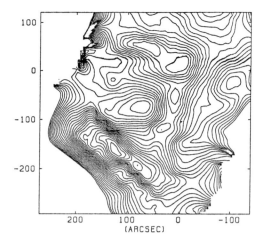

Figure 3 CO J=2-1 map of part of the Orion Nebula, where the 0,0 coordinate is RA (1950) = $05^h 32^m 46.8^s$, Dec (1950) = $-0° 24'23"$. The maps show ta map with a central velocity of 9 km s^{-1}.

a) *The Molecular Outflow*

The molecular outflow has been studied in many different molecular species, and with a wide range of angular resolutions. In the CO J=1-0 line, the highest angular resolution images have been obtained by Masson et al (1987) with the Owens Valley interferometer. These observations show two lobes of emission separated by ~ 15 arc seconds. The CO J=3-2 observations of Erickson et al (1982) show a similar distribution for the high velocity gas, however, the centre of their outflow lies ~ 10 arc seconds north of our CO J=1-0 (0,0) position, and also that of the Owens Valley observations. There is only marginal separation of the red and blue high velocity wings seen in our J=2-1 data, although differences to the appearance of the individual spectra can be seen over the range -2 to 20 km s^{-1}. The major new result from the present observations is the large extent of the outflow seen in the maps in the range 12 - 17 km s^{-1}. At least one HH- object (Jones & Williams 1985) is associated with the jet -like structure, and which has a proper motion lying along the lobe, away from the centre of the Orion Nebula, and is situated close to a peak in the molecular hydrogen emission seen in the maps of Gatley & Kaifu (1987).

b) *The Bright Bar*

The Orion Bright Bar marks the site of a shock front which results from expansion of the main Orion HII region into surrounding neutral material. The favorable geometry of this system, which lies ~ 2 arc minutes SE of the Trapezium stars, is almost edge-on. For the J=2-1 CO observations, the structure of the bar is confined to a small velocity range extending from 5.2 to 14.7 km s^{-1}. In the bar region, the lines generally appear singly peaked, with half-power widths ~ 4-6 km s^{-1}. With the present spectral and spatial resolution, no clear trend in velocity can be seen across the Bright Bar (observations to be published elsewhere obtained with 0.1 km s^{-1} resolution in the CO J=1-0 line show extremely complex line-shapes, but again no clear trend Sanderson et al 1988)). The temperature of the material in the Bar is extremely hot, being similar to that of gas close to the molecular outflow source. As discussed by Phillips & White (1982) and by Omodaka et al (1988), it is unlikely that with the present generation of telescopes, that appreciable antenna temperature enhancements will be observable, since post-shock radiative cooling is extremely efficient, and in this case would be expected to lead to a hot shock heated zone of less than one arc second width. A more likely source of heating for the gas in this bar is the photoelectric ejection mechanism of Tielens & Hollenbach (1985).

c) *The Molecular Streamers*

There is evidence for a number of 'streamers' or filamentary structures in the CO maps. The most obvious of these are the apparent extension to the compact high velocity flow, and the narrow filament extending northwards of the core. In this latter case, a narrow ridge extends more than two arcminutes northwards from the core of the cloud, seen most prominently between 5.2 and 8.1 km s^{-1}). A similar feature is also present in the CO J=1-0 map of Hasegawa (1987), who shows that the streamer extends northwards for almost 200 arc seconds. An Herbig-Haro object lies coincident with the streamer (close to (10,105)), but *not* at the tip of the feature. This particular streamer is also visible in other lines such as HCO$^+$ and HCN, and appears to be connected to the chemically peculiar 'Radical Position' (also know as the 3'N 1'E position - see for example White et al (1986). The origin and chemistry of this peculiar region will be discussed in a separate paper. As originally proposed by Hasegawa (1987), these streamers may represent interactions between highly collimated jets which are being ejected into the surrounding interstellar medium from the cloud core. At the interface, a shock interaction then leads to the optically visible HH-objects. The nature of these streamers remains unclear, although the reality of their existence is confirmed.

Acknowledgements

The observations of Orion reported in this paper were collected during the Scientific Comissioning phase of the James Clerk Maxwell Telescope.The data will be fully reported in a more extensive paper at a later time. The comissioning data are a tribute to the many people who have collaborated over the last decade to construct this very fine telescope. All though it is impossible to acknowledge the role of everybody here, I would like to mention the dedicated work and efforts of people at ROE, RAL, and the Universities of Cambridge, Kent, London (QMC), Manchester (Jodrell Bank), Utrecht and Groningen, as well as the Dwingeloo Laboratories of the Netherlands Foundation for Radio Astronomy and Lancashire Polytrechnic. In particular, the roles of Dr. Richard Hills, the Project Scientist, and Dr. Ron Newport, the Project Manager, were important to the success of the project. The James Clerk Maxwell Telescope is now operated by The Royal Observatory, Edinburgh on behalf of the partner nations: Holland, Canada and the United Kingdom. Mm and Submm Astronomy in the UK is supported by The Science & Engineering Research Council.

References

Bally,J. et al.,1987. *Astrophys. J.*, **312**, L45.
Batrla et al.,1983. *Astr. Astrophys.*, **128**, 279.
Becklin, E.E., Beckwith, S., Gatley,I., Matthews, K., Neugebauer, G., Sarazin., Werner,M.W. 1976. *Astrophys.J.*, **207**, 770.
Erickson, N.R., Goldsmith, P.F., Snell, R.L., Berson, R.L., Huguenin, G.R., Ulich, B.L and Lada, C.J., 1982. *Astrophys.J. (Letters)*, **261**, L103.
Gatley, I. & Kaifu, N., 1987. *IAU Symposium 120, Astrochemistry*, p 161, D.Reidel Press.
Gillespie A.R. & White, G.J., 1980. *Astr. Astrophys.*, **91**, 257.
Hasegawa, T., 1987. *IAU Symposium*, Star Formation Regions, p 123.
Kutner et al., 1987. *Astrophys. J.*, **215**, 521.
Loren, R.B., 1979. *Astrophys.J. (Letters)*, **234**, L207.
Masson, C.R., Berge, G.L.,Claussen, M.J., Heiligman, G.M., Leighton, R.B., Lo, K.Y., Moffet, A.T., Phillips, T.G., Sargent, A.I., Scott, S.L., Wannier, P.G. & Woody, D.P. 1984. *Astrophys. J. (Letters)*, **283**, L37.
Omodaka et al., 1988. *In preparation*.
Phillips,J & White,G.J. 1982. *Mon. Not. R. astr. Soc.*, **199**, 1033.
Tielens, A. & Hollenbach, D. *Ap.J.* **291**, 722, 1985a
Tielens,A. & Hollenbach,D. *Ap.J.* **291**, 747, 1985b
White, G.J. et al, 1986. *Astr. Astrophys.*, **162**, 153.
White, G.J., Rainey, R., Hayashi, S.S. & Kaifu, N., 1987. *Astr. Astrophys.*, **173**, 337.
White, G.J., 1988. *'Millimetre & Submillimetre Astronomy '*, p 27, Kluwer Academic Publishers, ed. R.D.Wolstencroft & W.B.Burton.
White,G.J & Phillips, J.P. 1988. *Astr.Astrophys.* **197**,253

DETECTION OF FAR-INFRARED ^{13}CO LINE EMISSION

R. Genzel, A. Poglitsch
Max-Planck-Institut für Physik und Astrophysik, D8046 Garching

G. Stacey
University of California, Berkeley CA 94720

ABSTRACT. The first detection of far-infrared ^{13}CO line emission toward the core of Orion-KL is reported. About 10 to 30 $M_\odot$ of dense and warm (T ≥ 200K) gas are required for the 151μm J=18->17 ^{13}CO line flux.

OBSERVATIONS AND RESULTS

The data were taken on the NASA Kuiper Airborne Observatory in January 1988 with the UCB tandem Fabry-Perot spectrometer. A linear 3-element detector array with a spatial resolution of 55" was employed. The spectral resolution was 32 km s^{-1} FWHM for the central detector. In addition to the ^{13}CO J=18->17 transition at 151.4315 μm, we also observed the nearby ^{12}CO J=17->16 line at 153.2669 μm.
    The results for the central detector are shown in the figure. The J=18->17 ^{13}CO line flux is 2.3±0.7x10^{-18} W cm^{-2}. The ratio of J=17->16 ^{12}CO to J=18->17 ^{13}CO line fluxes is 37±8. The ^{13}CO flux can be used directly for an estimate of the mass of warm molecular gas at the core of Orion-KL. For optically thin emission and a ^{13}CO/H$_2$ abundance ratio of 10^{-6}, total gas masses of 5 $M_\odot$ at a gas temperature of 500K and 50 $M_\odot$

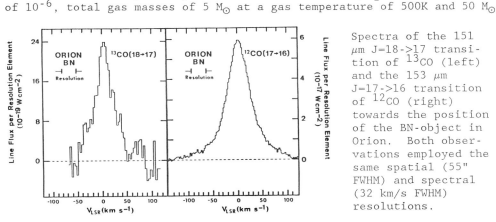

Spectra of the 151 μm J=18->17 transition of ^{13}CO (left) and the 153 μm J=17->16 transition of ^{12}CO (right) towards the position of the BN-object in Orion. Both observations employed the same spatial (55" FWHM) and spectral (32 km/s FWHM) resolutions.

at 200K are required to account for the ^{13}CO flux. Significant ^{13}CO emission was also detected in the 55" north-west off-center detector.
The ^{13}CO profile is narrower than the ^{12}CO profile. The intrinsic width of the ^{13}CO line is $\leq$ 17km s^{-1}, while the ^{12}CO line has a width near 27km s^{-1} (FWHM) as determined by MEM deconvolution.

DISCUSSION

Our measurements show that the ^{13}CO and ^{12}CO far-infrared lines have different line profiles and that the ^{13}CO line flux is significantly larger than expected for a single, optically thin ^{12}CO far-infrared emission component assuming the standard abundance ratio.

For quantitative evaluation of possible interpretations we computed level populations and line intensities as function of density, temperature, and column density in an escape probability radiative transfer formalism. We assumed a CO/H$_2$ abundance ratio of 10^{-4} (Watson et al. 1985) and a ^{12}CO/^{13}CO abundance ratio of 90 (Scoville et al. 1982).

The first model we considered assumes a single emission component with optically thick ^{12}CO far-infrared lines. With an area filling factor of $\approx$ 0.3 we can describe the far-infrared measurements adequately; the ^{12}CO lines also have enough optical depth to account for the difference in ^{12}CO and ^{13}CO line width. The model fails, however, to predict the observed submm brightness temperatures.

Our second model is based on three emission components: the high velocity plateau emission accounts for the mm and submm ^{12}CO lines; it represents warm gas (T $\approx$ 300K) in the outflow from the infrared cluster. The shock region is hot (T $\approx$ 700K) and optically thin gas which accounts for the far-infrared ^{12}CO lines and the high-velocity components of the ^{13}CO line. The hot core is a region of high column density but low velocity dispersion which is exhibited by the narrow peak in the ^{13}CO line profile. The ^{12}CO far-infrared emission from this region is optically thick. From the decomposition of the ^{13}CO line in a narrow component (hot core) and a wide component (shock) we infer a hydrogen mass of $\approx$ 10 M$_\odot$ in the hot core and a ^{12}CO/^{13}CO abundance ratio of 85 $\pm$ 20. For the 55" NW position, however, we find a ^{12}CO/^{13}CO flux ratio of 50 $\pm$ 20 which is either not consistent with the small size (10") of the hot core, or requires a lower ^{12}CO/^{13}CO ratio (cf. Blake et al. 1986). A decomposition of the center position line based on a lower abundance ratio would still be consistent with our data.

An explanation of the flux ratio based on only a very low ^{12}CO/^{13}CO abundance ratio is precluded by the clear difference in line shape between the two isotopic lines.

REFERENCES

Blake, G.A., Sutton, E.C., Masson, C.R., and Phillips, T.G. 1986, Ap.J. Suppl. 60, 357.
Scoville, N.Z., Hall, D.N.B., Kleinmann, S.G., and Ridgway, S.T. 1982, Ap.J. 253, 136.
Watson, D.M., Genzel, R., Townes, C.H., and Storey, J.W.V. 1985, Ap.J. 298, 316.

THE ROTATION CURVE OF THE S106 MOLECULAR DISC

RACHAEL PADMAN & JOHN RICHER
Cavendish Laboratory
Madingley Rd.
Cambridge CB3 0HE
England

ABSTRACT. We have observed the "disc" of the bipolar nebula S106 in the $J = 3 \to 2$ lines of HCN and HCO$^+$. MEM restoration of the position-velocity diagram shows evidence for differential rotation consistent with a central point mass of $\geq 10 M_\odot$

Discs are often invoked to explain the collimation into oppositely directed jets and flows of winds from pre-main sequence stars and young stellar objects. However despite intensive searches, almost all of the evidence pointing to the existence of discs is circumstantial. Only in a few sources is there direct *kinematic* evidence for the differential rotation curve that is the signature of Keplerian rotation about a massive central object [1 – 3] (but see [4] for a depressing recantation).

We have chosen to search for rotation by observing a well-known "disc" source in the relatively high density tracers HCN and HCO$^+$ in their $J = 3 \to 2$ transitions (which are typically excited at H$_2$ densities of between 10^7 and $10^8 \mathrm{cm}^{-3}$), and examining the position-velocity diagram along the major axis. Although the optically thick line core is expected to be confused by the ambient cloud material the more optically thin line wings should be a good indicator of the overall velocity field. By sampling at twice the minimum Nyquist rate we are in a position to apply a maximum-entropy (MEM) algorithm to obtain enhanced angular resolution.

The source, the bipolar nebula S106, is at first sight somewhat more evolved than most "classical" outflow sources. The absence of a strong molecular flow however should make it easier to detect the rotation signature of any disc that might be present. The observations were made at the JCMT [1], in 1988 April, using a 19-arcsec beam and sampling at 5-arcsec. intervals along an axis 1-arcminute long passing through IRS-4 at position angle $-54°$ (coincident with the dark lane in the 5GHz radio images in [5]).

The raw HCO$^+$ position-velocity diagram is shown in Fig.1, along with the MEM restoration. Also shown is the MEM restoration of the HCN data (which extends only to ±20arcsec). In both the HCO$^+$ and HCN data there is evidence for greatly increased linewidths near the position of IRS4, with strong asymmetry on the two sides of the source. Superimposed

---

[1] The James Clerk Maxwell Telescope is operated by the Royal Observatory Edinburgh on behalf of the Science and Engineering Research Council of the United Kingdom, the Netherlands Organization for Scientific Research and the National Research Council of Canada.

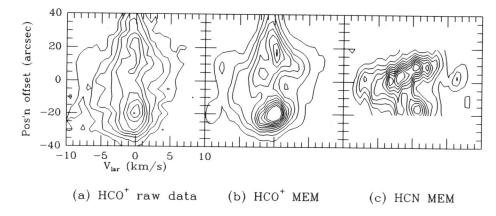

(a) HCO$^+$ raw data   (b) HCO$^+$ MEM   (c) HCN MEM

Figure 1: $HCO^+ J = 3 \rightarrow 2$ position velocity diagram (a) and MEM restoration (b). The HCN MEM restoration is shown in (c)

on top of the relatively smooth variations in the optically thin line wings we see "clumps" of emission (in l-v space), which may or may not be associated with the putative remnant disc. The intense emission at an offset of -20-arcsec from IRS4 is thought to be due to a separate clump.

The apparent differential rotation evident especially in the HCO$^+$ data is consistent in its inner parts with a central point mass of about $10 M_\odot$, although this is probably uncertain by a factor of 2 either way. There is some suggestion that the velocity falls off *faster* than Keplerian - this may be an indication that the disc has not yet relaxed in its outer parts, and the observed clumpy nature of the emission supports this view. The fact that the HCN emission peaks much closer to IRS-4 is probably due to the higher density required to excite this species, and to the fact that HCN is likely to be less optically thick than HCO$^+$. Higher resolution observations are necessary to decide whether the apparently linear velocity gradient across IRS4 in HCN is due to lack of resolution – if real it would indicate that the mass in the inner disc is significant when compared with the central point mass.

REFERENCES

[1] Vogel, S.N., Bieging, J.H., Plambeck, R.L. and Wright, M.C.H. (1985) *Astrophys.J.* **296** 600
[2] Bieging, J.H. (1984) *Astrophys.J.* **286** 591
[3] Jackson, J.M., Ho, P.T.P and Haschick, A.D. (1988) *Astrophys.J.* (in press)
[4] Loushin, R., Crutcher, R. and Bieging, J. (1988) in *Molecular Clouds in the Milky Way and External Galaxies*, J.Young and R.Dickman (eds) Riedel, Dordtrecht.
[5] Bally, J., Snell, R.L. and Predmore, R. (1983) *Astrophys.J.* **272** 154

ACKNOWLEDGEMENTS

It is a pleasure to acknowledge the assistance of P.F.Scott, R.E.Hills, N.D.Parker and A.P.G.Russell who helped with the observations described here. The maximum entropy restorations used the MEMSYS package written by S.F.Gull and J.Skilling.

A SEARCH FOR DENSE GAS AROUND YOUNG STARS

C. R. MASSON, J. B. KEENE, L. G. MUNDY, G. A. BLAKE
Caltech 320-47, Pasadena, CA USA 91125
E. C. SUTTON, W. DANCHI, P. JAMINET
University of California, Berkeley, CA USA 94720

ABSTRACT. We searched for CS(7-6) emission in a range of young stars, to test for the presence of dense gas. CS emission was detected only towards high-luminosity objects. We suggest that this is because the low-luminosity objects cannot heat a sufficient volume of gas to produce detectable CS emission.

1. Observations and Results

Many observations of millimeterwave dust emission have shown the presence of 0.01 to 1 $M_\odot$ of gas in regions of only a few hundred AU around young stars (e.g. Keene and Masson 1987). Even if the gas is uniformly distributed, the implied space densities are high. However, the dust properties are uncertain, and no direct measurement of the density has been obtained. CS(7-6) has a critical density of 2 10^7 cm^{-3}, an upper level energy of 65 K, and a rest frequency of 343 GHz. It is therefore an ideal probe for the presence of dense gas around young stars.

Using the UC Berkeley 345 GHz SIS receiver (Sutton *et al.* 1988) on the CSO, we measured CS(7-6) in 17 objects to a typical noise level of 0.075 K (1 σ). The objects were all young stars with evidence for nearby dense gas and they were chosen to cover a wide range of luminosity. Out of our sample, 10 objects were detected and 7 were not detected (2 σ). Of the detected objects, 5 were mapped and the emission was found to be quite strongly peaked. IRAS 16293-2422 was unresolved by our 24″ beam, which is consistent with the emission coming from the $\sim$ 6″ dense region mapped by Mundy, Wilking and Myers (1986) in this source.

2. Discussion

To find the difference between the non-detections and detections, we plotted the peak CS(7-6) antenna temperature against 100 μm flux density for the 16 objects for which we were able to find 100 μm data (mainly from IRAS). In this plot (Fig. 1) there is a striking correlation in that *all* of the objects with flux densities greater than 500 Jy are detected, while *none* of the weaker ones are.

There are two possible reasons for this correlation, since the 100 μm flux density is related both to the presence of dense gas and to the luminosity of an object. Indeed, we detected

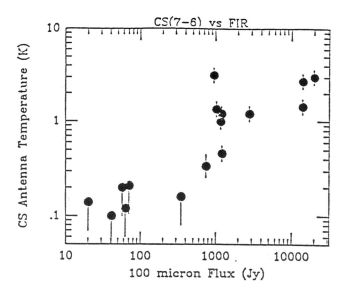

all objects with luminosities greater than 20 $L_\odot$ and none of the weaker ones. The cutoff of 500 Jy corresponds roughly to 0.5 $(d/160pc)^2$ $M_\odot$, which is substantially more mass than is required to produce the observed emission. For this reason, we believe that the crucial discriminator is luminosity, since low-luminosity stars can heat up only a small region around them to the 65 K required to excite the CS(7-6) transition, and beam-dilution would reduce the antenna temperatures below our detection limit. To obtain an crude estimate, we have calculated the volume heated to 65 K in an optically thick region and we find:

$T_a^* \sim 2.2\ (L/1L_\odot)\ (160pc/d)^2$ mK

This assumes that sufficiently dense gas exists in the vicinity of the star, which is likely since our objects all have evidence of circumstellar material. The magnitude of emission predicted agrees fairly well with the observations since, in a nearby cloud at 160 pc, our 2 σ limit of 0.15 K corresponds to a predicted threshold luminosity of 70 $L_\odot$. The diameter of the emission region at our detection threshold is 2″. More detailed calculations of CS emission from model accretion disks give similar results. With expected improvements in sensitivity, and increased integration times, it should be possible to detect CS(7-6) emission from the vicinity of stars down to a few $L_\odot$. The emission regions are predicted to be very small and a submillimeterwave interferometer will be required to study their structures. This research was supported by NSF grant AST 83-11849.

3. References

Keene, J. B., and Masson, C. R. 1987, B.A.A.S., **18**, 293.
Mundy, L. G., Wilking, B. A., and Myers, S. T. 1986, Ap.J.(Letters), **311**, L75.
Sutton, E. C., Danchi, W., and Jaminet, P., 1988, in preparation.

AN UNBIASED SURVEY FOR DENSE CORES AND STAR FORMATION IN THE ORION B MOLECULAR CLOUD

Elizabeth A. Lada
Department of Astronomy
The University of Texas
Austin, Texas 78712

This paper presents the results of a study of dense cores and of star formation in the Orion B Molecular Cloud. Over the last ten years, observations at radio and infrared wavelengths have revealed that stars form in molecular clouds. Little is known about the actual process of star formation, but it is becoming clear that knowledge of the density structure of molecular clouds is critical to understanding the evolution of the clouds and the formation of stars within them. Although the density structure of a number of individual molecular cores has been studied, very little is known about the large scale gas density distribution of star forming clouds.

Molecular line studies of regions with current star formation have shown that young stars are invariably associated with molecular gas. Is the converse true? Is star formation always present in dense gas? Until recently, this question has not been adequately addressed since the dense regions studied were selected because they had signposts of star formation.

For my thesis, I have completed an unbiased systematic search for density condensations in the Orion B Molecular Cloud and an unbiased search for embedded infrared sources in these condensations in an attempt to understand the relationship between dense cores and young stars.

To identify the dense cores, the Orion B cloud was surveyed in the J=2-1 transition of CS using the AT&T Bell Labs 7-m telescope (Lada, Bally and Stark 1988, in preparation). Thirteen thousand points were surveyed with one arc minute spacing to an rms noise level of 0.2 K. The total area covered by the survey was approximately 3.6 square degrees. Figure 1 presents the results of this unbiased survey in the form of an integrated intensity map. Emission is seen throughout much of the region surveyed and this emission is not uniform but clumpy. Individual velocity channel maps indicate an even higher degree of clumping than that present in figure 1.

In order to quantitatively study the clumping properties of the molecular gas, I have attempted to identify individual clumps using channel maps, position velocity maps and individual spectra. Thirty four clumps were identified above a five sigma noise level. Clump radii range from ≤ 0.13 to 0.68 pc. Twelve clumps are unresolved and only five have radii ≥ 0.50 pc. The FWHM linewidths for the peak intensity position of each clump range from 0.5 to 2.7 km/s. Virial masses were calculated for each clump and range from 13 M$\odot$ to ~2000M$\odot$. Most clumps have masses between 100 to 200 M$\odot$ and five have masses greater than 500 M$\odot$.

A near infrared survey of all CS clumps and some non-emission regions was carried out using the NOAO Infrared Array Camera on the KPNO 1.3 m telescope (Lada, Evans,

DePoy and Gatley 1988, in preparation). Three thousand 1 x 1 arc minute fields were surveyed at 2.2 microns, covering an area of approximately 0.8 square degrees. The sensitivity of this survey is estimated to be 14 magnitude at K, assuming no extinction.

Preliminary results indicate the presence of four stellar clusters. The locations of these young clusters follow the CS gas distribution and furthermore, the clusters are located near the most massive CS clumps (Mvirial > 500 M☉). Only one clump with Mvirial > 500 M☉ does not seem to be associated with a stellar cluster.

This research was supported in part by NASA Trainee Grant, NGT50320 and NSF Grant AST86-11784 to the University of Texas, at Austin. The author also acknowledges support from a Zonta Amelia Earhart Fellowship.

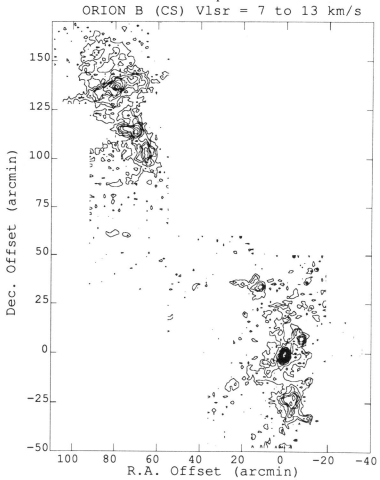

Figure 1. Distribution of CS(2-1) emission (integrated intensity) for the Orion B Cloud. The (0,0) position corresponds to R.A. = $5^h 39^m 12^s$, Dec. = -01° 55' 42". The lowest contour level is at 0.6 K km/s with subsequent levels at 1.2, 1.8, 2.4, 3, 4, 5,...15.

Physical Conditions in Molecular Clouds derived
from Sub-mm and far-IR Spectroscopic Observations

J. Stutzki[1]
and
G.J. Stacey[2], R. Genzel[1], U.U. Graf[1], A.I. Harris[1],
D.T. Jaffe[3], J.B. Lugten[4], and A. Poglitsch[1]

1) Max Planck Institut für Physik und Astrophysik, Institut für
Extraterrestrische Physik, D-8046 Garching b. München, F.R.G.
2) Physics Department, University of California, Berkeley, U.S.A.
3) Astronomy Department, University of Texas at Austin, U.S.A.
4) Institute for Astronomy, University of Hawaii, U.S.A.

Sub-mm and far-IR spectroscopic observations probe a very different regime of temperature and density of molecular clouds compared to mm-wave observations. In addition, transitions of new species, e.g. rotational transitions of light hydrides (e.g. HCl (Blake, Keene and Phillips 1986) or fine structure lines ([CI], [CII]) become observable only at these higher frequencies.

Exploration of the new regime of densities possible to trace with heavy rotor transitions in the submm has just started with the recent detection of the HCN J=9→8 (Stutzki et al. 1988a, Harris et al. in this volume), HCO^+ J=9→8 (Jaffe et al. 1989), SO_2 $7_{71} \rightarrow 6_{60}$, CH_3OH $13_1 \rightarrow 12_0$ E and one so far unidentified line (Stutzki et al. 1989) from the Orion Core. All lines have profiles characteristic for the low velocity outflow. Many new, important results from the sub-mm spectroscopic study of very high density cores can be expected in the near future with the high angular resolution now achievable with the new, large submm dishes (CalTech 10m, JCMT 15m).

The first half of this article summarizes our present knowledge about the warm, dense component recently discovered in all reasonably high UV luminosity sources in the galaxy. The second part discusses recent observations in the fine structure lines of neutral and ionized atomic carbon. The best explanation for the observed extended emission in these lines is its origin in photodissociation regions on clump surfaces in a clumpy, UV penetrated cloud.

I. Abundant Warm and Dense Gas in Molecular Cloud Cores

In Table 1 we summarize the present body of data on short-submm and far-IR CO observations.

Many of the energetic outflow sources are very bright in the submm- and far-IR CO lines, implying that large amounts of very warm, dense material participate in the outflow (Jaffe, Genzel, and Harris, 1986; Jaffe et al. 1988, Stacey et al. 1989). Due to the higher opacity of the warm gas in the submm lines, the outflow spectra often extend to much higher velocities than in the mm-wave lines.

The most striking result from table 1, however, is the presence of quiescent, warm gas in almost all galactic sources with high UV luminosities. The observed high brightness temperatures often imply a minimum kinetic temperature of about 100 K, substantially warmer than the dust. The increase of main beam brightness temperature with smaller beam sizes indicates that the source intrinsic temperatures are even higher. Indeed, in all cases were far-IR CO lines are also observed,

Table 1: Summary of Short-Submm and Far-IR CO Observations

Source	6-5 T_{mb}[K]	J=7→6 Δv[km s-1]	14-13	16-15	17-16	>20	Source	6-5	J=7→6 T_{mb}[K] (plus Quiescent Warm Gas) Δv[km s-1]	14-13	16-15	17-16	>20
I. Cold Clouds							III. Outflow Sources						
HH9	7	9					Ori BN/KL	§	175 50	x	x	x	>20
L1551	3	2.6					NGC 2071		7 25‡	x			
B35	<5								7 20†				
ρ Oph	<6								(25) (13)‡				
IC 1396	<6†						W33		(10) (9)‡				
V645	<10						GL490		50 15				
L134	<5								<6‡				
	<3†								<2‡				
II. Quiescent Warm Gas							W49		45 25	x	x		
W3	100§	7					W51	§	30 25				
		10‡							(45) (8)				
Ori θ¹C	26‡	7.1					G12591		4 15				
	80	6.8					DR21	§	65 25	x	x		
Ori Bar	65	4.8	x						(25) (8)				
OMC1, S3	67	1.8	x	x			W75 N		self absorbed				
OMC2	20	3.5					Ceph A		30 13				
NGC 2023	18	1.3							12 30‡				
NGC 2024	55 §	1.4							6‡ 30†				
NGC 2068	80 §	1.4‡							(30) (8)				
Mon R2	30†	2.3							(12) (9)‡				
S255	40	5.5†					G333.6-0.7			x	x		
G12.41+0.50	16†	7					NGC 7538		<2‡		x		x
W43	20	11					IV. Galactic Center						
M17 SW	15	7	x	x			20km/s cloud		15 30				
M17 KW	90	9†					2 pc ring		8 100		x		
G34.3+0.1	45†	9					Sgr B2	§	20 45				
S140	50	6‡					V. SN Remnants						
S106	14‡	9					IC 443 core		25 20				
	50	6†					IC 443 shock		6 70				
W75S	16†	7					VI. Late Type Stars						
	45						IRC+10216	§	13 21				
									9† 20†				
							CIT 6		<4				
							R Leo		<4				

¶ Boreiko, Betz, and Zmuidzinas, 1989, Ap.J., submitted
§ Koepf et al., 1982 (IRTF, 35")
† Krügel et al., 1988 (KAO, 100")
‡ Krügel et al., 1988 (UH88", 45")
$ UCB/MPE receiver at JCMT, 7"
- other submm observations: UCB/MPE receiver at IRTF (30") and UKIRT (25")
- other far-IR observations: UCB Tandem Fabry-Perot, KAO (55")

the relative intensity of the submm and far-IR lines indicates temperatures of a few 100 K and densities around $10^{4.5}$ cm^{-3} (e.g. Harris et al. 1987). At these high temperatures the distribution of the population over many rotational states results in mostly optically thin emission. Typically, the optical depth in the highest opacity lines near J=7→6 is barely above unity, assuming that about a tenth of the total CO column density is in warm gas. Even if the sources were resolved the brightness temperatures in the submm lines would thus be about a factor of two below the kinetic temperature.

The width of the submm lines is often narrower than in the low-J lines. Comparison of the line profiles shows that the lines match well in shape and intensity in the line wings, whereas the bright line core visible in the sub-mm transitions seems to be absorbed out in the lower-J lines, possibly by small amounts of cooler foreground material with a similar velocity dispersion (Stutzki et al. 1988b). As an extreme example the 80 K bright CO 7→6 lines from NGC 2024 have FWHM widths as low as 0.7 km s^{-1} (including the instrumental resolution of about 0.25 km ds^{-1}). This is only a factor of about 2 wider than the thermal width corresponding to the the minimum kinetic temperature

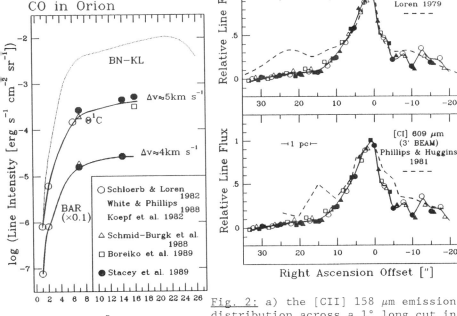

Fig. 1: Excitation curves for CO in Orion toward the bright bar and the region of Θ^1C.

Fig. 2: a) the [CII] 158 μm emission distribution across a 1° long cut in R.A. (Stacey et al. 1989) compared with CO J=1→0 emission, b) compared with [CI] 609μm emission.

derived from the line brightness under the assumption that the source fills the beam and is thermalized.

Fig. 1 compares the intensity in different J lines for the CO emission from Orion. Radiative transfer modelling shows that a range of temperatures up to 500 K and very high densities (up to 10^6 cm^{-3}) are needed to explain the increase of intensities up to J=17→16. The narrow linewidths found in the submm CO emission from the bright bar and the Θ^1 C region in Orion was also confirmed for the far-IR 17→16 line by recent heterodyne observations (Boreiko, Betz and Zmuidzinas 1989).

The narrow line width rules out shocks as a possible heating source. The correlation with UV intensity argues for a heating mechanism directly related to the UV radiation. However, all theoretical photodissociation region (PDR) models fail by about an order of magnitude to explain the amount of warm gas observed.

II. Ionized Carbon as a Tracer of the UV Radiation Field throughout Molecular Clouds

[CII] 158 μm fine structure emission is very extended. Modelling of the distribution in the M17 SW interface region (Stutzki et al. 1988b) showed that both the observed intensity and extended spatial distribution are very will fitted by the model of a clumpy molecular cloud, allowing the UV photons to penetrate deep into the cloud core and create many PDR's on the clump surfaces. In addition very extended [CII] emission is found to be well correlated with the distribution of the large scale molecular material. In the case of M17, the UV photons necessary to create the CII throughout the cloud can be provided by B stars embedded in the cloud, if the cloud is also clumpy on the large scale to allow their UV radiation to penetrate through the whole volume. Large scale extended [CII] emission following the distribution of molecular material has also been mapped recently in Orion (Fig. 2a and Stacey et al. 1989).

III. The Origin of Neutral Atomic Carbon

The first observations of the [CI] $^3P_1 \rightarrow ^3P_0$ 609 μm line showed that neutral atomic carbon is distributed throughout molecular clouds, whith line profiles and intensities very similar to ^{13}CO J=1→0, suggesting similar opacities in both lines and an abundance ratio CI/CO of ≈10% (Phillips and Huggins 1981, Keene et al. 1985). This was taken as an indication for a chemical origin of CI in the volume of the cloud. Several theoretical explanations were developed and are summarized in Keene et al. (1985).

The origin of CI in PDRs was at that time rejected as a possible explanation because the average visual extinction over the observed extend of [CI] emission is often 10 mag or higher. The new [CII] results dicussed in the previous section, showing that molecular clouds are indeed penetrated by UV radiation, can naturally explain the abundance of neutral atomic carbon throughout the cloud by its origin in PDRs.

This picture is nicely confirmed by the similar distribution of [CI] and [CII] emission along the extended cut through the Orion cloud (Fig. 2b), and the high angular resolution mapping in the [CI] $^3P_2 \rightarrow ^3P_1$ 370 μm line of the M17 SW interface region showing that the [CI] emission closely follows the emission from [CII], warm CO and the CO column

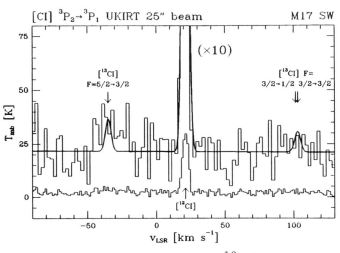

Fig. 3: Deep integration on the [^{12}CI] emission peak in M17 SW. The position of the [^{13}CI] hyperfine components, matching the position and width of the [^{12}CI] line, are indicated. The observed feature near the strongest [^{13}CI] hyperfine line is much too wide and may be another, yet unidentified line. With an ^{12}C/^{13}C abundance ratio of 60 the nondetection of [^{13}CI] to a level of <1.5 K T_{mb} emission gives an upper limit of $\tau([^{12}CI])\leq 5$.

density distribution in the clumpy interface (Genzel et al. 1988). The relation between integrated $C^{18}O$ 2→1 and [CI] 370 μm intensity indicates that the [CI] gets slightly optically thick. A recent deep integration at the CI peak in M17 SW (Fig. 3) gave an upper limit of $\tau([^{12}CI]\leq 5$. With the substantially lower beam dilution expected for a clumpy interface, the new, large submm telescopes will help to improve this limit in the near future.

References
Blake, G.A., Keene, J., and Phillips, T.G., 1986, Ap.J. 295, 501
Boreiko, R.T., Betz, A.L., and Zmuidzinas, J., 1989, Ap.J. submitted
Genzel, R., et al., 1988, Ap.J. 332, 1049
Harris, A.I., et al., 1987, Ap.J. (Letters) 322, L49
Jaffe, D.T., et al., 1988, Ap.J. submitted
Jaffe, D.T., et al., 1989, in prep.
Jaffe, D.T., Harris, A.I., and Genzel, R., 1987, Ap.J. 316, 231
Koepf, G.A., et al. 1982, Ap.J. 260, 584
Krügel, E., et al., 1988, Astr.Ap. submitted
Phillips, T.G., and Huggins, P.J., 1981, Ap.J. 251, 533
Keene, J., et al., 1985, Ap.J. 299, 967
Schmid-Burgk, J., et al., 1988, Astr.Ap. submitted
Stacey, G.J., et al., 1989, in prep.
Stutzki, J., et al., 1988a, Ap.J. (Letters) 330, L125
Stutzki, J., et al., 1988b, Ap.J. 332, 379
Stutzki, J., et al., 1989, in prep.

VIBRATIONALLY EXCITED H$_2$O IN ORION A

S. J. PETUCHOWSKI and C. L. BENNETT
Infrared Astrophysics Branch – Code 685
Laboratory for Astronomy and Solar Physics
NASA Goddard Space Flight Center
Greenbelt, MD 20771, U.S.A.

ABSTRACT. Various models of water maser emission in star forming regions, based both on collisional and 6.3 μm radiative pumping mechanisms, require appreciable rates of excitation of water into its first vibrationally excited bending mode. We report the detection in Orion, using the 12–meter NRAO radiotelescope at Kitt Peak, of a line coincident in frequency with the $4_{14} \rightarrow 3_{21}$ rotational transition of the ν_2 vibrational manifold of *ortho*–H$_2$O at v_{LSR} ~18 km s^{-1}. If the detected line is indeed due to this water transition, a beam–averaged column density of ~10^{13} cm^{-2} in the vibrationally excited upper level is implied.

1. Introduction

Water has been shown to constitute a significant coolant in warm molecular clouds and in shocked material in particular. Water has been detected previously in star–forming regions in its 22 GHz masing transition, in pure rotational transitions from an airborne platform (Frerking, this conference), and in the isotopic variants, H$_2^{18}$O (at 203 GHz, Phillips *et al.* 1978) and HDO, in which 8 lines have been detected to date (Petuchowski & Bennett 1988). Detected emission lines, besides that at 22 GHz, include spectral features associated with the hot core, compact molecular ridge and diffuse plateau components of the BN–KL region.

An understanding of the energetic role of water requires information about excitation conditions prevailing in the various regions. Of particular interest is the region, ~1' (~5x10^{17} cm) in diameter, surrounding BN, associated with shocked gas which is characterized by H$_2$ vibrational temperatures in excess of 2000 K. Shock models predict the liberation of water from grain mantles and the conversion to gaseous water of all oxygen not tied up in CO. These processes could account for water column densities in Orion A as high as 10^{18} cm^{-2} (Elitzur 1979).

The highly excited state of the shocked molecular material as well as the fact that various maser excitation models invoke vibrational pumping suggest a new observational approach to the detection of water: the detection of emission in the $4_{14} \rightarrow 3_{21}$ transition of the ν_2 (bending) mode, 1908 cm^{-1} above the ground state.

2. Observations

The rest frequency of $\nu = 67.80396(10)$ GHz has been reported by Belov *et al.* (1987). The contribution of telluric H_2O to zenith opacity at that frequency is calculated to be less than 0.001. Observations were conducted during April–May and October, 1988, using the 12–meter NRAO radiotelescope at Kitt Peak. Atmospheric opacity at this frequency is dominated by O_2 and was determined to be 0.5 at zenith. Effective system temperatures were in the range of 1500–2000 K for an air mass < 2. Switching the 1.5' beam by 4' in azimuth was found to produce reasonable baselines.

After 144 minutes on–source integration time in Spring, 1988, the rms (1σ) noise, based on the difference of left and right chopping scans, was 0.014 K, while a total of >10 hours has been spent on–source to date. A 3σ feature corresponds, if the H_2O identification is correct, to emission at a redshift ~18 km s^{-1}.

3. Discussion

On the basis of the Spring observations, we are able to derive a 5σ upper limit on the column density of water in the (010) 4_{14} state, assuming $\Delta v < 30$ km s^{-1}, of N< 1×10^{14} (1.5'/θ)^2 cm^{-2}, where θ^2 is the angular extent of the emitting region. It is unlikely that molecular densities are high enough to account for collisional thermalization of the (010) manifold since radiative cooling of the manifold is fast. It is possible, however, that at least some of the infrared transitions are optically thick and that photon trapping retards the cooling of the vibrational manifold. A typical 6.3 µm transition in water, with A~1 s^{-1}, becomes optically thick for column densities ~10^{17} cm^{-2} (for reasonable linewidths, in an LVG model). A confirming rotational line in the ν_2 manifold would allow us to determine to what degree H_2O is vibrationally thermalized or whether there is evidence for non–thermal population of selected levels of the manifold, via radiation, collision or electron impact, thereby yielding some insight into the mechanism pumping maser emission in the ground state.

4. References

Belov, S. P., Kozin, I. N., Polyansky, O. L., Tret´yakov, M. Yu., and Zobov, N. F. 1987, *J.Mol.Spec.*, **126**, 113.
Elitzur, M. 1979, *Ap.J.*, **229**, 560.
Petuchowski, S. J. and Bennett, C. L. 1988, *Ap.J.*, **326**, 376.
Phillips, T. G., Scoville, N. Z., Kwan, J., Huggins, P. J., and Wannier, P. G. 1978, *Ap.J.(Lett.)*, **222**, L59.

HCN J=9-8 Emission in the Orion Core and W49

A.I. Harris[1], R. Genzel[1], U.U. Graf[1], D.T. Jaffe[2], and J. Stutzki[1]

[1] Max-Planck-Institut für extraterrestrische Physik, 8046 Garching, FRG
[2] Astronomy Department, Univ. Texas, Austin, TX 78712

Submillimeter lines of heavy linear rotors with large dipole moments, such as HCN, are especially clean probes of warm and very dense cloud cores for three reasons. First, since the submillimeter rotational transitions are from collisionally excited states with energies 100 to 200 K above ground state, they probe only warm material. Second, submillimeter lines have high critical densities because of the Einstein A coefficient's $\nu^3 \mu_o^2$ dependence. Once the collision rate greatly exceeds the radiation rate, the line intensity becomes independent of density; the critical density is thus only a lower limit to density. Submillimeter lines therefore probe densities several hundred times higher than millimeter lines from the same molecules. For example, the HCN 9-8 transition's critical density is ~2×10^9 cm^{-3} in optically thin and thermalized gas. The gas is not optically thin, however, as these rapid radiative rates are accompanied by large absorption cross sections. Radiative trapping can reduce the effective critical density between one and two orders of magnitude. The comparatively simple internal energy structure of these simple rotors gives them their third advantage: one can reliably calculate radiative transfer effects to derive accurate values for the molecule's physical conditions.

Figure 1 is a high signal-to-noise HCN J=9-8 spectrum of the Orion Core, obtained with the MPE receiver on the UKIRT in Feb. 1988. The line is very bright and wide, and its complex lineshape indicates that there is structure still unresolved in our 25" beam. A decomposition of the line into three Gaussian components shows velocities and widths characteristic of millimeter line emission (see e.g. Blake, this volume) from the low and high velocity outflows (Δv_{FWHM} = 16 and 41 km s^{-1}, respectively), and of the "hot core" (v_{LSR} = 6 km s^{-1}, Δv_{FWHM} = 6 km s^{-1}). The simple detection of these components indicates that the bulk of the densest material is associated with the outflows, and not with the ambient cloud or "hot core."

Precise estimates of the physical conditions require knowledge of the source filling factor within the beam. Figure 2 is a map of the Orion Core in our 25" FWHM UKIRT beam. Most of the emission is contained in a barely resolved peak centered on IRc2; lower level emission extends to the south, west, and east, somewhat reminiscent of the H_2O

maser distribution. These observations confirm that the central source size, as well as lineshape, are consistent with those measured in millimeter wave interferometric observations (Plambeck et al. 1982). On the assumption that the submillimeter and millimeter sources had the same size, Stutzki et al. (1988) derived densities of 10^8 to 10^7 cm^{-3} for temperatures ranging from 200 to 300 K. These densities are at least an order of magnitude higher than had been derived from previous observations of thermal line emission. The column density gave a depth scale of ~10^{15} cm, about two orders of magnitude smaller than the beam's linear extent. This suggests that the gas fills only a fraction of the beam area, and has a filamentary, clumpy, or sheetlike structure. The high densities and temperatures, and lack of structure or graininess in the millimeter interferometry maps, indicates that the material is shock-compressed sheets or small filaments in the outflow from IRc2.

Weak HCN J=9-8 emission has also been detected toward W49 (T_{MB} = 5 K, v_{LSR} = 6 km s^{-1}, $\Delta v_{FWHM} \approx$ 6 km s^{-1}). Intrinsically, the line must be very bright, given the large distance to W49 and the consequent beam dilution. The detection indicates that this warm and very dense material is a common feature of the active interstellar medium. Observations of these densest cores, and possibly protostellar disks, with the new large submillimeter telescopes will immediately yield new and interesting results.

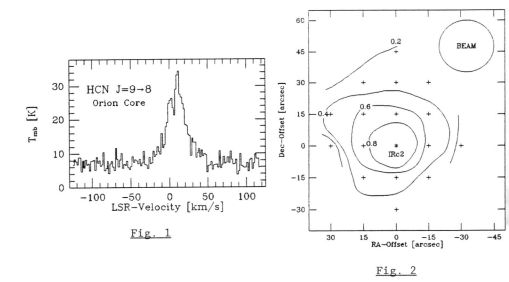

Fig. 1

Fig. 2

Plambeck, R.L., Wright, M.C.H., Bieging, J.H., Baud, B., Ho, P.T.P., and Vogel, S.N. 1982, Ap. J. 259, 617.
Stutzki, J., Genzel, R., Harris, A.I., Herman, J., and Jaffe, D.T. 1988, Ap. J. (Lett.) 330, L125

MOLECULAR LINE STUDIES OF W49A

R. MIYAWAKI
Fukuoka University of Education, Munakata-city, Fukuoka 811-41, Japan
M. HAYASHI
Department of Astronomy, University of Tokyo, Bunkyo-ku, Tokyo 113, Japan
and
T. HASEGAWA
Institute of Astronomy, University of Tokyo, Mitaka, Tokyo 181, Japan

ABSTRACT. We have carried out high S/N observations of various molecular species toward W49A. Line profiles of optically thin lines clearly show two peaks at $V_{LSR} = 4$ kms^{-1} and 12 kms^{-1}, suggesting that there are two velocity components inside the W49A core. This can be interpreted into two ways: there are two interacting clouds with 10^5 $M_\odot$ inside the core or there is a rotating ring of molecular gas.

W49A is one of the most active star forming regions in the Galaxy. Its luminosity is of order 10^7 $L_\odot$. The mass of the W49A core is $\sim 10^5$ $M_\odot$, which is closely packed in a compact region 3 pc in diameter. The origin of such activities has not been well understood because it lies far away from the sun (12 kpc). Miyawaki *et al.* (1986) have suggested the interaction of two clouds inside the W49A core as an origin of the high activities. However the presence of the two interacting clouds has not directly been confirmed and there are other interpretations (Welch *et al.* 1987).

We have observed various molecular lines with high S/N ratios, angular resolutions (15"–20"), and velocity resolutions (~ 0.6 kms^{-1}) using the 45–m telescope of Nobeyama Radio Observatory. The observed molecular lines are classified into three types: optically thick lines (^{12}CO ($J =1-0$), HCO$^+$ ($J =1-0$), and HCN ($J =1-0$)), optically thin lines (^{13}CO ($J =1-0$), $C^{18}O$ ($J =1-0$), $H^{13}CO^+$ ($J =1-0$), $H^{13}CN$ ($J =1-0$), and $HC^{15}N$ ($J =1-0$)), and molecular lines like SiO (v=0, $J =2-1$), SO, and SO$_2$, which are often detected in highly excited regions around newly formed stars such as in hot cores in Orion KL. We call these highly excited lines as "hot core" lines.

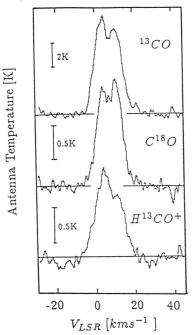

Figure 1. Line profiles of optically thin lines. The optical depth of ^{13}CO is <0.35.

Characteristics of the observed line profiles are summarized as follows:

1. *Optically thick lines*: Sharp and deep dips at $V_{LSR} = 7\ kms^{-1}$ are clearly caused by the absorption of foreground gas.

2. *Optically thin lines*: Figure 1 shows that there are two peaks at $4\ kms^{-1}$ and $12\ kms^{-1}$ with a shallow dip whose velocity is determined by the relative intensities of the two velocity features, *i.e.* the HCO$^+$ dip is found at a redder velocity than the ^{13}CO and C^{18}O dips because the bluer velocity feature of H^{13}CO$^+$ is stronger than the redder one. If ^{13}CO were optically thick, its dip would be much deeper than the C^{18}O dip.

3. *Hot core lines*: SO and SiO show single peaks at $6\ kms^{-1}$, while SO$_2$ shows a single peak at $13\ kms^{-1}$.

We conclude that (1) there are two velocity components inside the W49A core suggested by the presence of the two velocity features seen in the optically thin lines, (2) there is a compact and highly excited region inside the W49A core, and (3) there is less excited absorbing gas in front of the W49A core.

These facts can be interpreted into two ways:

Case 1: There are two clouds with $\sim 10^5\ M_\odot$ in the W49A core. Extremely active star formation occurs in the active region formed through the interaction of the two clouds.

Case 2: There is a rotating molecular ring of $\sim 10^5\ M_\odot$ inside the W49A core. Extremely active star forming region is located at the center of the ring.

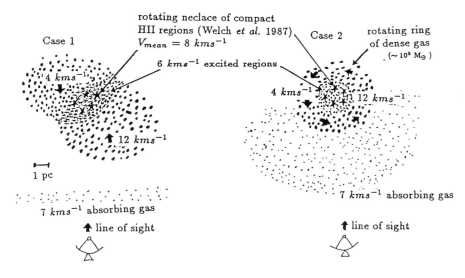

Figure 2. Two interpretations inferred from the presence of the two velocity components inside the W49A core. Case 1 is the interacting cloud model, and Case 2 is the rotating ring model.

References

Miyawaki, R., Hayashi, M., and Hasegawa, T. 1986, *Ap. J.*, **305**, 353.
Welch, W. J., Dreher, J. W., Jackson, J. M., Terebey, S., and Vogel, S. N. 1987, *Science*, **238**, 1550.

ATOMIC AND MOLECULAR OUTFLOW IN DR21

Adrian P. G. Russell
Joint Astronomy Centre
665 Komohana Street
Hilo
Hawaii 96720
USA

ABSTRACT. 21 cm VLA line observations are presented of DR21. These reveal a massive ($24 M_\odot$) HI jet which is red-shifted by up to 90 km/s. These data are compared with HCO^+ J=3-2 data taken at UKIRT. The momentum flux of the HI is sufficient to drive the molecular outflow in a momentum conserving interaction.

1. INTRODUCTION

We have made 21 cm observations of the HI line in DR21. The data were taken with the VLA "D" array and have spatial and velocity resolutions of 44 arcsec and 10 km/s respectively. The molecular line HCO^+ J=3-2 data were taken at UKIRT and have a spatial resolution of 60 arcsec.

2. RESULTS

Fig. 1 shows the HI data and reveals a massive red-shifted jet extending to the east from the DR21 HII region. This jet is visible out to +90 km/s and has a mass of $24 M_\odot$ and a momentum flux of 0.05 $M_\odot$ km/s/yr. Unfortunately confusion with Galactic HI prevents us from seeing any blue-shifted counter jet. The Galactic HI is also visible in the +30 and +40 km/s frames and gives the misleading impression that the 30 km/s jet bends to the south. Also visible in the HI data is a small knot of red-shifted emission about 80 arcsec to the west of the

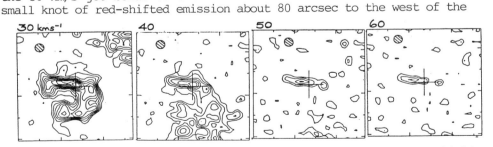

Figure 1. Red-shifted HI jet in DR21. Individual frames are 10x10 arcmin and cover 10 km/s velocity intervals.

outflow centre. This red-shifted knot is coincident with a secondary peak in the integrated HCO$^+$ data (fig. 2) and appears to be the site of a dissociative interaction between an unseen blue-shifted counter jet and the ambient molecular cloud. In fact this is also the point at which the S(1) line of shocked molecular hydrogen shifts from blue- to red-shifted (Garden et al. 1988) - again indicating an interaction. The HCO$^+$ data indicate a molecular outflow with a total (red+blue) mass of 170M_o and a momentum flux of 0.05 $M_o$km/s/yr.

3. DISCUSSION

An obvious possible interpretation of the HI in DR21 is one of shock-dissociated H_2. However it is more attractive to interpret the the HI as a neutral wind: Fig. 3 shows the 30 km/s HI frame overlaid on the contours of S(1) emission by Garden et al. (1986) - at the east of the jet the HI lies beyond the H_2 as if it has burst out of the ambient molecular cloud. If we interpret the HI as a neutral wind, then the H_2 emission would represent the shocked interface between the wind and the ambient cloud. A comparison between the momentum fluxes of the atomic and molecular components shows that, if the HI is a neutral wind, then it is capable of driving the massive molecular flow in DR21.

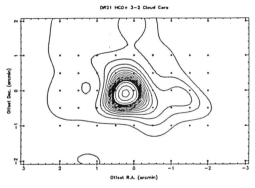
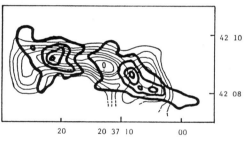

Figure 2. Total integrated intensity of HCO$^+$ J=3-2 emission.

Figure 3. Contours of H_2 S(1) emission (bold) overlaid on the 30 km/s HI frame.

4. ACKNOWLEDGEMENTS

My collaborators in this work were Richard Hills, Rachael Padman and John Bally (VLA work).

5. REFERENCES

Garden R. et al. 1986 Astrophys. J., MNRAS <u>220</u>, 203.
Garden et al. 1988 in prep.

FIRST HETERODYNE OBSERVATIONS AT 690 AND 800 GHz WITH THE JAMES CLERK MAXWELL TELESCOPE

A.S.WEBSTER, A.P.G.RUSSELL, H.E.MATTHEWS, G.D.WATT, S.S.HAYASHI, I.M.COULSON
Joint Astronomy Centre, Hilo, Hawaii, USA.

R.GENZEL, A.I.HARRIS, J.STUTZKI, U.U.GRAF
Max-Planck-Institut fuer Physik und Astrophysik, Garching, FRG.

R.PADMAN
Mullard Radio Astronomy Observatory, Cambridge, UK.

The first heterodyne observations on the James Clerk Maxwell Telescope in the far sub-millimetre bands were made in 1988 August with the laser-pumped cooled Schottky receiver of the Max-Planck-Institut (Harris et al., 1987) mounted at the Nasmyth focus. Observations were made at 691 GHz (CO 6-5), 797 GHz (HCN 9-8) and 806 GHz (CO 7-6) with the aims both of characterising the antenna and of carrying out astronomical research. At 691 GHz a scan of Mars showed the primary beam size to be consistent with the 7 arcsec FWHM expected from diffraction theory; the antenna efficiency was determined to be 14 % and the (lunar) beam efficiency 50 %. At 806 GHz, there was sufficient dry time available to measure only the antenna efficiency, which was 9 %. The rms accuracy of the surface of the telescope implied by these figures is about 40 microns, close to the value inferred from holographic measurements made at 3 mm wavelength.

The spectrum given in Fig. 1 is particularly interesting: it shows not only the quality of the data available from this combination of instrument and telescope but also the very different appearance that objects can present in the submillimetre bands compared with the millimetre. The spectrum is of Orion in the direction of IRc2, and was made at 806 GHz in the 7-6 line of CO. The diffraction beamsize of the telescope at this frequency is 5 arcsec FWHM. The observation was made in a beamswitching mode, making use of the chopping secondary mirror on the telescope, and the integration time was about 3 hours. The source is known to be extended compared to the beam, so the lack of a beam efficiency measurement prevents a proper calibration. A good idea of the calibration can, however, be obtained from a spectrum of the same position in the same line obtained earlier at the United Kingdom Infrared Telescope which indicates that a value of 150 - 200 K is appropriate for the peak temperature (main beam) in the present spectrum.

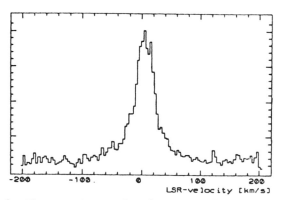

Figure 1. The spectrum of Orion towards IRc2 in CO 7-6.

At first sight this spectrum resembles those at lower frequencies, with a spike component from the general cloud sitting on top of the plateau component from the high-velocity outflow driven by IRc2. A second look, however, shows that this cannot be so because the velocity width is nearly 40 km/s FWHM and the spike is only about 6 km/s wide. This spectrum, therefore, is dominated by the plateau and there is little sign of any other component. The absence of any clear sign of emission from the spike is unusual, and cannot be explained in terms of an inability of the material in the general molecular cloud to radiate this high transition: Schmid-Burgk (1989) has mapped the Orion Molecular Cloud in CO 7-6 with a 98 arcsec beam and finds strong spike emission everywhere. The absence requires a different explanation, and we suppose it is simply that the gas in the outflow is optically thick in this transition and absorbs the spike emission from the molecular cloud behind it. The optical depth in CO of the plateau near its central velocity does not appear to have been determined before at values of J = 7 or smaller, although a recent determination of the relative fluxes of the 12 and 13 isotopes for J = 17 and 18 has also led to the conclusion that the gas is optically thick (Genzel 1989). In order to absorb the spike completely in 7-6 CO, not only must the outflowing gas be opaque but it must also be ubiquitous throughout the beam; the covering factor must be close to 100 %. This is a new inference which has a bearing on models of the outflow and constrains how clumpy the gas may be. The outflow source is usually reckoned to be embedded in the molecular cloud near its front face, so the question also arises as to why the spike component is not seen in absorption. A possible answer is that the molecular cloud in front of the outflow is heated, perhaps by the Trapezium stars and M42, and is no longer at its usual temperature of 60 - 80 K but is at say 170 K, at which it can be seen neither in emission nor absorption against the background source at the same radiation temperature.

Genzel R., Poglitsch A. & Stacey G. 1989. This volume, p261.
Harris A.I., Jaffe D.T., Stutzki J. & Genzel R. 1987. Internat. J. Infrared Millimetre Waves, vol. 8, 857.
Schmid-Burgk J. 1989. This volume, p11.

Extensive C+ 158 μm Line Mapping of W3 and NGC 1977

J. E. Howe,[1] N. Geis,[2] R. Genzel,[2] D. T. Jaffe,[1] A. Poglitsch,[2] and G. J. Stacey [3]

[1] Department of Astronomy, University of Texas at Austin
[2] Max Plank Institute for Extraterrestrial Physics
[3] Department of Physics, University of California, Berkeley

1. Introduction and Observations.

The 158 μm $^2P_{3/2} \rightarrow ^2P_{1/2}$ fine structure line of C+ is one of the dominant coolants in the photoionized interface region between a source of far ultra-violet (FUV) radiation and a molecular cloud. In the 1-D model of Tielens and Hollenbach (1985), C+ is the predominant species of carbon at the UV illuminated edge of a cloud, while neutral carbon and CO dominate farther into the cloud. In this model, at the position where the CO abundance levels off the C+ abundance has dropped by several orders of magnitude.

We mapped the 158 μm C II line in W3 and NGC 1977 with the UC Berkeley Mk II Tandem Fabry-Perot spectrometer on the NASA Kuiper Airborne Observatory in January 1988. The beam FWHM at 158 μm is 55" and subtends a solid angle of 8.6×10^{-8} sr on the sky. The spectral resolution was set at 67 km s^{-1} for the observations. The maps were generated by observing the integrated C II line intensity in each object with a 1' grid spacing. For the map of W3, the central 4'x 3.5' region was mapped with a 30" grid spacing. We estimate the absolute calibration uncertainty to be ± 25%.

2. Results

The region mapped in C II in the star formation region W3 (L~10^6 L$_O$, d~2.4 kpc) contains the compact H II regions W3B and W3C (IRS 3 and IRS 4, respectively) and the H II region W3A which encompasses IRS 2. The most luminous source in the far-infrared (FIR) is IRS 5, whose luminosity of ~ 2×10^5 L$_O$ implies a spectral class of O6.5 or earlier. However, IRS 5 has no apparent radio continuum emission. Figure 1 shows the C II emission peaking near IRS 5 with an extension toward IRS 4 to the northwest, similar to the distribution of 100 μm continuum emission mapped by Werner et al. (1980). There is also an extension of the C II emission to the south which is not seen in the 100 μm map. The C II region is > 4 pc in extent, and has a luminosity of ~ 600 L$_O$, ~3×10^{-3} of the total FIR luminosity.

NGC 1977 (d~450 pc) is an ionization front at the northern edge of the Orion molecular cloud that is excited by the B1.5 star 42 Ori to the northeast. The edge-on geometry of NGC 1977 makes it an ideal source to study the interaction of a FUV field with a molecular cloud. Makinen et al. (1985) conclude that 42 Ori is the dominant energy source for the region. The C II emission shown in Figure 1 closely follows the 100 μm continuum mapped by Makinen et al. The total flux of C II emission in the region mapped is ~ 5 x 10^{-16} W cm^{-2}, or ~ 8 x 10^{-3} of the FIR flux. Observations of CO J=3→2 with a similar beamsize obtained at the University of Texas Millimeter Wave Observatory show that the C II emission peaks within the molecular cloud boundary and extends > 0.8 pc into the molecular gas. Stutzki et al. (1988) also report C II emission extending several parsecs into the M 17 SW molecular cloud. The close morphological association of the partially ionized gas (as traced by the C II emission) and the warm dust in NGC 1977 and W3, as well as in M 17, and the large scales over which the two types of emission are coextensive, strongly support the picture of Stutzki et al. in which the C II emission arises in the FUV illuminated boundaries of dense clumps embedded in less dense gas. The Tielens and Hollenbach model would then refer to the photodissociation regions of the individual clumps rather than to the region as a whole.

References

Makinen, P., Harvey, P. M., Wilking, B. A., and Evans, N. J., II 1985, *Ap. J.*, **299**, 341.
Stutzki, J., Stacey, G. J., Genzel, R., Harris, A. I., Jaffe, D. T., and Lugten, J. B. 1988, *Ap. J.*, **332**, 379.
Tielens, A. G. G. M., and Hollenbach, D. 1985, *Ap. J.*, **291**, 722.
Werner, M. W., et al. 1980, *Ap. J.*, **242**, 601.

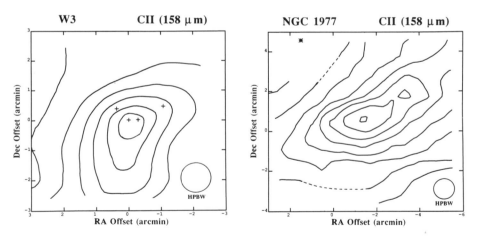

Figure 1: <u>Left:</u> Integrated C II emission in W3. (0,0) is at the position of IRS 5. The crosses mark the positions of IRS 2, IRS 5, IRS 3, and IRS 4 (E to W, respectively). The contours are from 30 to 80 percent of the peak flux of 2.1 x 10^{-17} W cm^{-2} in 10% intervals. <u>Right:</u> Integrated C II line and continuum emission in NGC 1977. (0,0) = 05^h $32^m 49^s$, -04° 57' 30". The contours are from 10 to 90 percent of the peak flux of 1.5 x 10^{-17} W cm^{-2} in 10% intervals. The asterisk marks the position of 42 Ori. The 158 μm continuum level is ~ 15% of the total intensity at the emission peak.

FIRST SEST OBSERVATIONS OF THE LARGE MAGELLANIC CLOUD

L.E.B. JOHANSSON[1,2], R.S. BOOTH[2], D.M.MURPHY[3], and M.OLBERG[1]
[1] *SEST project, ESO, Casilla 19001, Santiago 19, Chile*
[2] *Onsala Space Observatory, S-43900 Onsala, Sweden*
[3] *Université de Montréal, Montréal, Canada*

ABSTRACT. Using SEST, two smaller areas in LMC have been mapped in the ^{12}CO (1-0) line with a grid point spacing of 1'. A sample of 20 molecular clouds is identified and analysed with regard to CO luminosities, sizes, velocity dispersions and virial masses. The results suggest that the conversion factor from CO integrated intensities to H_2 column densities is five times higher in LMC compared to the Galactic value.

The present knowledge of the abundance of molecular clouds in galaxies is based on the extrapolation of CO emissivity to H_2 column density as defined by observations in the Galaxy. However, observations of dwarf galaxies and, specifically, the Magellanic Clouds show significantly weaker CO emission than late type spirals, indicative of a different conversion factor CO to H_2; a dependence on the ambient conditions like UV-field and metallicity has been suggested (see e.g. Israel, 1988).
The Magellanic Clouds offer a unique possibility to investigate the molecular cloud component in detail and its dependence on global properties (e.g. metallicity) significantly different from those found in the Galaxy. The large scale distribution of molecular clouds in LMC is known from the recent Colombia ^{12}CO (1-0) survey (Thaddeus, these proceedings; Cohen et al., 1988). Guided by the Colombia survey we have used the Swedish – ESO Submillimetre Telescope (SEST; for a technical description see Booth et al., 1988) in a pilot mapping of two smaller regions in the ^{12}CO (1-0) line: one area (14' × 11') centered on 30Dor; the second (11' × 27') includes the HII regions N159, N160 and N158. A rectangular grid with 1' spacing was used. At 115 GHz the beamwidth is 45"(corresponding to 11 pc linear resolution) and the main beam efficiency is 0.67. In the observed areas we found 20 well discernable molecular clouds. For this sample we have calculated CO luminosities, sizes, velocity dispersions and virial masses assuming spherical clouds. In Figure 1 we compare these molecular cloud parameters with those of a Galactic sample, analysed by Solomon et al. (1987). Statistical reasons (too few gridpoints within individual clouds) prevented us to calculate the parameters with the method used by Solomon et al.. However, tests, where statistically significant, indicate no systematic differences between the two methods used.

Figure 1a) as well as b) suggests that the CO luminosities in LMC are weaker by a factor of five relative to those observed in the Galaxy. On the other hand, the size - velocity dispersion relation is within the scatter identical for the two galaxies. These plots combined indicate that the CO to H_2 conversion factor is about five times higher in LMC than in the Galaxy, consistent with the result of Cohen et al..

We confirm previous observations which show that the CO emission is significantly weaker towards 30Dor than in the star formation regions southwards. From 30Dor to N159 there is a sequence of star formation regions of decreasing age. However, on the basis of our data (Figure 1) we can not distinguish the two areas observed with the exception of the CO line strength.

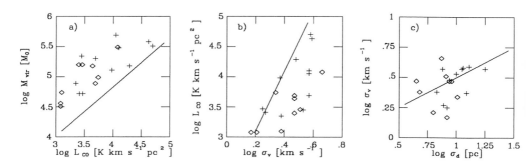

Figure 1. Observed relations of a) virial mass against CO luminosity, b) CO luminosity against velocity dispersion and c) velocity dispersion against size, for a sample of 20 molecular clouds in LMC. Separate symbols are used for the two areas observed (◊ 30Dor, + N159 to N158). The lines represent the Galactic relations according to Solomon et al. (1987).

REFERENCES

Booth, R.S., Delgado, G., Hagström, M., Johansson, L.E.B., Murphy, D.M., Olberg, M., Whyborn, N.D., Greve, A., Hansson, B., Lindström, C.O., and Rydberg, A. 1988, *Astr. Ap.*, , in press.
Cohen, R.S., Dame, T.M., Garay, G., Montani, J., Rubio M., and Thaddeus, P. 1988, *Ap. J. (Letters)*, **331**, L95.
Israel, F.P., 1988. in: 'Millimetre and Submillimetre Astronomy', Eds. R.D. Wollstencroft and W.B. Burton, Kluwer Academic Publishers, Dordrecht, p.281.
Solomon, P.M., Rivolo, A.R., Barret, J., and Yahil, A. 1987, *Ap. J.*, **319**, 730.

HIGHLY COLLIMATED, HIGH VELOCITY GAS IN THE ORION B MOLECULAR OUTFLOW

JOHN RICHER, RICHARD HILLS, RACHAEL PADMAN & PAUL SCOTT
Cavendish Laboratory
Madingley Rd.
Cambridge CB3 0HE
England

ADRIAN RUSSELL
Joint Astronomy Centre
Hilo
Hawaii

ABSTRACT. Observations of the Orion B molecular outflow reveal an unusually well-collimated high velocity component to the flow, lying inside a massive envelope of more slowly outflowing poorly collimated gas. This jet appears to accelerate as it moves away from the exciting star. Observations at this resolution do not suggest that the emission arises in a thin swept up shell of gas.

The molecular outflows from YSOs and protostars exhibit a variety of structures. The large mass of outflowing material [1], and the 'hollow shell' structure observed in some sources [2] has led to models where the CO emission arises in a thin shell of gas swept up by an undetected neutral wind. The poor collimation observed in most molecular flows is then attributed to a poorly collimated driving wind. In contrast, our observations of the outflow in Orion B reveal a fast jet-like component to the molecular flow, which appears to accelerate away from the driving star.

Previous observations revealed an energetic and bipolar outflow in the Orion B molecular cloud core [3]. Our JCMT[1] observations of the CO($2 \rightarrow 1$) line are shown below: on the left are maps of redshifted emission at three velocities, and to the right a position-velocity map taken down the jet axis. The map centre is RA=$05^h39^m12^s.6$, Dec.=$-01°57'00''$ (1950), which is close to the presumed (but undetected) object driving the flow. We note the following features:

(1) The highest velocity red-shifted gas, with a velocity of $30 \mathrm{kms}^{-1}/\sin(i)$ with respect to the cloud core, is found at the end of the outflow.

(2) The high velocity gas is extremely well collimated, lying apparently inside an envelope of extended emission.

(3) In the position-velocity plot we see an abrupt end to the flow, where gas at *all* velocities is brought to rest.

---

[1] The James Clerk Maxwell Telescope is operated by the Royal Observatory Edinburgh on behalf of the Science and Engineering Research Council of the United Kingdom, the Netherlands Organisation for Scientific Research and the National Research Council of Canada.

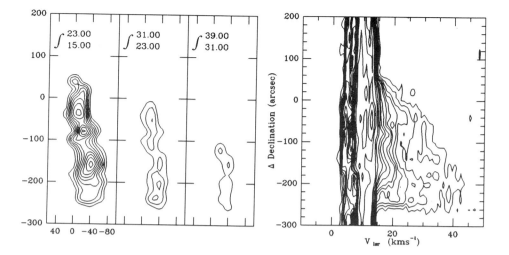

Figure 1: *(Left): Antenna temperature T_A^* averaged over 8km/s velocity intervals. The label indicates the velocity range of each map (V_{lsr}). Cloud core is at 11km/s. Contours start at 0.7K (4σ). (Right): Declination-Velocity plot down the jet. First contour is 3σ.*

These features are difficult to understand in terms of a thin swept up shell of emission. We could explain the 'acceleration' if the driving source was more energetic in the past than now, but it is unclear how this can account for the observed collimation structure. A possibility here is that we are seeing *in situ* acceleration of molecular gas: we note that the material detected at the highest velocities is very light (only a few percent of the total flow mass) yet contributes a major fraction to the flow kinetic energy. Thus, it is possible that this light energetic component is entraining ambient cloud material, leading to the observed velocity and spatial structure.

REFERENCES

[1] Snell, R.L. & Schloerb, F.P. (1985) *Astrophys.J.* **295**,490.
[2] Moriarty-Schieven, G.H., Snell, R.L., Strom, S.E., and Grasdalen, G.L. (1987) *Astrophys.J.* **319**,742.
[3] Sanders, D.B. & Willner, S.P. (1985) *Astrophys.J.* **293**,139.

Section VII

Solar and Solar System Studies

One-Millimeter Observations of Asteroids with the James Clerk Maxwell Telescope

Russell O. Redman, Paul A. Feldman, Ian Halliday
National Research Council of Canada
100 Sussex Drive, Ottawa, Ontario, Canada K1A 0R6

Henry E. Matthews
Joint Astronomy Center
665 Komohana Street, Hilo, Hawaii 96720, U. S. A.

ABSTRACT

Five asteroids (1 Ceres, 2 Pallas, 3 Juno, 18 Melpomene, and 444 Gyptis) have been observed at 1100μm, with an additional measurement of 1 Ceres at 800μm. These are the first radio-frequency observations ever reported for 3 Juno, 18 Melpomene, and 444 Gyptis, and the highest radio frequency observations reported for 1 Ceres and 2 Pallas.

OBSERVATIONS

The observations reported here were taken on the morning of 1988 August 13, using the 1100μm and 800μm filters on the UKT14 bolometer with the James Clerk Maxwell Telescope on Mauna Kea in Hawaii. The telescope pointing was checked repeatedly by observing Mars, 1 Ceres, and several nearby pointing sources, and was found to be within 3 arcseconds over the whole sky. The data were calibrated by making repeated observations of Mars. The principle systematic uncertainty in the calibration is the gain G at 1100μm, estimated to be 13.4 $Jy\ mV^{-1}$. It may be shown that the true flux F_* from an asteroid and the true gain G_*, are related to the values F_{est} and G_{est} estimated from the calibration data by expressions of the form $F_* = F_{est} \cdot (G_*/G_{est})^\epsilon$.

DISCUSSION

The simplest realistic model for the thermal emission of an asteroid is the "standard model" of Lebofsky [1]. For comparison with the data, flux densities F_ν were computed assuming an emissivity of 1 at mm wavelengths for the asteroids. The diameters and albedos used for the asteroids were taken from Table 17.1 of Cunningham [2][pages 147-164].

Table 1 lists the derived fluxes for the asteroids. The quantities shown are the asteroid's number and name, the Tholen type of the asteroid, the wavelength filter, the exponent ϵ described above, the predicted flux from the standard model, and the observed flux density in mJy with a 1σ estimate of the internal error of the measurement.

In general, the standard model predicts the observed fluxes quite well. Only 18 Melpomene and 444 Gyptis appear significantly different from the standard model. The measurement of 18 Melpomene should be the second most reliable in the data set and the difference is significant at the -3.7σ level. However, the optical light curve of 18 Melpomene has a large amplitude (0.33 mag) and no correction has yet been made for the rotational phase or viewing geometry. The flux from 444 Gyptis also appears to be significantly weaker than predicted ($-3.8\ \sigma$). In this case the rotational phase is irrelevant, since the discrepancy is much larger than the amplitude of the optical light curve. Instead, the emissivity of 444 Gyptis may be significantly less than 1.

Table 1: Observations of Thermal Radiation from Asteroids

Object	Type	Filter microns	ϵ	$F_\nu(pred)$	$F_\nu(obs)$
1 Ceres	G	1100	-0.09	2530	2522 ± 50
		800	-0.32	5402	5485 ± 300
2 Pallas	B	1100	-0.85	542	511 ± 43
3 Juno	S	1100	0.79	119	118 ± 24
18 Melpomene	S	1100	-0.24	431	383 ± 13
444 Gyptis	C	1100	-0.88	250	131 ± 31

Flux Density (mJy) column header spans $F_\nu(pred)$ and $F_\nu(obs)$.

CONCLUSIONS

The successful observation of these five asteroids demonstrates that it is now practical to study the thermal emission from asteroids with diameters down to 150 km. This represents a increase of 50% in the number of asteroids detected in the radio, over those listed in the recent survey by Webster and Johnston [3]. Most of these asteroids appear to be represented by a "standard model" as used in the IR. However, the low flux from 444 Gyptis suggests that the emissivity of some of the asteroids may be significantly less than 1. Detailed interpretation will require that rotational phase and viewing geometry be taken into account.

The authors would like to acknowledge the assistance of Dr. Göran Sandell who advised us on the correct approach for calibrating the data, and of Dr. Fokke Creutzberg who helped in the calculation of the positions of the asteroids. The James Clerk Maxwell Telescope is operated by the Royal Observatory Edinburgh on behalf of the Science and Engineering Research Council of the United Kingdom, the Netherlands Organization for Pure Research, and the National Research Council of Canada.

References

[1] Lebofsky, L. A., Sykes, M. V., Tedesco, E. F., Veeder, G. J., Matson, D. L., Brown, R. H., Gradie, J. C., Feierberg, M. A., and Rudy, R. J., (1986) 'A Refined "Standard" Model for Asteroids Based on Observations of 1 Ceres and 2 Pallas', Icarus **68**, 239-251

[2] Cunningham, C. J. (1987) Introduction to Asteroids, Willmann-Bell Inc., Richmond, VA

[3] Webster, W. J. Jr., and Johnston, K. J. (1988) 'Passive Microwave Observations of Asteroids', in T. Gehrels (ed.), Asteroids II, University of Arizona Press, Tucson, in press

PROSPECTS FOR SUBMILLIMETER OBSERVATIONS OF THE TOTAL SOLAR ECLIPSE OF 1991

Roellig, T.L. (*NASA/ARC*) and Lindsey, C.A. (*U.Hawaii*)

The Solar Eclipse of July 11, 1991

The solar eclipse that will take place early in the morning of July 11, 1991 will be total over the entire Island of Hawaii (i.e. the Big Island), lasting for up to 253 seconds. Over the rest of the State of Hawaii the eclipse will not reach totality, except for sections of the south coasts of Kahoolawe and Maui. A narrow strip along the south coast of Maui, centered about Kaupo Village, will probably experience a maximum of between 50 and 70 seconds of totality. The center of the path passes about 1.2 nautical miles south of the Mauna Kea summit. The width of the path, 122 nautical miles, will cover the Big Island completely and skim the south-eastern coast of Maui and Kahoolawe.

The edge of the Moon will first appear against the Sun's limb just slightly right of the top, and progress almost directly downward as the Sun rises, the tilt angles of 71.8, 76.4, and 81.1 degrees for the first contact, mid totality, and fourth contact, respectively. First contact of the lunar silhouette to the solar limb will occur at 06:30:41.6 AM, with the Sun approximately 8 degrees above the horizon; totality (i.e. second contact) will occur at about 07:28:08.5 AM, local time, and will last approximately 252 seconds. At mid totality the Sun will be at an elevation of approximately 20 degrees on an azimuth of 73 degrees (east of north); last contact will occur at 08:37:38.0 AM, at an elevation of about 37 degrees.

The Advantages of Submillimeter Observations of the Sun

Submillimeter observations of the sun offer unique diagnostic advantages in that 1) the mechanism of opacity is simple free-free interactions of electrons with neutral hydrogen atoms and protons, 2) the radiation emanates in Local Thermodynamic Equilibrium (LTE) from the solar chromosphere, and 3) the source function varies only linearly with temperature. The main opacity mechanism in the submillimeter continuum is free-free absorption due to the interaction of electrons with neutral atoms and ions. In the low chromosphere, this is almost entirely H⁻ free-free absorption, a well understood mechanism. Submillimeter continuum opacity increases strongly with wavelength (it is simply proportional to the square of the wavelength). It therefore provides a powerful and flexible tool for sampling any layer in a broad range of heights in the solar chromosphere. The Earth's atmosphere is opaque to radiation from 30 to 300 microns. However, radiation at 350 and 800 microns is accessible from the large telescopes on Mauna Kea. The weighting function for this radiation shows that it originates just at and just above the temperature minimum region of the solar chromosphere. Because of the dominance of the free-free processes, the submillimeter continuum is formed in LTE, providing a unique advantage for temperature diagnostics. This means that, much more than any visible spectral line, the submillimeter continuum is a direct and easily interpretable atmospheric thermometer. The submillimeter source function follows a simple Rayleigh-Jeans law, which maintains a simple direct proportionality of source function to temperature and allows much more direct modeling of the atmosphere as a multicomponent structure. In contrast, the emission at shorter wavelengths disproportionately weights hotter material.

Previous Submillimeter Eclipse Observations

Previous infrared and submillimeter observations of solar eclipses have generally been made from aircraft observatories. This has the advantages of getting above most of the atmospheric water vapor and also allows placement of the telescope directly in the path of totality. By observing the limb of the sun near the times of second and third contact, sub arc second resolution of the solar atmosphere is possible, much higher resolution than the diffraction limited beam sizes at these wavelengths. In the past, these observations have typically been made using a dichroic detector system, so that the observations were made at a number of wavelengths simultaneously. During the eclipse of July 31, 1981, the sun was observed at 30, 50, 100, and 200 microns from the Kuiper Airborne Observatory (Lindsey et al., *Ap.J.* **308**, 448, 1986). During the eclipse of March 18, 1988, the sun was observed at 400 and 800 microns wavelengths from the KAO in addition to the wavelengths listed above (Roellig et al., BAAS, **20**, 689, 1988). The results from these observations extended the sampling of the solar atmosphere to higher altitudes, allowing further refinement of the models of the chromosphere (Hermans and Lindsey, *Ap.J.* **310**, 907, 1986, also Braun and Lindsey, *Ap.J.* **320**, 898, 1987).

Submillimeter Eclipse Observations with the Mauna Kea Telescopes

The James Clerk Maxwell Telescope and the Caltech Submillimeter Observatory would be excellent instruments for submillimeter observations of the 1991 eclipse. With their alt-az mounts they are capable of pointing to the low elevations that would be needed to view the sun early in the morning. They will, however, need to protect their primary surfaces from the visible sunlight. The question of how to best do this will be addressed in more detail in the following section. The large apertures of the Mauna Kea telescopes relative to those of the airborne telescopes means that they will have significantly smaller diffraction limited beam sizes. This will not affect the spatial resolution of the observations directly, since the spatial resolution is provided by the limb of the moon, but will have a large affect on the signal to noise of the signal as the sun disappears. This will allow a more precise determination of the emission of the solar chromosphere as a function of height above the photosphere. The atmosphere above Mauna Kea is generally dry and excellent transmission can be expected at 800 microns wavelength. At 350 microns the atmospheric transmission is more variable, but will probably still be good enough for better quality data than that previously obtained at this wavelength.

Protection of the Mauna Kea Telescopes from Direct Sunlight

The entire dome of the JCMT is covered with a finely woven nylon-teflon fabric, Gortex, to allow the dome interior to be kept at a constant temperature. This covering serves as a highly effective shield against focused visible light for solar observations and has been previously tested in submillimeter solar observations. As far as the CSO is concerned, a similar system would work, although a great deal of tension would be needed to keep the covering material from sagging onto the secondary assembly. Covering only the primary mirror would use less material and would also allow a boresight telescope to be mounted on the edge of the reflector. The disadvantages of this scheme are that two passages of the light through the material would be necessary and that stretching the fabric over the reflector may put pohibitively large stresses on the support structure.

SOLAR ASTRONOMY ON THE NEW LARGE SUBMILLIMETER FACILITIES ON MAUNA KEA

CHARLES A. LINDSEY
Institute for Astronomy, University of Hawaii, Honolulu, HI

THOMAS L. ROELLIG
NASA, Ames Research Center, Moffett Field, CA

The submillimeter continuum represents an area of great potential interest in the study of the solar atmosphere. Submillimeter wavelengths accessible from Mauna Kea can provide a particularly valuable tool for thermal diagnostics of the solar chromosphere. Thus far, the submillimeter solar continuum remains relatively unexplored. Recent advances in submillimeter technology will open this new and powerful diagnostic tool to solar physics.

1. WHY WE WANT SUBMILLIMETER OBSERVATIONS OF THE SUN

The following list outlines briefly the most important reasons we know of for doing submillimeter observations of the sun:

1.1. The Solar Chromosphere

Submillimeter radiation comes from the solar chromosphere, the same very interesting region from which most of the sun's ultraviolet radiation also comes. We want to understand the processes that heat the chromosphere to thousands of degrees above the temperature of the underlying photosphere, giving rise to this ultraviolet radiation.

1.2. Thermal Diagnostics

Submillimeter radiation is perhaps the best known atmospheric thermometer. Unlike almost all strong spectal lines, the submillimeter continuum is emitted in local thermal equilibrium (LTE) with free electrons in the atmospheric medium.

The comparative illustration is given by Figures 1 and 2. Figure 1 shows an intensity map of the sun in the light of a strong chromospheric line, the K-line if singly ionized calcium. Note that the limb is considerably darker than disk center. This normally results from a decrease in temperature with height, in this case the population temperature of the calcium ion. However, the kinetic temperature of the chromospheric medium (i.e. the free electron kinetic temperature) increases with height in the chromosphere. Because calcium atoms are radiatively coupled efficiently to cold space, they cool off rapidly with height, as collisions decrease with decreasing density. They are, therefore, a poor thermometer, like all strong visible lines that radiate efficiently into space from the chromosphere.

Infrared continuum emission is due to electron collisions with atoms and ions. It is a far better thermometer than visible spectral lines. Figure 2 shows an image of the sum in 820 μm continuum radiation. This image shows clear limb brightening, directly consistent with the increase in temperature with height in the chromosphere.

1.3. Height Discrimination

Submillimeter opacity is strongly dependent on wavelength. Thus different wavelengths can be used to sample different heights. Figure 3 shows weighting functions for the solar atmosphere for 350 and 850 μm radiation, wavelengths uniquely accessible to the new large submillimeter telescopes on Mauna Kea. Radiation at 350 μm samples the low chromosphere, just above the temperature minimum, where the onset of anomalous heating causes a sudden reversal in the vertical temperature gradient. The 850 μm continuum samples the overlying hotter chromosphere.

1.4. Because It's There

However strong our justification for exploring into a new region, we are somewhat unfairly forced to omit those that remain to be discovered. The submillimeter continuum offers all of the mystery of a new and nearly totally unexplored domain. With the resources that have been invested in the new submillimeter facilities, we simply cannot allow this opportunity to escape us.

2. WHY HAVE SO FEW SUBMILLIMETER OBSERVATIONS OF THE SUN BEEN MADE?

Only recently has technology begun to make serious submillimeter solar observations practical. Solar scientists are accustomed to vast quantities of high-resolution high signal-to-noise data. Only the latest generation of large submillimeter telescopes offers discrimination surpassing that of the naked eye in visible light.

This problem is complicated by the fact that submillimeter radiation is heavily absorbed by water vapor. Only wavelengths longward of about 300 μm are visible from any ground-based sites, and those are accessible only to the very driest places. We are very fortunate to have at the summit of Mauna Kea easy access to one of the driest sites in the world. Figure 4 shows atmospheric transmission of the submillimeter band typical of Mauna Kea. The transmission windows centered at 350, 450 and 850 µm are very useful for solar chromospheric diagnostics.

The recent advent of large sumillimeter telescopes on Mauna Kea with new and powerful detectors will have a particularly powerful impact in solar research. We believe that these new telescopes can be used for solar observations with a modest investment in precautions to protect them from focused visible and near infrared radiation.

3. TOPICS OF PARTICULAR INTEREST

The following is a brief outline of solar opportunities that are wide open for exploitation by the new large submillimeter telescopes on Mauna Kea. A single definitive observation in any of these areas is bound to throw considerable light on the atmospheric structure and dynamics of the solar chromosphere.

3.1. Sunspots

The photospheres of sunspots are cooler than the quiet photosphere. However, at some height, the sunspot chromospheres become hotter. Where? At the temperature minimum, or well above it? What does this say about the role of strong magnetic fields in heating the low chromosphere? Submillimeter observations are the best way to answer these questions. No telescopes have been able to resolve sunspots at submillimeter wavelengths to date. The new large submillimeter dishes will be able to resolve them.

3.2. The Chromospheric Supergranular Network

This is a pattern of convection cells visible all over the sun in any K-line photograph. The centers of the cells are darker than the boundaries separating the cells. This seems to be associated with magnetic fields, which are concentrated at the boundaries. Until now, this network has not been resolved in submillimeter radiation. The new large submillimeter dishes can resolve it. Submillimeter observations of the supergranular network will provide thermal diagnostics important for refining models of the quiet solar chromosphere and of heated magnetic flux tubes.

3.3. Solar Oscillations

The new large submillimeter telescopes will be useful for studying the thermal response of the chromospheric medium to hydrodynamic waves propagating through it. The main mechanism causing thermal variations is work done by compression as the wave passes through the medium (see Figures 5 and 6). The high spatial resolution the new telescopes offer will show features previously unresolved, resulting in a greater signal to noise than has been possible before.

3.4. The Total Solar Eclipse of 1991

The total solar eclipse of July 11, 1991 will cover the entire Big Island, including the summit of Mauna Kea. The path of totality of the eclipse is shown in Figure 7. The center of the path will pass near the summit missing the new submillimeter facilities by only hundreds of meters. The eclipse offers us a once-in-a-lifetime opportunity to observe the occultation of the solar limb with sub-arcsecond resolution at submillimeter and millimeter wavelengths. This will provide important data for modeling the thermal state and height distribution of the middle chromosphere. The technique for determining the extreme solar limb profile of the sun by lunar occultation has been developed by Lindsey et al. 1986 and by Roellig et al. 1989 using the Kuiper Airborne Observatory. Figure 8 shows limb brightness profiles determined by the former using this technique in the total solar eclipse of 1981 July 31.

REFERENCES

Braun, D and Lindsey, C. 1987, Ap. J., 320, 898.
Hermans, L. M. and Lindsey, C. 1986, Ap. J., 310,907.
Lindsey, C., Becklin, E. E., Orrall, F. Q., Werner, M. W., Jefferies, J. T., and Gatley, I. 1986, Ap. J., 308, 448.
Lindsey, C., de Graauw, T., de Vries, C., and Lidholm, S., 1984, Ap. J., 277, 424.
Lindsey, C., Kopp, G., Becklin, E. E., Orrall, F. Q., Roellig, T. L., Werner, M. W., Jefferies, J. T. 1988, Ap. J. (in press).
Roellig, T. L., Werner, M. W., Kopp, G., Becklin, E. E., Lindsey, C., Orrall, F. Q. and Jefferies, J. T. 1988, Ap. J. (in preparation).
Traub, W. an Stier, M. T. 1976, Appl. Optics, 15, 364.
Vernazza, J. E., Avrett, E. H., and Loeser, R. 1981, Ap. J. Supp., 45, 635.

Figure 1: Intensity map of the full solar disk in light of the K-line of singly ionized calcium. Note the strong limb darkening. (Projected relief map provided by Mr. Matthew Penn from observations from the Mees Solar Observatory.)

Figure 2: A visible continuum intensity map of the sun shown in (a) is compared with projected relief map of solar intensity in the 820 μm continuum (b and c). The map in (c) is cut across the middle to show more clearly the brightening of the limbs. The 820 μm observations were made at the IRTF with T. de Graauw's heterodyne receiver in 1980 (see Lindsey et al. 1984). The angular resolution is about 90", poorer than the naked eye in visible light. The new large submillimetre telescopes will resolve ~15".

Figure 3: Atmospheric weighting functions for 350 and 850 μm radiation show the vertical distribution of radiation emanating from the chromosphere throughout the submillimeter continuum. These wavelengths come primarily from anomalously heated material just above the reversal of the vertical temperature gradient in the low chromosphere. Only wavelengths longward of 350 μm are accessible from Mauna Kea. These characterize the temperature minimum region, the height above which a sharp rise in temperature occurs, due to anomalous heating of the chromospheric medium. The weighting functions were computed from the Model C of Vernazza, Avrett and Loeser (1981).

Figure 4: Atmospheric transmission of submillimeter radiation from the summit of Mauna Kea with 1.2 mm of precipitable line of sight water vapor (condensed from the work of Traub and Stier 1976).

Figure 5: Chromospheric Doppler oscillations in the D_1 line of neutral sodium are shown over a period of 6 hours (lower plot). These observations were made with the Stokes Polarimeter at the Mees Solar Observatory in collaboration with D. L. Mickey, who built the instrument. The upper plot shows the velocity difference between both sodium D lines (there are two). The D_2 line is twice as strong, and, therefore, forms about 80 km higher in the chromosphere. The velocity difference plotted above gives a measure of compression rate, and work done by compression causes thermal variations in the chromospheric medium associated with oscillations. The above measurements were made with a resolution of ~40", finer than is possible at the IRTF at 800 μm. Poorer resolution than this results in very poor signal to noise in the compression signal. The greater resolution provided by the new large submillimeter telescopes will improve this considerably, making comparative measures of compression and thermal variations possible.

Figure 6: Submillimeter intensity oscillations (top box) observed from the Kuiper Airborne Observatory (Lindsey et al. 1988) are strongly correlated with Doppler oscillations made of the same region from the Mees Solar Observatory at Haleakala. The new large submillimeter telescopes, with their higher angular resolution will allow us to compare compression with thermal response with a much higher signal to noise than has been possible with the IRTF.

Figure 7: The total solar eclipse of 1991 July 11 covers the entire Island of Hawaii. The path of totality is the corridor between the oblique solid lines. The center of the path (dashed line) will pass almost directly over the centroid of the Mauna Kea science reserve, offering a unique opportunity to Mauna Kea facilities. Totality will occur at 7:28 A.M. local time with the sun about 20° above the horizon and will last for ~4.5 minutes.

Figure 8: Submillimeter limb brightness profiles computed from airborne limb-occultation observations in the total solar eclipse of 1981 July 31 (Linesey et al. 1986). The dashed and dash-dotted curves show how these observations have revised chromospheric models based on the assumption of gravitational-hydrostatic equilibrium (Hermans and Lindsey 1984). More sophisticated models incorporating spicular inhomogeneities have subsequently been published by Braun and Lindsey (1987). The total solar eclipse of 1991 July 11 will provide a unique opportunity to repeat this observation at millimeter wavelengths. With the large submillimeter dishes, it should be possible to resolve vertical detail in the solar limb profile to ~0.2", about 150 km in height.

303

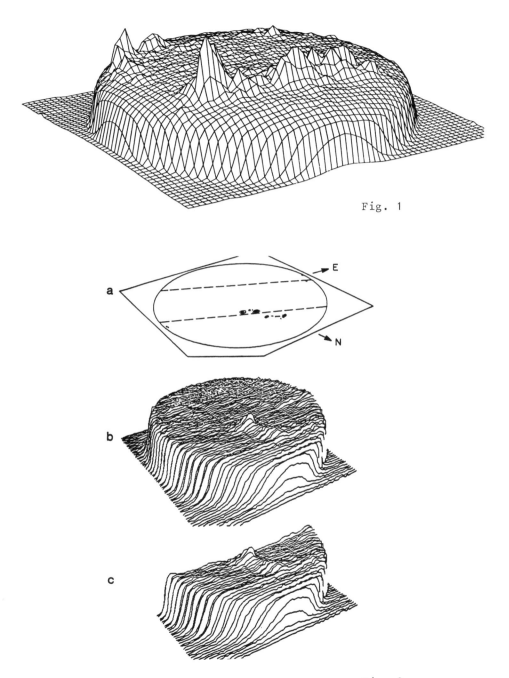

Fig. 1

Fig. 2

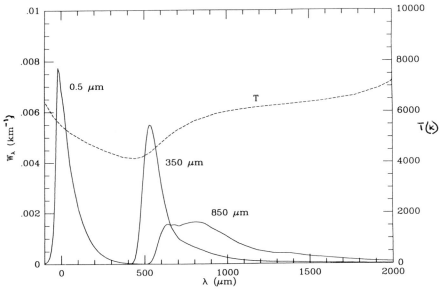

Fig. 3

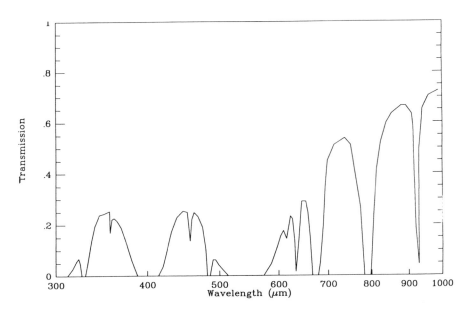

Fig. 4

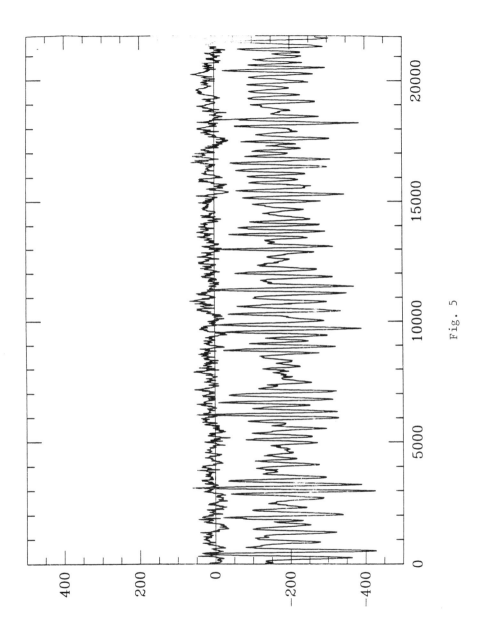

Fig. 5

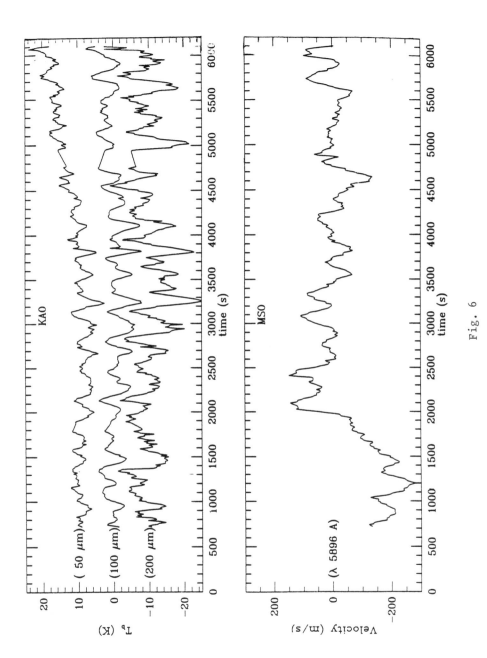

Fig. 6

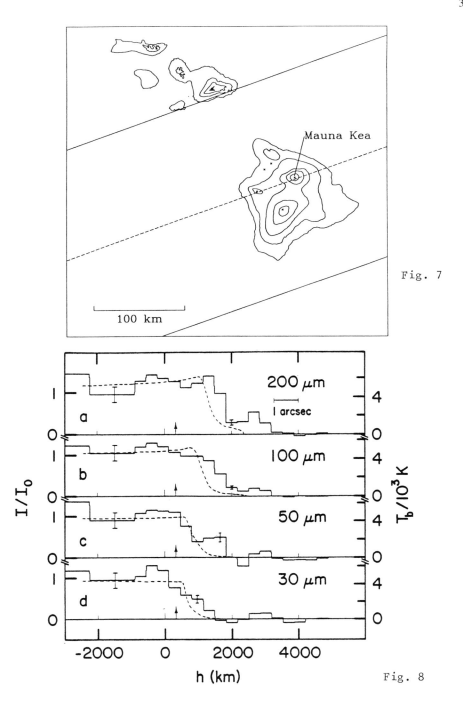

Fig. 7

Fig. 8

Indices

Appendix A: Authors
Appendix B: Molecules
Appendix C: Objects
Appendix D:General

The page numbers referred to within the indices reference the first page of the relevant paper and not to the actual page containing the index item.

AUTHOR INDEX

Alexander P	231
Allen M	187
Aspin C	29
Baars J W M	45
Baas F	225
Barsony M	193
Bash F N	227
Becklin E E	147
Belcourt K	187
Bennett C L	275
Benson P J	35
Bernstein G M	53, 79
Betz A L	117, 167
Black J H	103, 225
Blake G A	105, 265
Blitz L	173
Bloemhof E E	55
Bogey M	107
Booth R S	287
Boreiko R T	117, 167
Boulanger F	107
Brett M	75
Brown R L	87
Butner H M	129, 157
Byrom C N	77
Carico D P	209
Casey S C	35, 189
Casoli F	249
Chikada Y	89
Chin G	65
Chini R	19, 229
Combes F	107, 249
Coulson I M	283
Cowie L L	247
Cunningham C T	71
Danchi W C	105, 265
Dave H	65
Davidson J A	35, 189
Davies S R	71
Davis J H	227
De Graauw Th	225
Dent W R F	169
Dhawan V	55
Dickman R L	171
Duncan W D	51, 191, 217, 231
Dupraz C	249
Eales S A	211, 217, 231, 241
Ellison B N	77
Emrich A	61
Encrenaz P	107
Erickson N R	31
Evans N J	157
Fairclough J H	233
Falgarone E	3, 9
Feldman P A	293
Fischer M L	79
Frerking M A	77
Friberg P	115
Fuller G A	163
Gammie C F	33
Gautier T N	53
Gear W K	203
Geballe T R	29
Geis N	67, 285

Genzel R	67, 125, 127, 261, 269, 277, 283, 285	Iwashita H	89
Gerber R A	235	Jaffe D T	127, 227, 269, 277, 285
Gerin M	107, 249	Jaminet P A	105, 265
Gezari D	173	Jarrett T H	171
Goldsmith P F	31, 183	Jewell P R	111
Gordon M	191	Johansson L E B	287
Graf U U	269, 277, 283	Judge P G	179
Griffin M	49	Kaifu N	63
Grossman E N	57	Kameya O	181
Guelin M	97	Kanzawa T	89
Haggerty M	67	Kasuga T	73, 89
Halliday I	293	Kawabe R	73, 89
Handa K	89	Keene J B	209, 265
Harper D A	35, 189	Kenney J D P	251
Harris A I	127, 269, 277, 283	Knapp G R	33
Hasegawa T	63, 279	Kobayashi H	89
Hasegawa T I	175	Koornneef J	225
Hauschildt M	233	Krügel E	45
Hayashi M	63, 279	Kulkarni S R	239
Hayashi S S	159, 175, 283	Kwan J	31
Herbst W	171	Lada E A	267
Hildebrand R H	25	Ladd E F	35
Hills R E	183, 289	Lamb S A	235
Hjalmarson Å	115	Lan-Ping X	131
Ho P T P	219	Langer W D	123
Hollis J M	111	Lasenby J	183
Howe J E	127, 157, 285	Laycock S C	157
Hughes V A	37, 177	Leduc H G	77
Inatani J	73, 89	Levreault R M	35
Irvine W M	115	Lindsey C A	295, 297
Ishiguro M	89	Linsky J L	179
Ishizuki S	89	Lis D C	183
Israel F P	225	Little L T	71

Lo K Y	69, 211, 241	Peterson J B	79
Lovas F J	111	Petuchowski S J	275
Lucas R	97	Phillips T G	9, 33, 41, 103, 213, 221, 223
Lugten J B	125, 269	Poglitsch A	67, 261, 269, 285
Madden S C	115	Pudritz R E	135
Maillard J P	187	Puget J L	3
Mangum J G	185	Rand R J	239
Martin R N	45, 47, 219	Redman R O	293
Masson C R	41, 105, 265	Richards P L	53, 79
Matheson D N	71	Richer J	263, 289
Matthews H E	115, 283, 293	Rickard L J	131
Mayer C E	157	Rieke G H	53
McGrath W R	77	Robson E I	191, 215
Melnick	93	Roellig T L	295, 297
Miller R E	47, 77	Rogers C	175
Miller R H	235	Rosenthal E	211, 241
Mitchell G F	187	Routledge D	75
Miyawaki R	279	Russell A P G	281, 283, 289
Moriarty-Schieven G H	37, 177	Sage L J	245
Morita K -I	89	Sakamoto A	73
Mundy L G	155, 185, 265	Sandell G	29, 141
Murata Y	89	Sanders D B	213, 221, 223
Murphy D M	287	Sargent A I	155, 213, 221, 223
Myers P C	35, 189	Schmid-Burgk J	11
Nakai N	237	Scott P F	165, 289
Neugebauer G	209	Scoville N Z	155, 197, 213
Ohishi M	113	Serabyn E	41
Okumura S K	89	Sharifi F	69
Olberg M	287	Smith B F	235
Padin S	155	Smith M G	29
Padman R	165, 263, 283, 289	Snell R L	31, 37
Parker N D	165	Snyder L E	111
Pendleton Y	189	Soifer B T	209
Pernic R	189		

Stacey G J	67, 125, 127, 261, 269, 285
Stencel R E	179
Stephens S	211, 241
Stern J A	77
Stutzki J	127, 269, 277, 283
Sunada K	63
Sutton E C	105, 227, 265
Tacconi L	243
Takahashi T	89
Tauber J A	31
Tielens A G G M	13
Timbie P T	53
Timusk T	79
Townes C H	67, 125
Tsuboi M	73
Turner B E	115, 131
Turner J L	219
Van Dishoeck E F	103, 225
Van Harlingen D J	69
Vaneldik J F	75
Walker C E	47
Walker C K	47
Walker T M	233
Wall W F	157, 227
Walther D	191
Wang Z	247
Ward-Thompson D	191
Watazawa K	73
Watt G D	283
Wattenbach R	59
Webster A S	283
Welch Wm J	81
Werner M W	53
White G J	255
Wilson C D	161
Woody D P	43, 155
Wootten A	107, 185
Wright G S	233
Wright M C H	91
Wrobel J M	245
Wynn-Williams C G	211, 217
Yamamoto M	73
Young K	33
Zhou S	107, 157
Ziurys L M	109, 115
Zmuidzinas J	69, 117, 167
Zuckerman B	147

MOLECULE INDEX

21cm HI	281
atomic carbon	93, 269
C_3H_2 $1_{1,0}$-$1_{0,1}$	131
carbon	123
CH_2CN	113
CO	269
CO 1-0	81, 93, 161, 197, 213, 223, 225, 237, 239, 243, 245, 289
CO 2-1	165, 197, 213, 221, 255
CO 3-2	29, 31, 103, 219, 241, 255
CO 6-5	283
CO 7-6	11, 283
CS	193
CS 2-1	103
H_2 emission	225
H_2CO	131
H_2O	93, 275
H_2O masers	181
H_3O^+	107
HC_2CHO	113
HC_3N 1-0	131
HC_3N 2-1	131
HCN 3-2	263
HCN 9-8	283
HCO^+ 3-2	263, 281
MgH	109
O_2	93
OCS	107
PAHs	13
SH^+	109
SiO 2-1	109

OBJECT INDEX

0420-014	215
0736+017	215
0851+202	215
1 Ceres	293
1253-055	215
1308+326	215
13349+2438	203
16293-2422	141
1641+399	215
18 Melpomene	293
1921-293	91, 215
2 Pallas	293
2145+067	91
3 Juno	293
30Dor	287
3C273	91
3C279	91
3C345	91
3C84	91, 203
NGC1275	203
444 Gyptis	293
Arp 220	19, 155, 197, 209, 219
Arp 243	213
Arp 55	197
Asteroids	293
B335	109, 141, 175
B35	269
B5	189
B5-IRS1	163
Beta Pictoris	147
BL Lac	91
Centaurus A	**223, 225**
Cep A	157, 159, 177, 269
CIT6	269
CPD-56 8032	33
CRL2688	33
CRL618	33, 97
Cyg OB7	9
Cygnus A	19
DR21	269, 281
DR21(OH)	155, 157, 185
Eta Eradini	147
Formalhaut	147
Frosty Leo Nebula	97
FU Orionis	141
G12.41+0.50	269
G333.6-0.7	269
G34.3+0.1	269
G35.2N	169
Galactic Centre	19
GL 2591	269
GL 490	157, 269
HD169454	103
He2-113	33
Herbig Ae/Be stars	141
HH7-11	155
HH9	269
HL Tau	135
HR451	147
IC10	233
IC1396	269
IC342	93, 219
IC443	269
IC443B	109

IC443G	109	NGC404	245
IRAS 09371+1212	97	NGC520	197
IRAS 16235-2416	191	NGC958	209
IRC+10216	31, 97, 269	NGC1055	209
L1262	165	NGC1068	197, 219, 237
L134	269	NGC1097	237
L134N	109	NGC1143/3	209
L1489	35	NGC1275	91
L1551	109, 269	NGC1977	285
L1551-IRS5	155	NGC2023	127, 269
L43	165	NGC2024	19, 117, 157, 159, 269
L483	165		
L588	165	NGC2068	269
LkHa101	193	NGC2071	37, 141, 159, 269
LkHa234	141, 169	NGC2146	125
LMC	287	NGC2346	97
M1-92	33	NGC2623	213
M17	157, 255	NGC2903	219
M17(SW)	19, 157, 269	NGC3245	245
M17KW	269	NGC3256	221
M2-56	33	NGC3665	245
M33	161	NGC3690	155, 209, 219
M51	81, 219, 239	NGC4151	203
M82	89, 203, 211, 219	NGC4293	251
M83	19, 219, 237	NGC4419	251
Maffei 2	81, 89, 219, 237	NGC4429	245
Magellanic Clouds	93, 287	NGC4435	245
Mon R2	157, 269	NGC4459	245
Mrk 231	209, 213	NGC4526	245
N158	287	NGC4710	245, 251
N159	287	NGC4945	19
N160	287	NGC5195	245
NGC205	245	NGC5363	245
NGC253	19, 219, 227	NGC5907	125

NGC6090	213	T Tauris	141
NGC6240	219	Tau Eradini	147
NGC6334	173	Taurus	9, 93
NGC6946	93, 219, 241, 243	TMC-1	109, 113, 115
NGC7027	33, 97	TMC1 (Heiles 2)	171
NGC7252	249	UGC2982	209
NGC7538	109, 181, 269	V Hya	33
NGC7538-IRS1	155	V645	269
NGC7541	209	Vega	147
NGC7771	209	Virgo	251
NRAO530	91	Virgo A	19
OH 0.9+1.3	97	W3	269, 59
OH 231.8+4.2	33, 97	W3(OH)	157
OJ287	91	W3-IRS3	285
OMC-1/Orion A	11, 59, 93, 107, 109, 111, 157, 167, 255, 261, 269, 277, 275, 283	W3-IRS4	285
		W3-IRS5	187
		W33	269
OMC-2	269	W43	269
Orion B	267, 289	W49	81, 269, 277
Perseus	93	W49A	279
R Leo	269	W51	157, 269
Rho Oph	93, 171, 191, 269	W51-IRS2	117
S87	193	W75N	157, 269
S106	193, 263, 269	W75S	269
S140	157, 159, 269	Zw 049.057	209
S235	157		
S255	157, 269		
S88	157		
Sgr A	269		
Sgr A (East)	19		
Sgr B2	105, 113, 183, 269		
Sgr B2(OH)	115		
SSV13	141		
Sun	295, 297		

GENERAL INDEX

Accretion disk
- luminosity — 197
- structure and evolution — 135

Alfven
- velocity — 3
- waves — 135

Ambipolar diffusion — 135

Antenna Structure
- antenna/beam efficiencies — 283
- far-field measurements — 41
- holographic techniques — 41
- homology measures — 43
- pyrex glass — 45
- reflector panels — 45
- surface measurement — 41, 283
- survey telescope — 63

Asteroids — 293

Bell Labs Telescope — 103, 267

BL Lacs — 203

Black hole — 197

Blazars — 203, 215

Bolometers - see Instrumentation

Carbon
- column densities — 117
- graphite — 13
- isotope ratio — 97, 117
- optical depths — 117

Canada-France-Hawaii Telescope (CFHT) — 161, 187

Chemistry
- composition — 97, 105, 109, 123, 129, 131
- deuterium fractionation — 129
- differentiation — 105
- endothermic reactions — 109
- high temperatures — 109
- ion-molecule chemistry — 109, 123
- metal deficient chemistry — 131
- modelling — 103, 123, 255

Circumstellar Material — 31, 33, 97, 141, 163, 179, 193

Calcium Lines — 297

Carbon Monoxide
- column density — 167
- distribution — 247

Comet clouds — 147

Compact HII regions — 141

Continuum Observations
- 2 micron H_2 lines — 29
- 12-100 microns — 147
- 100 microns — 25

- 158 micron CII	125, 269, 285	- condensation	191
- 160 and 360 micron	189	- density	35
- 100-300 microns	59	- dust/gas conversion ratio	155
- 350 microns	183, 217, 297	- emissivity	13, 147, 203
- 450 microns	141, 169, 177, 183, 203, 215, 217	- mantles	191, 275
		- size distribution	13, 147, 131
- 600 microns	215	- temperatures	13, 35
- 800 microns	141, 147, 183, 215, 293	**Embedded young stars**	141
- 850 microns	231, 297	**Extinction Measures**	13
- 870 microns	211	**Far-Infrared**	
- 1100 microns	141, 169, 173, 183, 215, 231, 293	- observations	35
		- spectroscopy	261, 269
- 1300 microns	19, 197, 209, 215, 229		
- 1.4 mm	155	**Five College Radio Astronomy Observatory (FCRAO)**	37, 91, 109, 115, 163, 243
- 2.0 mm	215		
- 2.6 mm	197		
		Fluorescent H_2 emission	127
Cosmic Background Radiation	53, 79	**Galactic Centre**	25
		Galaxies(types)	
Cosmology	53	- AGNs	197, 203, 215
Caltech Submillimeter Observatory (CSO)	9, 33, 41, 43, 103, 105, 109, 209, 213, 221, 223, 227, 265	- disk galaxies	223, 225, 247, 249
		- ellipticals	223, 225, 249
		- interacting galaxies	197, 221, 235, 249
Cerro-Tololo International Observatory (CTIO)	171, 173	- IRAS galaxies	19, 197, 209, 213, 221
		- irregular	233, 287
Differential Mass Spectrum	135	- Seyferts	91, 203, 215, 237
Dust grains	13, 147, 173, 217	- spirals	19, 81, 89, 125, 161, 217, 219, 227, 237, 239, 241, 243, 245, 251
Dust Grain Properties			

- starburst galaxies	19, 125, 197, 203, 211, 221, 249	Herzberg Institute for Astronomy (HIA)	71
- Virgo galaxies	19, 251		
		Hydromagnetic waves	135
Galaxies(structure)		**Infrared**	
- CO/HI ratio	251	- cirrus clouds	13, 131
- CO-H2 conversion factor	287	- emission features	13
- dust lane mapping	223, 239		
- evolution	197, 247	**Instrumentation**	
- flaring	91, 203, 215	- bolometers	19, 45, 47, 49, 51, 53
- gas distribution	197, 225, 243, 247, 251	- corner-cube antennas	57
- HI stripping	251	- Gaussian beams	57
- initial mass function	197	- imaging photometer(MIPS)	53
- interarm molecular clouds	239, 243		
- jet formation	91	- INSU CIRCUS-IR camera	135
- luminosities	287	- IRCAM	29
- nuclear regions	197, 223, 225, 237, 241	- laser local oscillator	59, 65
		- photoconductive detectors	67
- physical modelling	235	- radiative coupling efficiency	57
- radial distribution	247		
- ring	197	- Schottky mixers	45, 47, 59, 63, 81, 93, 111, 117, 157, 249, 255, 283
- spiral arm maps	239, 243		
- spiral density wave	81, 239, 243		
- superluminal motion	91	- SCUBA	49, 51
- synchrotron emission	91, 203, 215	- SIS junctions	55, 69, 75
- variability	177, 215	- SIS - junctions (Niobium)	55, 69, 73, 75, 77
		- SIS - junctions (Pb-alloy)	71, 77
Green Bank Telescope	113	- SIS mixers/receivers	47, 55, 61, 69, 71, 73, 75, 77, 81, 89, 105, 113
Harvard-Smithsonian Center for Astrophysics (CFA)	55		
		- UCB 32 ch. bolometer	189
Hat Creek Telescope	81, 91, 163	- UCB/MPIfR receiver	11, 269, 283

Interferometry	43, 55, 81, 87, 89, 91, 155, 161, 163, 181, 185, 197	Magnetic fields	3, 25, 135
		Mass loss studies	29, 31, 33, 165, 179, 197
IRAM	9, 19, 97, 109, 229	Max Planck Institut für Radioastronomie (MPIfR)	45, 283
Infrared Astronomical Satellite (IRAS)	147, 171, 173, 191, 197, 209, 213, 217, 221, 233, 245	Maynoorth College, Ireland	51
Infrared Telescope Facility (IRTF)	35, 127, 189, 297	Maximum Entropy Method (MEM) restoration	263
Interstellar Dust - see Dust Grains		Mees Solar Observatory (MSO)	297
James Clerk Maxwell Telescope (JCMT)	29, 49, 51, 71, 109, 141, 147, 159, 165, 169, 175, 181, 183, 203, 211, 227, 231, 233, 241, 255, 263, 269, 283, 289, 293	Millimeter Array (MMA)	87
		Mixers - see Instrumentation	
		Molecular Clouds	9, 19, 37, 81, 93, 109, 115, 127, 141, 155, 157, 161, 163, 165, 167, 169, 171, 175, 181, 183, 187, 189, 191, 193, 261, 263, 267, 269, 279, 281, 283, 289
Junctions - see Instrumentation			
Kuiper Airborne Observatory (KAO)	11, 35, 59, 117, 125, 127, 189, 261, 269, 285, 297		
Kitt Peak National Observatory (KPNO)	91, 171, 267	Molecular Cloud Structure - bars	255
Leighton telescopes	43	- clumps	163, 189, 255, 263, 267, 269, 283
LTE Modelling - see Radiative Transfer		- cores	35, 129, 131, 135, 141, 165, 189, 267, 277, 283, 289
LVG Modelling - see Radiative Transfer		- dense gas	265, 269

- density condensations	267
- disks	125, 135, 263
- filaments	255
- fragmentation	3
- physical modelling	3, 37, 135, 227, 261
- self-similar structure	9
- streamers	255
- supersonic turbulence	3
- temperature structure	35
- translucent clouds	103
- velocity dispersion	3, 287
- virial mass	131, 267
Mount Graham	45
Multicolour observations	171
Millimeter Wave Observatory (MWO)	107, 127, 129, 157
Near-Infrared survey	267
Nobeyama Radio Observatory (NRO)	73, 89, 91, 113, 159, 181, 237, 279
Nobeyama Millimeter Array	73, 81, 89, 181
National Radio Astronomy Observatory (NRAO)	107, 109, 111, 115, 127, 129, 131, 161, 173, 219, 245, 275
Onsala Telescope	61, 91
Outflow Sources	29, 33, 37, 109, 141, 159, 165, 169, 177, 181, 193, 233, 255, 263, 269, 281, 283, 289
Outflow Structure	
- atomic gas	281
- backflow	37
- high velocity outflows	187, 283
- high velocity wings	11, 29, 283
- neutral wind	281
- winds/jets	29, 33, 289
Owens Valley Radio Observatory (OVRO)	43, 81, 91, 155, 161, 185, 193, 197, 239
Polycyclic Aromatic Hydrocarbons (PAHs)	13
Photodissociation regions	117, 123, 125, 127, 255, 269, 285
Planetary nebulae	33
Plateau de Burre (IRAM)	81
Polarization	25
Poynting-Robertson drag	147
Precipitable water vapour	45, 297
Protostars	35, 185, 187, 191, 193, 289
Protostellar Disks	159, 181
Protoplanetary Nebula	29
Queen Mary College (QMC)	51
Quasars	91, 203, 229, 237

Radiative Pressure	33	Submillimeter Telescope (SMT)	45, 47
Radiative Transfer			
- LVG specific	107, 129, 131, 175, 181	Solar Eclipse 1991	295, 297
- LTE specific	107, 131, 147, 167, 295, 297	Spectral Lines	
		- absorption lines	225
		- non-Gaussian shapes	9
- modelling	31, 183, 269, 277	- vibrationally excited species	97, 275
- non-LTE specific	227		
		Spectrometers	
		- acousto-Optical	59, 93
Radio continuum observations	177	- autocorrelation	47, 61, 81
Rutherford-Appleton Laboratories (RAL)	71	- Fabry-Perot	67, 125, 269, 285
		- Fourier Transform	41, 113
Royal Observatory Edinburgh (ROE)	51		
		Stars	
Submillimeter Continuum Bolometer Array - see Instrumentation		- Asymptotic Giant Branch	33
		- chromospheric heating rates	179
		- evolved	33, 179, 233
Submillimeter Wave Astronomy Satellite (SWAS)	93	- FU Orionis	141
		- giant	29, 33
		- Herbig Ae/Be	141
Sky emissivity	49	- T Tauri	141
Sunyaev-Zel'dovich effect	53	- Wolf-Rayet	33
Southern European Submillimeter Telescope (SEST)	19, 225, 249, 287	Star Formation	3, 29, 37, 93, 125, 131, 135, 141, 159, 161, 165, 169, 171, 177, 187, 191, 197, 213, 221, 233, 235, 239, 245, 247, 255, 267, 279, 285, 287
Shock Regions	181, 193, 235, 255, 275, 281		
Submillimeter/Infrared Telescope Facility (SIRTF)	53		

Star Formation Efficiencies	161, 239, 243, 245	Wyoming Infrared Observatory	31

Stellar
- evolution   33
- winds   31

Steward Observatory   45, 47

Submillimeter Continuum   203, 295

Sun
- chromosphere   295, 297
- continuum   295, 297
- oscillations   297
- sunspots   297
- supergranulation 297

Survey Observations   97, 105, 111, 113, 157

University of California, Berkeley (UCB)   125, 189, 269

University of Hawaii (UH88)   11

University of Kent, Canterbury (UKC)   71

United Kingdom Infrared Telescope (UKIRT)   29, 215, 217, 269, 277, 281

Very Large Array (VLA) data   177, 281

Very Long Baseline Interferometry (VLBI)   91